Biological and Medical Physics, Biomedical Engineering

This series is intended to be comprehensive, covering a broad range of topics important to the study of the physical, chemical and biological sciences. Its goal is to provide scientists and engineers with monographs- both, authored and contributed volumes to address the growing need for information. The fields of biological and medical physics and biomedical engineering are broad, multidisciplinary and dynamic. They lie at the crossroads of frontier research in physics, biology, chemistry, and medicine.

Books in the series emphasize established and emergent areas of science including molecular, membrane, and mathematical biophysics; photosynthetic energy harvesting and conversion; information processing; physical principles of genetics; sensory communications; automata networks, neural networks, and cellular automata. Equally important is coverage of applied aspects of biological and medical physics and biomedical engineering such as molecular electronic components and devices, biosensors, medicine, imaging, physical principles of renewable energy production, advanced prostheses, and environmental control and engineering.

Nirmal Mazumder • Hemanth Noothalapati
T. M. Sampath Udeni Gunathilake
Editors

Vibrational Spectroscopy for Biopolymer Characterization

Springer

Editors
Nirmal Mazumder
Department of Biophysics
Manipal Academy of Higher Education
Manipal, Karnataka, India

Hemanth Noothalapati
Faculty of Life and Environmental Sciences
Shimane University
Matsue, Japan

T. M. Sampath Udeni Gunathilake
Department of Polymer Science
University of Sri Jayewardenepura
Nugegoda, Sri Lanka

Editor-in-Chief
Bernard S. Gerstman

ISSN 1618-7210 ISSN 2197-5647 (electronic)
Biological and Medical Physics, Biomedical Engineering
ISBN 978-3-032-15305-0 ISBN 978-3-032-15306-7 (eBook)
https://doi.org/10.1007/978-3-032-15306-7

This Springer imprint is published by the registered company Springer Nature Switzerland AG
The registered company address is: Gewerbestrasse 11, 6330 Cham, Switzerland

Contents

Vibrational Spectroscopic Characterization of Chitosan

Medard Rukshika Fernando and T. M. Sampath Udeni Gunathilake

Abstract Chitosan, derived from chitin—the second most abundant natural biopolymer after cellulose—stands out as a sustainable and multifunctional material with wide-ranging industrial and biomedical applications. Characterized by biodegradability, biocompatibility, and non-toxicity, chitosan serves as an eco-friendly alternative to synthetic polymers. Its precursor, chitin, is mainly obtained from crustacean shells and fungal cell walls, consisting of N-acetyl glucosamine monomers arranged in α, β, and γ polymorphic forms. Through deacetylation, chitin is converted into chitosan, a β-(1,4)-linked D-glucosamine polymer with improved solubility and reactivity. This transformation enhances its usability in fields such as drug delivery, agriculture, wound healing, and water purification. Chemical modifications—such as N-substitution for antimicrobial properties, O-substitution for solubility enhancement, crosslinking for hydrogel formation, and grafting for improved adsorption—further expand its potential. To optimize such modifications, precise characterization is achieved through vibrational spectroscopy techniques, including infrared (IR) and Raman spectroscopy, which provide insights into structural and chemical features. Owing to its exceptional tunability, chitosan is increasingly recognized as a transformative biopolymer suited for biomedical, environmental, and food applications. Continued research combining chemical modification and spectroscopic analysis can further optimize chitosan's structure–function relationships, establishing it as a cornerstone in sustainable material science and biotechnological innovation.

Keywords Chitosan · Biodegradability · Spectroscopy

M. R. Fernando · T. M. S. U. Gunathilake (✉)
Faculty of Applied Science, Department of Polymer Science, University of Sri Jayewardenepura, Nugegoda, Sri Lanka
e-mail: sgunathilake@sjp.ac.lk

N. Mazumder et al. (eds.), *Vibrational Spectroscopy for Biopolymer Characterization*, Biological and Medical Physics, Biomedical Engineering,
https://doi.org/10.1007/978-3-032-15306-7_1

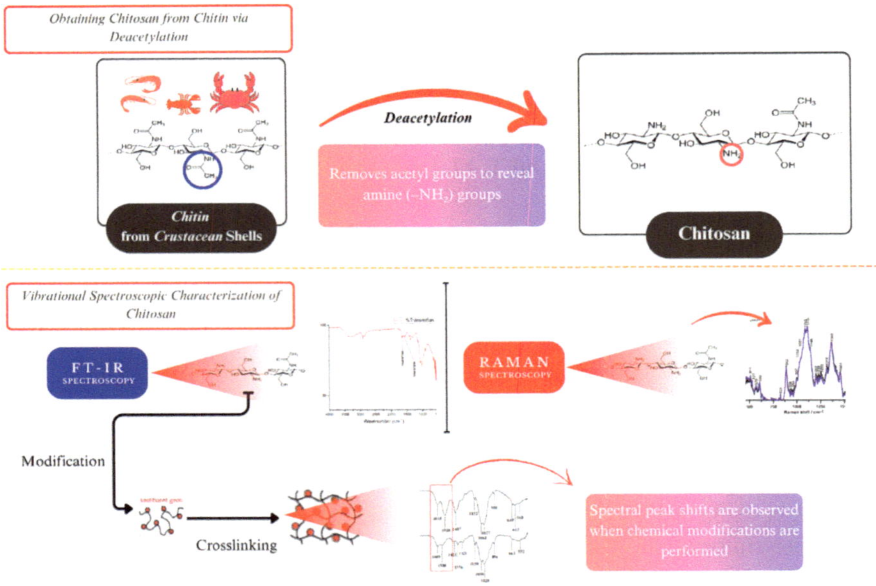

1 Introduction

Chitosan, a derivative from Chitin, the second most abundant natural biopolymer sitting right after cellulose, is a possible cornerstone of scientific innovation across various fields of application. The unique combination of biodegradability, biocompatibility and nontoxic nature helps Chitosan stand out as an attractive resource in addressing modern challenges faced in critical fields such as biomedicine, agriculture and food technology. Unlike existing synthetic polymer alternatives, which often pose environmental as well as health hazards, Chitosan stands out as an environmentally sustainable polymer derived from natural sources. Through research studies it has been proven that this remarkable polymer is not just another passive material but a transformative polymer whose applications extend from biopesticides to hydrogels which help in sustainable agriculture and wound healing respectively, thus embodying the potential of natural biopolymers [1]. To understand and appreciate the versatility of chitosan, it is essential to understand the biological background from which its precursor chitin is acquired.

Chitin is a natural biopolymer which is widely distributed across a plethora of biological systems. It is primarily found in the exoskeletons of arthropods such as crustaceans in which chitin is accounted for approximately 20–30% of the total tissue composition [2]. Chitin can also commonly be found in cell walls of certain

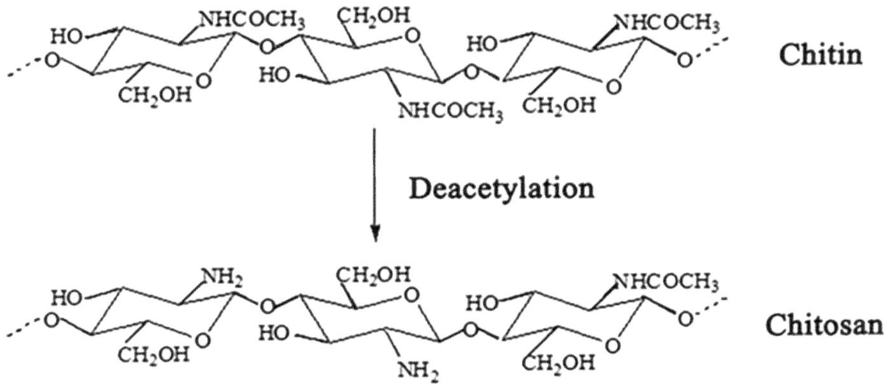

Fig. 1 Illustrates the deacetylation process, where acetyl groups in chitin are replaced with hydroxyl and amino groups, producing chitosan [5]

fungi species (*Zygomycetes, Basidiomycetes*, etc.) [3], insect cuticles and the structural frameworks of mollusks [4]. When considering chitin on a molecular level, it consists of repeating N-acetyl glucosamine (NAG) monomer units. Chitin as a polymer exists in three known polymorphic forms in nature, categorized as α, β and γ. Each polymorphic form exists in unique arrangements of the molecular chains thus resulting in unique structural features depending on the source of acquired chitin [2]. Chitin can be transformed into chitosan through the process of deacetylation (Fig. 1). Chitosan consists of β-(1,4)-linked D-glucosamine and N-acetyl-D-glucosamine units. While chitin itself is a well-structured linear polysaccharide that offers robust properties, however its direct applications are very limited due to the presence of acetyl groups. Thus, via the process of deacetylation, these acetyl groups are converted into hydroxyl and amino functional groups. This product is known as Chitosan which offers better solubility, reactivity and other features thus outshining chitin. This specific transformation helps broaden its overall usability as a functional biopolymer [2, 6].

While chitosan's intrinsic properties make it an exceptional polymer, by chemical modifications and extensive testing, the true hidden potential of this polymer can be unlocked. Chemical modifications such as N-substitution (e.g., quaternary ammonium chitosan for antimicrobial activity), O-substitution (e.g., carboxymethylation to improve solubility), crosslinking for applications such as hydrogels for controlled drug release and grafting for applications such enhancing absorptivity in water treatment proves to significantly enhance chitosan's structural and functional properties in unique manner for each process [7, 8].

To ensure the overall success of such modifications and to replicate reliable results, precise and accurate analysis of this material is crucial. Thus, highlighting the need for precise characterization techniques such as spectroscopy. Vibrational spectroscopy circumscribing Infrared (IR) and Raman Spectroscopy stands out as a complementary set of valuable tools for unraveling chitosan's structural and chemical properties. The above-mentioned techniques offer a well-balanced advantage in detecting functional groups and analyzing chitosan's molecular vibrations [9, 10].

The combination of unique properties such as biodegradability and biocompatibility accentuates the superiority of chitosan as an alternative to synthetic polymers with wide ranging applications, highlighting its remarkable potential of natural biopolymers as versatile and sustainable materials. Thus, the pressing need for further research on this remarkable polymer persists where the crucial factors focus on understanding its inherent material properties. As research continues, the application of advanced analytical techniques, such as vibrational spectroscopy, will play a crucial role in optimizing chitosan for targeted uses.

This paper aims to discuss and explore the structural and molecular characteristics of chitosan, its derivation from natural sources and its diversified applications in different fields with a particular focus on how vibrational spectroscopic methods can be employed to understand and optimize this biopolymer in applications.

1.1 Core IR Characteristic Bands of Chitosan

When considering analysis of Chitin and Chitosan natural polymers, Infrared(IR) spectroscopy is a an essential tool [11]. IR spectroscopy provides essential insights into these polymers which are crucial, such as degree of deacetylation, molecular structures and intermolecular interactions. When specifically considering applications of IR spectroscopy for characterizing Chitosan and Chitin, key structural properties are reflected by Amide groups, amine and hydroxyl vibrations along with polysaccharide backbone. These features give rise to characteristic absorption bands (Fig. 2).

The Amide I band which appears within the range of 1650 cm^{-1} [12], is referenced to the C=O stretching found in the acetyl groups of N-acetylglucosamine units. But this exact position of the band can shift depending on the source of chitin according to the study conducted by *Kumirska* [11]. It is shown that in alpha Chitin samples the Amide I band was found at a slightly higher wavenumber ~1660 cm^{-1}. This is attributed to the higher crystallinity which is found in alpha chitin when compared to beta chitin. Beta chitin has more open and less closely bonded structures (less H-Bonds) Thus, beta chitin shows the above band within the range of ~1620–1640 cm^{-1}. The intensity of these bands is directly correlated with the degree of deacetylation (DD), weakening as Chitin undergoes deacetylation. As shown, variations and shifts of this band indicate changes in strong intermolecular interactions such as hydrogen bonding. When considering N–H bending and C–H stretching in these polymers, the characteristic band Amide II is found near the range of ~1550 cm^{-1} [13]. This band is also subjected to variations and shifts based on the

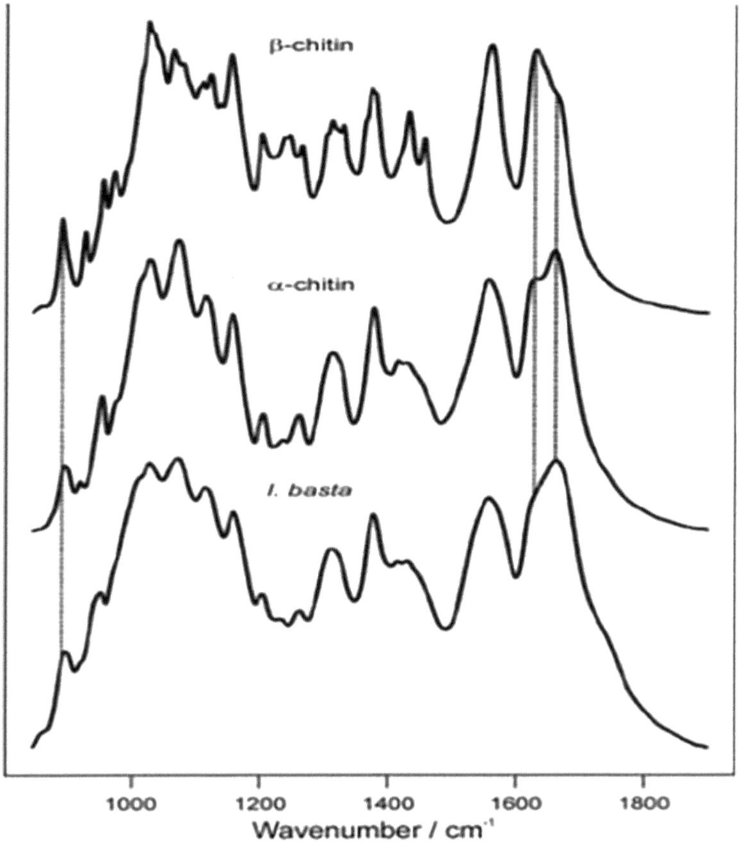

Fig. 2 Variation and shifts of Amide I and II bands in various sources of Chitin [11]

source of chitin. Due to the strong inter sheet H-bonding found in alpha chitin compared to beta chitin the Amide band II is found closer to 1565 cm^{-1} while in beta chitin it is shifted to ~1550 cm^{-1}. This observation reflects a much more relaxed polymeric arrangement in beta chitin networks [11]. Unlike the Amide I band, the Amide II band is more persistent and noticeable in more deacetylated chitosan due to the presence of unconsumed Amide groups and other amine functionalities.

It is found that the NH_2 scissoring motion which occurs between the range of 1590 to 1570 cm^{-1} overlaps with the Amide II band thus making it difficult to distinguish [14]. However, this spectral feature remains important in analyzing partially deacetylated chitosan. Since the presence of primary amines marks deacetylated chain segments. The broad N–H and O–H stretching which is noticeable in the range of 3200–3400 cm^{-1} range signifies hydrogen bonding interactions within the polymers [14]. Shifts of these peaks can indicate intra and intermolecular interactions where stronger H-bonding generally results in lower wavenumber shifts. Thus, it affects the ability of chitosan to retain moisture and its film forming properties. These interactions are key players in determining the hydrophilicity and solubility of chitosan-based materials. Greatly influencing its potential applications.

In practical applications of Chitosan, its ability to interact with different substances through interactions such as H-bonding and electrostatic forces has been demonstrated particularly in membrane technologies in water filtration and scaffolds used in biomedical applications [15]. When considering the composite forms of chitosan in cases such as incorporating fillers (clay, cellulose etc.) or synthetic polymers, modifications in the inherent spectral features are observed. These modifications are especially quite evident in Amide I and II bands where new bonding interactions give rise to structural rearrangements. When changes occur, intensity variations and shifts of the peaks are observed. These changes can essentially indicate specific bonding changes and alterations or electrostatic interactions with composite fillers. These alterations influence mechanical strength, moisture retention, and degradability.

When fine tuning properties which are to be expected, FTIR analysis has thus proven its worth by providing valuable insight. Furthermore by analyzing IR spectral features, researchers have successfully determined the impact of processing conditions such as pH level, thermal and solvent system effects on the polymer thus providing means to delicately cater properties for specific usage scenarios [15].

Overall, the IR spectral characteristics of chitosan serves an important role as an analytical too for assessing crucial properties [11, 12]. Observed differences in spectral features in Amide I and II bands between alpha and beta chitin samples highlights how FTIR can provide insight into structural variations within sources of chitin. Vital data such as the source of raw materials can be specified using the above discussed methods so that industrial optimization can be done effectively for specific usage scenarios. Chitosan continues to be a subject of extensive research, aiming to enhance its efficacy in diverse scientific and industrial domains.

1.2 Spectral Features and Characteristics of Raman Spectroscopy

Raman spectroscopy is a vibrational spectroscopic technique which is a powerful tool in analyzing spectral features of Chitosan. It stands along with IR spectroscopic techniques (FTIR) providing valuable insights. This spectroscopic technique excels in providing molecular level insights into structural and functional modifications of chitosan, which have been demonstrated in several studies. These studies were focused on vibrational modes, crystallinity and molecular interactions of the chitosan polymer [16]. Considering the unique advantages of using Raman spectroscopy over (IR), its lower interference from water is a major factor. This makes Raman spectroscopy more suitable for studying hydrated chitosan or biological gel matrices over conventional IR spectroscopic techniques [17]. Due to the inherent nature of Raman spectroscopy being based on polarizability of molecules rather than being based on the change of dipole moment like FTIR [18], Raman spectra often exhibit sharper peaks in certain areas. Particularly in non-polar or weakly hydrogen bonded systems [19]. Thus the ability to differentiate chitosan's chemical modification in various applications are enhanced [16].

Raman spectroscopy can be instrumental in applications such as monitoring cross linked chitosan systems due to its higher sensitivity to molecular interactions and structural changes. Currently Raman spectroscopy is widely employed in distinguishing ionic and covalent crosslinking mechanisms as discussed by Souza et al. The findings were based on distinct spectral shifts that were observed when Oxo-Carbon derivatives were introduced to modify the backbone of Chitosan [16]. Several vibrational bands are fundamental when the key spectral features of Raman spectra of Chitosan are considered. Amide I (~1650–1630 cm^{-1}) which corelates with C=O stretching vibrations, commonly employed to analyze the secondary structure of Chitosan. Shifts which occur in this region can indicate hydrogen bonding and molecular interactions in Chitosan based samples. The Amide III (~1260–1240 cm^{-1}) corresponds to chitosan's conformations as well as its secondary structure as reported by [20]. This region is often used to analyze and evaluate modifications involving amide groups or functionalized derivatives. Ring breathing modes serve as a unique fingerprint for polysaccharides, representing chitosan's backbone structure and are essential in distinguishing chitosan from other biopolymers [21]. When interactions with composites and other materials such as nanoparticles are observed and assessed, CH2/CH3 deformations which are typically found within (~1450 cm^{-1}) provides valuable insight [19].

Although there are notable advantages of employing Raman spectroscopy, Fluorescence interference and low scattering efficiency poses a significant challenge. Several techniques have been developed to minimize these effects thus improving sensitivity and overall applicability in Chitosan research (Tables 1 and 2). One such significant advancement is Surface-Enhanced Raman Scattering (SERS). This method employs metallic nanoparticles to amplify weak Raman signals [17].

Table 1 Spectroscopic techniques for chitosan analysis: a comparative overview of raman and FTIR

Aspect	Raman spectroscopy	FTIR spectroscopy
Primary focus	Structural characterization, functional group identification, crosslinking, crystallinity analysis	Functional group verification, polymer modifications, deacetylation analysis
Key spectral features	Amide I (~1650 cm^{-1}) (protein structure, crosslinking) [16]	C=O stretching (~1650 cm^{-1}) (amide presence) [22]
	Ring breathing modes (724–736 cm^{-1}) (chitosan backbone) [16]	NH bending (~1550 cm^{-1}) (deacetylation monitoring) [23]
	C=O stretching (1700–1800 cm^{-1}) (crosslinking confirmation) [16]	Glycosidic bonds (1100–1200 cm^{-1}) (polysaccharide structure) [20]
	OH (3362 cm^{-1}), CH3 (2932 cm^{-1}) (hydroxyl and methyl groups)	OH groups (~1350 cm^{-1}) (hydroxyl presence)
Advantages	Highly specific to molecular vibrations [24]	Easier peak assignments for functional groups [25]
	Provides detailed structural information [20]	More quantitative (e.g., degree of deacetylation) [14]
	Effective for analyzing crystalline and amorphous forms [26]	Works well for bulk material analysis [27]
	Suitable for aqueous environments [17]	Less affected by fluorescence interference [22]
Limitations	Fluorescence interference affects weak signals [19]	Limited structural resolution compared to Raman [25].
	Lower sensitivity to some functional groups	Can be challenging for overlapping peaks [22]
	More expensive and complex instrumentation [24]	Requires complementary methods (XRD, UV-vis) for full characterization [14]
	Requires special sample preparation (e.g., SERS for enhanced signals) [17]	–
Special techniques used	SERS (surface-enhanced Raman spectroscopy): Increases sensitivity for trace analysis [28]	FTIR combined with XRD and UV-vis: For enhanced structural insights [14]
	SVD-based background correction: Reduces fluorescence interference [19]	Regression models: Used for quantitative chitosan deacetylation analysis [27]
	PCA & PLS regression: Enhances spectral reproducibility [19]	–
Application examples	Crosslinking studies (ionic vs. covalent bonding) [16]	Identification of chitosan purity and modifications [25]
	Chitosan-based composites and biomaterials [26]	Deacetylation quantification for biomedical applications [14]
	Rapid, non-destructive analysis for food safety (SERS for pesticide detection) [17]	Used for composite material analysis [27]

Table 2 Literature-based confidence assessment of FTIR and Raman applications in chitosan characterization

Aspect	Confidence level	Reasoning
FTIR is effective for functional group identification	High	Multiple studies consistently used FTIR for identifying chitosan's key functional groups [14, 25, 27]
Raman provides better structural insights	High	Studies such as Souza et al. [16] and Babatunde et al. [26] show that Raman captures molecular interactions and crystallinity better than FTIR
FTIR is useful for quantifying deacetylation	High	Multiple studies show deacetylation quantification via FTIR, often validated using XRD [14, 22]
Fluorescence interference is a major issue in Raman	Medium-high	Beattie et al. discuss fluorescence, but some papers mitigate this using computational corrections (SVD, PCA, PLS regression, SERS) [19]
SERS significantly enhances Raman sensitivity	High	Zhang et al. provides strong evidence that SERS lowers the detection limit to ppt levels, proving its effectiveness [28]
FTIR and Raman are complementary techniques	High	Many studies use both methods to validate chitosan modifications and structure [16, 20, 26]
Raman is more sensitive to crosslinking and molecular interactions	Medium-high	Studies show Raman detects crosslinking shifts, but not all modifications show significant Raman changes [16, 26]
Computational corrections (SVD, PCA) improve Raman reproducibility	Medium	Beattie et al. presents compelling evidence, but more studies are needed to validate its broad applicability in Chitosan research [19]
FTIR struggles with overlapping peaks, requiring complementary techniques	High	Studies emphasize that overlapping bands require XRD or other methods for complete interpretation [14, 22]
Raman is becoming more widely used in recent research	Medium-high	The dataset shows a rise in Raman-based studies post-2020, but FTIR remains dominant in broader chitosan research

2 Quantification of Degree of Deacetylation Via IR/RAMAN Spectroscopic Methods

The degree of deacetylation is a crucial characteristic determiner when considering Chitosan. The degree of deacetylation (DD) defines the balance between N-acetylglucosamine and D-glucosamine units within the polymer backbone. Physiochemical properties of Chitosan ranging from aqueous solubility, viscosity, crystallinity and many other factors are directly governed by the balance of above-mentioned units within the polymer. A higher DD corresponds to an increased density of free amino groups thus enhancing the cationic character of the overall polymer. Thus, precise and reproducible quantification of DD is essential as it governs how the polymer perform across various applications in a variety of fields such as biomedical, environmental as well as food related applications.

Commonly employed traditional techniques in quantifying DD include potentiometric and conductometric titrations, elemental analysis and nuclear magnetic resonance spectroscopy. Even though these methods are reliable and analytically robust, they are either destructive, time intensive or even heavily reliant on sample solubility and sample purity. Vibrational spectroscopic analysis on the other hand offers faster nondestructive alternative techniques which can be employed on both solid and solution states with minimal sample preparation. However, these techniques are not without challenges. Common reported issues are spectral overlaps, baseline drifts and hydration effects along with instrumental variations. These factors can induce uncertainties if not thoroughly controlled. The study by Weißpflog et al., 2021 stands out as an extensive study done on this subject in a methodologically rigorous and comprehensive way [29].

In their study, they characterized three main commercially available samples of DD values of 75%, 85% and 90%. These samples are labeled Ch75, Ch85 and Ch90 respectively. To extensively study the DD as the key variable they kept the viscosities of all samples set at 200 mPa.s thus excluding molecular weight as a potential source of confusion when studying the spectral data. FTIR spectra were recorded in ATR mode using solid state flakes and dissolved chitosan, allowing comparison between spectral behavior in different phases.

Three sophisticated FTIR methods were developed to estimate DD each offering several advantages and disadvantages over each other. Peak Height Ratio Method which is the most trivial out of the three compares intensity of the amide-I band (νC=O at ~1652 cm^{-1} in solids and ~1647 cm^{-1} in solution) to a relatively stable C–H stretch band (~2848–2875 cm^{-1}). This was done after preprocessing the data (normalizing to the saccharide ring stretch at 1024 cm^{-1} (solid) or 1033 cm^{-1} (dissolved)). This method helped to minimize variations across different thicknesses or concentrations. DD was calculated via the below empirical formula.

$$DD_{FTIR}(\%) = 100 - \left(\frac{A_{\nu(C=O)}}{A_{\nu(C-H)}} \right) \times \frac{100}{1.33} \tag{1}$$

The key here is that the Amide-I band diminishes as the polymer increasingly becomes deacetylated thus reflecting the decrement of N-acetyl groups in the polymer backbone. The C–H band which is tied to the glucopyranose backbone remains unaffected throughout the process of deacetylation and thus serves as a reliable reference. While straightforward and rapid, this method has a major drawback which is hydration and baseline shifts making it prone to unreliable results if not controlled and monitored closely. It was found that the sensitivity to these factors were amplified in solid samples due to distortions in peak shapes caused by hydrogen bonding and residual moisture.

The second method introduced by the authors was the Peak Maxima Matrix Approach, which employs a multivariate regression model incorporating the normalized heights of three key spectral features: the amide I band (νC=O), the δ(NH) bending vibration (~1548 cm^{-1}), and the ρ(CH) wagging vibration (~1374 cm^{-1}). Each peak was normalized relative to the C–O–C polysaccharide band (~1024 cm^{-1}) due to this bond being practically unaffected by deacetylation, and their values were inserted into a linear equation derived through Gauss-Jordan matrix inversion

$$DD = -0.67 \bullet x_{\nu(C=O)} + 4.85 \bullet y_{\delta(NH)} - 1.36 \bullet z_{\rho(CH)} \quad (2)$$

The reason for the higher reliability and robustness of this method is due to the use of multiple informative bands unlike in the peak height ratio method. When deacetylation is proceeded with δ(NH) of the spectra will increase due to higher primary ammine content while νC=O reflects the residual acetyl content. Here ρ(CH) acts as an independent variable due to it being less influenced by deacetylation. However, its vibration environment can reflect some shifts due to changes in ring conformation, hydrogen bonding or crystalline order due to deacetylation. Thus, including ρ(CH) can reduce overreliance on a single given band and it will help in averaging out spectral noise in practical applications. The given regression coefficients are empirically fitted to referenced DD data. This method tends to be more labor intensive when compared to the previous method, but it actively enhances reliability by compensating for any local distortions or noise in any peak.

The most advanced method introduced by the author was the Peak Deconvolution Method. Here a pseudo-Voight profile was employed which is a linear combination of Gaussian and Lorentzian components. The reason for using this model was to separate the Amide I, II and III signals which typically exhibit superposition thus broadening in FTIR spectra, especially in solid state (~1647, 1548, and 1374 cm^{-1}, respectively). Once Deconvolution was performed the peak areas of the deconvoluted bands were then employed to calculate intensity ratios which were correlated with independently measured HNMR and elemental analysis values of DD. The models yielded R^2 values of more than 0.98, thus demonstrating considerably high accuracy. This method allowed precise DD estimation.

Even though Weißpflog et al. [29] did not specifically mentioned about Raman spectral analysis, Raman based DD estimation is increasingly validated in literature. Stephan et al. and Zając et al. have demonstrated that Raman spectroscopy when combined with first derivative spectra or deconvoluted spectra can produce predictions for DD values using ratios such as ν(C=O)/ν(C–O–C), 1591/896 cm^{-1}, and $\delta NH_2/\rho CH_2$. Reported R^2 value for this experimental method was 0.9962 when calibrated against certified DD values obtained via elemental analysis [30, 31]. Here a distinct advantage in using Raman can be mentioned as its insensitivity towards moisture thus reducing interference. However, Raman method is not without its own drawbacks such as susceptibility to fluorescence and weak Amide scattering.

Considering estimation of DD, several spectroscopic challenges persist thus limiting its overall potential in applications. Band overlap within the region of 1500–1700 cm^{-1} region, particularly between amide I and water OH/NH vibrations, remains a critical obstacle. Baseline drift, especially in ATR mode, can significantly skew peak height or area calculations if not corrected. Hydration effects further complicated interpretation. As absorbed moisture essentially broadens bands and alters relative intensities. However, among the available methods, the Peak

Deconvolution method introduced by Weißpflog et al. stands out as a practical and an applicable solution to some of the above discussed limitations [29].

Notable comprehensive and methodological study was also conducted by Rehman et al. [9] discussing both Raman and FTIR approaches. Their work introduces a fully automated vibrational spectroscopic platform with a high throughput. This platform integrates ATR-FTIR and FT-Raman techniques. Authors developed and validated robust partial least squares regression (PLSR) models which were calibrated against proton NMR values, thus ensuring a higher degree of analytical robustness. The study was also focused on how practical factors such as the physical state of a sample and the choice of normalization method can affect the accuracy of the overall measurement. By addressing these practical factors, the study provides a strong methodological foundation for the synergistic use of Raman and FTIR spectroscopy in high-throughput, standardized chitosan analysis workflows.

A foundational study by Brugnerotto et al. [32] complements this work by validating FTIR-based DD quantification. The specialty is that the study was conducted across a wide range of chitin and chitosan sources, including α- and β-polymorphs from different origins. Their analysis compared transmission, DRIFTS, and ATR modes. The study concluded that the A1320/A1420 ratio offered superior linearity ($r = 0.99$) and moisture resistance compared to amide-based ratios. Their work is still standard for method selection and instrument-specific restrictions, even though it predates more sophisticated deconvolution techniques. In particular, its variable penetration depth discourages the use of ATR mode for film analysis.

In conclusion, FTIR and Raman spectral methods offer accessible yet robust methods for estimation of DD in practical settings. While simpler methods can offer quick approximations, regression-based matrices and deconvoluted peak fitting provides higher accuracy. When considering solid state formulations Raman spectroscopy proves to be a valuable tool even though it is currently underutilized if paired with rigorous calibration and preprocessing. Ultimately a hybrid spectroscopic approach proves to be the most promising path for standardized and reproducible results in quantitative analysis of DD in Chitosan based polymers.

3 Analysis of Chemical Modifications in Chitosan

Property enhancement through chemical and physical modifications in Chitosan proves its immense potential as a structurally versatile biopolymer. Modifications can be tailored to alter solubility, bioactivity, mechanical properties and functional versatility thus widening applicability of Chitosan and Chitosan incorporated materials in biomedical, environmental as well as other industrial fields. Confirming and characterizing these modifications are crucial since the repeatability and authenticity of modifications are a must have feature. Vibrational spectroscopic techniques such as FTIR and Raman spectroscopy serve as pivotal tools in non-destructive analysis of such modifications. As discussed earlier, FTIR remains central for detecting polar functional groups, thus monitoring deacetylation and monitoring molecular interactions in modified chitosan while Raman spectroscopy complements FTIR along with its heightened sensitivity towards carbon backbone vibrations and resistance to interferences by hydration. Enhanced techniques such as Surface Enhanced Raman Scattering (SERS) further amplifies weaker Raman signals thus maximizing its applicability. Especially in applications where nanostructured composites are utilized.

Enhancing water solubility and bioactivity of Chitosan by introducing sulfate groups is such a way of modification. This was studied by Zhang [28] by employing FT-Raman spectroscopy to quantify the Degree of Sulfation (DS) in Chitosan sulfate derivatives. Material was observed for newly formed Raman bands at 1070 cm^{-1}, and 823–834 cm^{-1} which represents (O=S=O) and (C–O–S) stretching vibrations respectively. Alongside with disappearance of glycosidic backbone signals and achieving a quantitative DS estimation through intensity ratios. Karthik et al. [33] further confirmed monitoring sulfation via vibrational spectroscopy by employing FTIR and FT-Raman by detecting S–O stretching 1262 cm^{-1} and C–O–S modes at 838 cm^{-1} and 1110.1 cm^{-1}, respectively.

Phosphorylation of Chitosan represents a chemical modification of introducing phosphate groups onto the backbone of the Chitosan polymer, thus enhancing its hydrophilicity, chelation characteristics and potentially biomedical applications. Phosphorylation occurs at the amino or the hydroxyl groups of the Chitosan chain. The pattern of substitution directly affects and influences the physicochemical behavior of the polymer as a result. This was studied by Wang and Liu [34] through Diffuse Reflectance Infrared Fourier Transform Spectroscopy (DRIFT). The findings indicated that synthesis routes targeting amino groups produced N-P bonds which could be characterized by new FTIR peaks formed at 1216 and 1234 cm^{-1}. However, phosphorylation of hydroxyl groups yielded C–O–P linkages which was indicated by a distinctive peak at 1084 cm^{-1}. The widespread substitution was confirmed by red shifts in the amide bands at 1533 cm^{-1} and 1415 cm^{-1} as well as the elimination of N–H bending vibrations. Thus, these experimental results confirm the ability of vibrational spectroscopy to be employed as a tool in characterizing and quantifying phosphorylation at nitrogen and oxygen sites in a distinct way.

Crosslinking is a very practical method when considering strengthening polymers via three dimensional networks. This can be effectively identified and characterized via vibrational spectroscopy, particularly FTIR analysis of bond evolution. Sampath et al. confirmed the formation of covalent bonds in chitosan (C=N) crosslinked with glutaraldehyde. An increased absorption at 1548 cm^{-1}. significant shoulders were discovered at 1548 and 1635 cm^{-1} indicating C=C and aldehyde functional groups respectively [35]. Additionally, an increase near 2910 cm^{-1} was reported which was associated with C–H stretching. This indicated alterations in hydrophobicity associated with crosslinking. This understanding was further analyzed and extended by [36]. He analyzed several chitosan hydrogels crosslinked with various aldehydes thus reporting a systematic decrease in the NH_2 band intensity (1560–1580 cm^{-1}) and subtle shifts around 1656 cm^{-1}. These results were in line with the formation of Schiff bases. Their results concluded that glutaraldehyde promoted more robust crosslinking when compared to formaldehyde or acetaldehyde due to its dual carbonyl functionality. These studies clearly demonstrate that vibrational spectroscopy provides a clear and straightforward approach in verifying

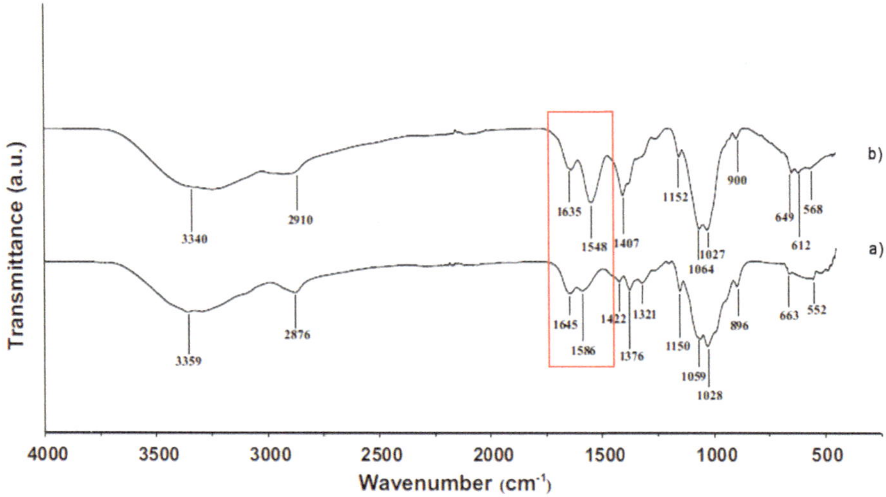

Fig. 3 FTIR spectrum of (**a**) pristine chitosan and (**b**) chitosan crosslinked with glutaraldehyde [35]

the occurrence and degree of chemical crosslinking in Chitosan matrixes. By monitoring the position and intensities of imine, amine and carbonyl bands. However, [37] applied FTIR to prove that physical crosslinking in Chi-TEOS (chitosan and tetraethyl orthosilicate) hydrogels yields no new covalent bonds thus emphasizing the role of vibrational spectroscopy in distinguishing between physical and chemical crosslinking processes (Fig. 3).

Chemical modifications such as sulfation, phosphorylation and crosslinking significantly expand the functionality of Chitosan by tailoring its properties for diverse applications. When considering tools for the verification of the above-mentioned chemical processes, vibrational spectroscopy proves to be indispensable by providing direct molecular level evidence of bond formation, dissociation and substitute patterns. Further, these tools not only allow the quantification of degrees of substitution, ensuring the reproducibility of results, but also play a central role in the rational design and validation of chemically engineered chitosan derivatives as the demand for advanced biopolymers continues to grow.

4 Conclusion and Future Directions

Vibrational spectroscopy has proven to be essential when considering chitosan characterization. While FTIR offers higher sensitivity to polar groups and deacetylation, Raman offers structural insights especially in hydrated or other complex systems. Spectral markers such as Amide I/II, OH/NH stretches, and glycosidic bonds are key features which enable reliable and reproducible structural assessments. DD quantification has evolved from simple ratios to robust multivariate models which are validated against NMR and other reliable analytical methods. With emerging modern tools and innovations like SERS for trace detection and FTIR in 3D-print scaffolds enhance functionality, while aspects of machine learning boosts spectral interpretation. Vibrational spectroscopy was found to be useful in verifying modifications (e.g., sulfation, phosphorylation and crosslinking) through analyzing characteristic peaks and their shifts. However, challenges persist, especially considering spectral overlaps, hydration effects, and baseline drifts, which affects reproducibility and consistency. Addressing these requires a rigorous framework that employs spectral normalization, deconvolution, internal referencing, and cross-validation with complementary techniques like NMR and titration.

Future directions lie in the way of combining vibrational spectroscopy and machine learning based tools to achieve standardized and application specific characterization. In conclusion, vibrational spectroscopy stands not just as a method of analysis but as a guiding tool for the future of polymer science, enabling more targeted, sustainable, and intelligent use of this versatile biopolymer.

References

1. Pillai CKS, Paul W, Sharma CP (2009) Chitin and chitosan polymers: chemistry, solubility and fiber formation. Prog Polym Sci 34(7):641–678
2. Kumari S, Kishor R (2020) Chitin and chitosan: origin, properties, and applications. In: Handbook of chitin and chitosan: volume 1: preparation and properties. Elsevier, Amsterdam, pp 1–33
3. Abo Elsoud MM, El Kady EM (2019) Current trends in fungal biosynthesis of chitin and chitosan. Bull Natl Res Cent 43(1):59
4. Saenz-Mendoza AI et al (2023) Insects as a potential source of chitin and chitosan: physicochemical, morphological and structural characterization. A review. Emirates J Food Agric 35(5):388–407
5. AlMohammed HI et al (2021) Chitosan-based nanomaterials as valuable sources of anti-leishmanial agents: a systematic review. Nano 11(3):689
6. Aranaz I et al (2021) Chitosan: an overview of its properties and applications. Polymers 13(19):3256
7. Ji J et al (2014) Chemical modifications of chitosan and its applications. Polym-Plast Technol Eng 53(14):1494–1505
8. Madera-Santana TJ, Herrera-Méndez CH, Rodríguez-Núñez JR (2018) An overview of the chemical modifications of chitosan and their advantages. Green Mater 6(4):131–142
9. Rehman HU et al (2023) High-throughput vibrational spectroscopy methods for determination of degree of acetylation for chitin and chitosan. Carbohydr Polym 302:120428
10. Bumgardner JD et al (2017) Characterization of chitosan matters. Chitosan Based Biomater Fundamentals 1:81–114
11. Kumirska J et al (2010) Application of spectroscopic methods for structural analysis of chitin and chitosan. Mar Drugs 8:1567–1636
12. Duarte ML et al (2002) An optimised method to determine the degree of acetylation of chitin and chitosan by FTIR spectroscopy. Int J Biol Macromol 31:1–8
13. Branca C et al (2016) Role of the OH and NH vibrational groups in polysaccharide-nanocomposite interactions: a FTIR-ATR study on chitosan and chitosan/clay films. Polymer 99:614–622

14. Negrea P et al (2015) The study of infrared spectrum of chitin and chitosan extract as potential sources of biomass. Dig J Nanomater Biostruct 10(4):1129–1138
15. Trung TS et al (2006) Functional characteristics of shrimp chitosan and its membranes as affected by the degree of deacetylation. Bioresour Technol 97(4):659–663
16. Souza NLGD et al (2015) Synthesis, vibrational spectroscopic and thermal properties of oxocarbon cross-linked chitosan. J Braz Chem Soc. https://doi.org/10.5935/0103-5053.20150090
17. Zhang Q et al (2022) Synthesis of a 3D Ag-decorated chitosan film as a simple and stable flexible SERS substrate for the detection of pesticides in food. ACS Agric Sci Technol 2(2):323–329
18. Smith B (2011) Fundamentals of Fourier transform infrared spectroscopy, 2nd edn. CRC Press, Boca Raton
19. Beattie JR, McGarvey JJ (2012) Estimation of signal backgrounds on multivariate loadings improves model generation in face of complex variation in backgrounds and constituents. J Raman Spectrosc 44(2):329–338
20. Varma R et al (2021) Synthesis and physicochemical characteristics of chitosan extracted from pinna deltoides. Lett Appl NanoBioSci 11:4061–4070
21. Gieroba B et al (2023) Application of vibrational spectroscopic techniques in the study of the natural polysaccharides and their cross-linking process. Int J Mol Sci 24(3):2630
22. Furuhashi T et al (2009) Molluscan shell evolution with review of shell calcification hypothesis. Comp Biochem Physiol B: Biochem Mol Biol 154(3):351–371
23. Barroso-Solares S et al (2025) Chitin and chitosan quantification in fungal cell wall via Raman spectroscopy. Spectrochim Acta A Mol Biomol Spectrosc 334:125928
24. Priya G et al (2024) Modern analytical approach in biopolymer characterization. Phys Sci Rev 9(3):1149–1170
25. Yasmeen S et al (2016) Chromium (VI) ions removal from tannery effluent using chitosan-microcrystalline cellulose composite as adsorbent. Int Res J Pure Appl Chem 10(4):1–14
26. Babatunde EO et al (2020) Market structure and economic analysis of timber business in oluyole local government of Oyo state. Nigerian J Agric Food Environ 16(2):42–50
27. Abdel Aziz M et al (2014) FTIR spectroscopy of natural bio-polymers blends. Middle East J Appl Sci 4:816–824
28. Zhang K et al (2011) FT Raman investigation of novel chitosan sulfates exhibiting osteogenic capacity. Carbohydr Polym 83(1):60–65
29. Weißpflog J et al (2021) Characterization of chitosan with different degree of deacetylation and equal viscosity in dissolved and solid state—insights by various complimentary methods. Int J Biol Macromol 171:242–261
30. Stephan B et al (2012) Determination of the degree of N-acetylation (DA) of chitin and chitosan in the presence of water by first derivative ATR FTIR spectroscopy. Carbohydr Polym 87:117–122
31. Zajac A et al (2015) Determination of N-acetylation degree in chitosan using Raman spectroscopy. Spectrochim Acta Part A Mol Biomol Spectrosc 134:114–120
32. Brugnerotto J et al (2001) An infrared investigation in relation with chitin and chitosan characterization. Polymer 42:3569–3580
33. Karthik R et al (2016) Structural characterization and in vitro biomedical activities of sulfated chitosan from Sepia pharaonis. Int J Biol Macromol 84:319–328
34. Wang K, Liu Q (2014) Chemical structure analyses of phosphorylated chitosan. Carbohydr Res 386:48–56
35. Sampath UGTM et al (2017) Preparation and characterization of nanocellulose reinforced semi-interpenetrating polymer network of chitosan hydrogel. Cellulose 24(5):2215–2228
36. Emil B, Noverra Mardhatillah N, Siti Prilia M (2015) Effect of crosslinking agents, pH and temperature on swelling behavior of cross-linked chitosan hydrogel. Asian J Appl Sci 3:2321
37. Ryan C et al (2017) Synthesis and characterisation of cross-linked chitosan composites functionalised with silver and gold nanoparticles for antimicrobial applications. Sci Technol Adv Mater 18(1):528–540

Vibrational Spectroscopic Characterisation of Cellulose

Thisal Induwara Vitharana Denagama Vitharanage and Hemanth Noothalapati

Abstract As the planet's most abundant biopolymer, cellulose is the foundation of industries from textiles and construction to pharmaceuticals and biomedicine. In this chapter, we dive deep into how modern vibrational spectroscopy, particularly infrared (IR) and Raman spectroscopy, can unravel the intricate details of cellulose's structure. We demonstrate how these tools can identify cellulose's molecular conformations and distinguish its various crystalline forms (cellulose I, II, III, and IV). They also allow us to gauge how crystalline a sample is (the crystallinity index), reveal the orientation of cellulose fibers, and detect subtle chemical modifications in its structure. Along the way, we highlight key spectral fingerprints and employ robust analytical methods such as polarized spectroscopy for orientation analysis, crystallinity indexing for quantifying order, and ATR-FTIR to extract meaningful insights. Finally, we explore emerging trends like performing spectroscopy in situ and in real time, harnessing machine learning to enhance spectral interpretation, and combining spectroscopic techniques with other analytical tools for a more comprehensive understanding. By bridging foundational theory with real-world applications, this chapter serves as an engaging, thorough guide that provides researchers with valuable spectroscopic insights into cellulose, inspiring further exploration and innovation in cellulose-based materials.

Keywords Cellulose · Spectroscopy · Raman bands · FTIR spectroscopy

---

T. I. V. D. Vitharanage
Faculty of Applied Science, Department of Polymer Science, University of Sri Jayewardenepura, Nugegoda, Sri Lanka

H. Noothalapati (✉)
Department of Biomedical Engineering, Chennai Institute of Technology, Chennai, Tamil Nadu, India

Department of Chemical Engineering, Indian Institute of Technology Hyderabad, Kandi, Sangareddy, Telangana, India

Faculty of Life and Environmental Sciences, Shimane University, Matsue, Japan
e-mail: nvhnag@life.shimane-u.ac.jp

N. Mazumder et al. (eds.), *Vibrational Spectroscopy for Biopolymer Characterization*, Biological and Medical Physics, Biomedical Engineering,
https://doi.org/10.1007/978-3-032-15306-7_2

1 Introduction

Cellulose is the most ubiquitous among the biopolymers found on earth, with a global annual production of approximately 1.5×10^{12} tons [1]. Cellulose serves as the major structural component in plant cell walls, contributing to a substantial 40% of the carbon fraction in plants (Fig. 1). Cellulose polysaccharide is a linear biopolymer with D-glucopyranose ring units linked by β-1,4-glycosidic bonds in its cellobiose repeat unit of length approximately 1.3 nm [1]. The glucopyranose ring units occur in the lowest-energy $^4C_1$-chair conformation, where all the hydroxyl groups are located on equatorial positions and methine groups on the axial positions [3].

Due to its abundance and the convenience of extraction as a raw material, cellulose plays a vital role in many industrial sectors such as food, textiles, and construction and in cutting-edge medical and pharmaceutical applications such as drug delivery and wound healing [2, 4, 5]. Unraveling and comprehending the intricate architecture of cellulose is essential to unleashing its true potential in these domains. The effectiveness of vibrational spectroscopic analytical techniques such as infrared (IR), Raman, and sum frequency generation (SFG) spectroscopy can be harnessed to elucidate the complex structures of different polymorphisms of cellulose from the molecular level to microfibril organization while also yielding useful insights into the chemical functionalities and molecular conformations [3].

Infrared (IR) spectroscopy uses and quantifies the absorption of infrared radiation by molecular vibrations, providing useful insights on the presence of functional groups and hydrogen bonding configurations, which is especially useful in the determination of the crystallinity of cellulose (Fig. 2). Crystalline and amorphous regions display unique IR spectra, facilitating the determination of crystallinity indices [3]. For instance, the absorbance ratio between 1420–1430 cm^{-1} (crystalline band) and 893–898 cm^{-1} (amorphous band) [7]. is frequently employed to assess the relative quantities of crystalline and amorphous cellulose. IR spectroscopy can assist in the identification of cellulose crystalline polymorphs as well [3]. Raman spectroscopy is a vibrational spectroscopic technique that is based on the phenomenon of inelastic Stokes and anti-Stokes scattering between molecular vibrational

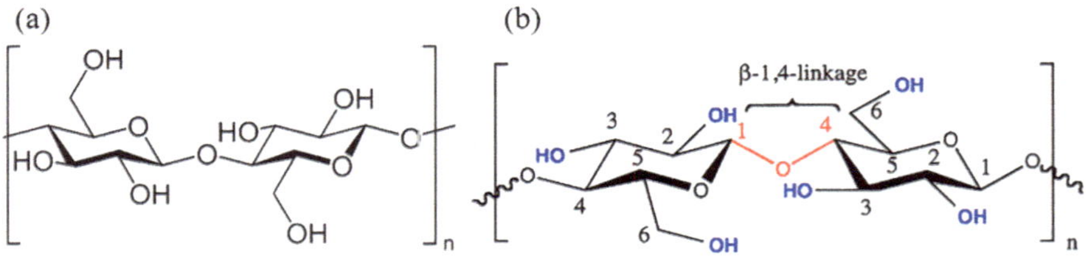

Fig. 1 Representation of the (**a**) cellulose molecule and the (**b**) formation of the β-1,4-glycosidic bonds in its cellobiose repeat unit of cellulose. Adapted from Chaka, K. T. Extraction of cellulose nanocrystals from agricultural by-products: a review. Green Chemistry Letters and Reviews 2022, 15 (3), 582–597 [2]

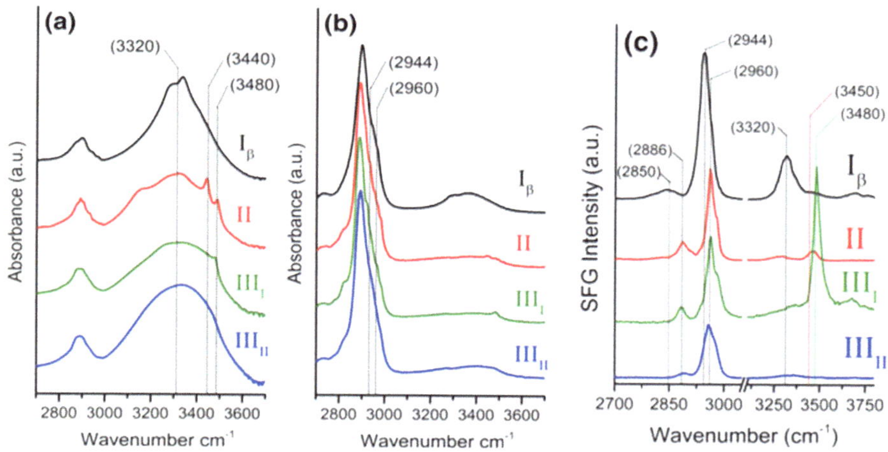

Fig. 2 Comparison of the (**a**) IR (**b**) Raman and (**c**) SFG spectra of cellulose I_β, II, III_I, and III_{II} [6]

modes. Utilization of Raman spectroscopy is also quite essential in cellulose analysis, as Raman spectra display distinct characteristics for various cellulose polymorphs and are highly responsive to alterations in crystallinity [8]. Water, a principal constituent of plant cell walls, exhibits weak Raman scattering, facilitating direct examination of hydrated specimens [7]. Raman spectroscopy can elucidate molecular conformation and disorder within amorphous regions, thereby providing a more thorough understanding of cellulose architecture [9].

Six interconvertible crystalline polymorphic forms of cellulose (Cellulose I, II, III_I, III_{II}, IV_I, and IV_{II}) have been found [10]. Natural cellulose exists as cellulose I, comprising two allomorphs: cellulose Iα and cellulose Iβ [11, 12]. Cellulose Iα predominates in basic species such as bacteria and algae, but cellulose Iβ is prevalent in higher plants. The two forms have similar chain conformations but can be distinguished from crystalline unit cell parameters and hydrogen-bonding patterns, resulting in different vibrational spectroscopic signatures [7]. The presence of these two forms was confirmed by spectroscopic methods, while their lattice patterns were elucidated via diffraction techniques [10]. Cellulose II can be obtained via regeneration or mercerization processes; cellulose II possesses a distinct crystal structure compared to cellulose I. For example, a distinctive band at 577 cm^{-1} in the Raman spectra differentiates cellulose II from cellulose I and amorphous cellulose [13]. Cellulose III and IV polymorphs are quite uncommon and can be synthesized through modifications of cellulose I or II. They exhibit unique spectroscopic characteristics that facilitate their identification and discrimination [8].

This review will focus on comprehensively examining the principles, applications, and current advancements in the utilization of two significant and widely used forms of vibrational spectroscopic methods, IR and Raman spectroscopy, for cellulose characterization. By comprehending the relationship between cellulose's structure, properties, and spectroscopic characteristics, researchers can realize its complete potential and facilitate creative applications across several domains.

2 Theoretical Background

2.1 *Raman Spectroscopy*

Raman spectroscopy is based on inelastic light scattering arising from molecular vibrations. When monochromatic photons (usually from a laser) interact with a material, most scatter elastically (Rayleigh scattering) without energy change. A small fraction, however, exchanges energy with vibrational modes of the molecule, yielding shifted photons: lower-energy Stokes-shifted photons (if the molecule gains vibrational energy) or higher-energy anti-Stokes-shifted photons (if the molecule loses vibrational energy). The energy difference between the incident and scattered photons corresponds to a specific vibrational mode, known as the Raman shift, typically reported in wavenumbers (cm^{-1}). For a vibrational mode to be Raman-active, it must involve a change in the molecule's polarizability (the ease of distorting its electron cloud) during the vibration. Symmetric vibrational modes and skeletal deformations often meet this criterion, making Raman spectroscopy well-suited to detecting C–C and C–O backbone vibrations and ring deformations in polymers like cellulose. Raman selection rules thus complement IR selection rules: for example, a symmetric stretch of a nonpolar bond can give a strong Raman signal even if it is IR-inactive. Because water has very weak Raman scattering, Raman spectroscopy can analyze wet samples or *in situ* processes in aqueous media with minimal interference, an advantage for biological and plant materials that contain moisture.

2.2 *Infrared Spectroscopy*

By contrast, IR spectroscopy is based on the absorption of infrared radiation as molecules undergo transitions between vibrational energy levels. A vibrational mode is IR-active if it involves a change in the molecular dipole moment. Thus, polar bonds and asymmetric vibrations (especially involving heteroatoms like O–H, C–O, and C=O) produce strong IR absorption bands. In an IR spectrum (often obtained by Fourier-transform IR, FT-IR), the positions of absorption peaks reveal the vibrational frequencies of functional groups, and their intensities relate to bond dipole changes. Attenuated Total Reflectance (ATR)-FTIR is a convenient mode that probes the sample's surface by evanescent-wave absorption, allowing rapid analysis of solids like cellulose films or powders without extensive sample preparation. ATR-FTIR spectra of cellulose are very similar to transmission IR spectra, except for some intensity differences due to the shallow penetration depth (on the order of a few microns) of the evanescent wave into the sample.

3 Overview of Raman Spectroscopic Characterization of Cellulose

3.1 Characteristic Raman Bands of Cellulose

Cellulose has a rich Raman spectrum with distinct bands arising from its molecular vibrations. Notably, strong Raman bands appear in the 100–1500 cm^{-1} region (the "fingerprint" region) corresponding to ring deformation and backbone vibrations. For example, cellulose I (native cellulose) shows prominent bands near 380 cm^{-1} (assigned to coupled CCC ring bending and skeletal deformations), around 433–436 cm^{-1} (skeletal motions), and at 1095 cm^{-1} (C–O–C stretching of the glycosidic linkage). A closely spaced band at 1120 cm^{-1} also originates from C-O–C stretching modes and is characteristic of the β-1,4-glycosidic bond. The region around 895–900 cm^{-1} is associated with C–H bending and ring "breathing" of the glucose unit (this band's relative intensity is sometimes used as an indicator of amorphous content, as discussed later). In the high-frequency region, cellulose shows Raman bands for C–H stretching at 2890–2900 cm^{-1} (CH and CH_2 stretches) and weaker, broad features for O–H stretching around 3400 cm^{-1} (hydrogen-bonded O–H stretches, which are weak in Raman due to minimal polarizability change). Overall, the positions, widths, and relative intensities of these bands provide a spectral fingerprint of cellulose's molecular structure. For instance, the band around 1460–1480 cm^{-1} is sensitive to the conformation of the exocyclic CH_2OH group: cellulose I (with predominantly *tg* (trans–gauche) conformation) shows a band near 1480 cm^{-1}, whereas cellulose II (with *gt* conformation) shows a band at 1460 cm^{-1}. Such differences help distinguish cellulose polymorphs, as illustrated by comparative spectra (e.g., cellulose II lacks the 1480 cm^{-1} band of cellulose I but shows a strong 1460 cm^{-1} band, among other differences).

3.2 Crystallinity and Ordering

Raman spectroscopy is a valuable tool for assessing cellulose crystallinity and chain ordering. Crystalline and amorphous cellulose exhibit differences in certain Raman band intensities and positions. A common metric is the *Raman crystallinity index*, defined as the intensity ratio of a band representative of crystalline order to one representative of disordered structure. One widely used measure is the ratio I_{380}/I_{1096}—the intensity of the 380 cm^{-1} band to that of the 1096 cm^{-1} band. The 380 cm^{-1} band is associated with collective ring deformation vibrations that are enhanced in ordered (crystalline) regions, whereas the 1096 cm^{-1} C–O stretch is present in all cellulose and is often taken as an internal reference. A higher I_{380}/I_{1096} ratio indicates higher relative crystallinity. Alternative Raman indices have been proposed, including one based on a lattice mode near 93 cm^{-1} (accessible with specialized low-frequency Raman setups), which directly probes the crystal lattice vibrations. Such low-frequency Raman measurements (sometimes called terahertz Raman, since 3 THz =100 cm^{-1}) correlate with the presence of crystalline domains and can complement traditional X-ray diffraction methods for crystallinity assessment. Recent studies have highlighted that even disordered (non-crystalline) cellulose can contribute intensity to ostensibly "crystalline" Raman bands. For example, Agarwal et al. [13] showed that ball-milled amorphous cellulose still produces a weak 380 cm^{-1} band. Therefore, careful analysis (often via multivariate methods or by combining Raman with IR data) is needed for accurate crystallinity assessment, a point revisited under combined analysis. Raman-based crystallinity metrics continue to be refined, and protocols have been developed to minimize analyst and instrument variability for such measurements.

3.3 Molecular Orientation

Polarized Raman spectroscopy can probe the orientation of cellulose chains and functional groups. By measuring Raman spectra with different polarization configurations of incident and scattered light, one can determine the orientation of specific vibrational transition moments relative to the sample's axis. In highly oriented cellulose samples (e.g., drawn fibers or aligned tunicate cellulose), the intensity of certain bands (like the 1095 cm^{-1} C–O–C stretch or the 1120 cm^{-1} band) [14] will vary with the polarization direction. Vibrational modes aligned along the chain axis (for example, the C–O–C glycosidic linkage stretch, which lies roughly along the chain) scatter light more strongly when the incident polarization is parallel to that axis. Polarized Raman analysis has been used to determine the orientation distribution of cellulose microfibrils in plant cell walls and to monitor alignment in manufactured fibers. It provides insight into how processing (stretching, shearing, etc.) affects molecular orientation in cellulose materials [8]. For instance, the Raman band at 1095 cm^{-1} typically shows higher intensity when the polarization is parallel to the fiber axis in a highly oriented sample, [15] reflecting alignment of the glycosidic linkages with the fiber direction.

3.4 Chemical Modifications and Interactions

Raman spectroscopy is very useful for monitoring chemical modifications of cellulose, as the introduction of new functional groups or changes in structure often leads to identifiable changes in the Raman spectrum [16]. For example, partial oxidation of cellulose (to yield carbonyl or carboxyl groups) can be detected by the appearance of new Raman bands or shifts in existing bands. An oxidation that introduces C=O groups (such as TEMPO-mediated oxidation that yields C_6 carboxylate groups) may result in new bands around 1760 cm^{-1} (for unconjugated carbonyls; although in practice IR is more sensitive to C=O) and changes in the C–O stretching region. Hydrolysis or acid treatment that removes amorphous regions (to produce microcrystalline cellulose or nanocrystals) tends to sharpen the Raman bands and increase the relative intensity of crystallinity-sensitive bands (e.g., the 380 cm^{-1} band) [17]. Functionalization (such as silylation, etherification, grafting of polymers, etc.) produces combinatorial spectral features: the Raman spectrum will show both the cellulose backbone bands and new bands from the substituent. For instance, grafting poly(ethylene glycol) (PEG) chains onto cellulose would introduce PEG's C–O–C stretching band (1100 cm^{-1}) and C–H stretching bands (2880 cm^{-1}) on top of the cellulose spectrum. Raman can detect such changes provided the new bands fall within the spectral window and are not completely masked by cellulose's own bands. Moreover, intensity changes in cellulose's characteristic bands can indicate successful modification; for example, a decrease in the O–H stretching region intensity might accompany extensive substitution of hydroxyls in a cellulose derivatization reaction [18].

In summary, Raman spectroscopy provides a detailed molecular fingerprint of cellulose, reflecting its backbone conformation, degree of crystallinity, and even subtle differences between polymorphs. For instance, it can readily distinguish native cellulose I from regenerated cellulose II by the presence of the 577 cm^{-1} band and shift of certain CH_2 bending modes in the latter, as well as differentiate these from cellulose III by its unique pattern of lattice mode frequencies. In the next section, we turn to IR spectroscopy, which offers complementary information, particularly on functional groups and hydrogen bonding [13] to build a fuller picture of cellulose structure.

4 Overview of IR Spectroscopic Characterization of Cellulose (Fig. 3)

4.1 Characteristic IR Bands of Cellulose

The IR spectrum of cellulose is dominated by absorptions from O–H, C–H, and C–O bonds in the sugar rings. In the O–H stretching region (3600–3000 cm^{-1}) [20] cellulose exhibits a broad, intense band centred around 3300–3400 cm^{-1} [20]. This broad envelope results from the myriad hydroxyl groups engaged in hydrogen bonding; strong hydrogen bonds shift O–H stretches to lower frequencies and broaden

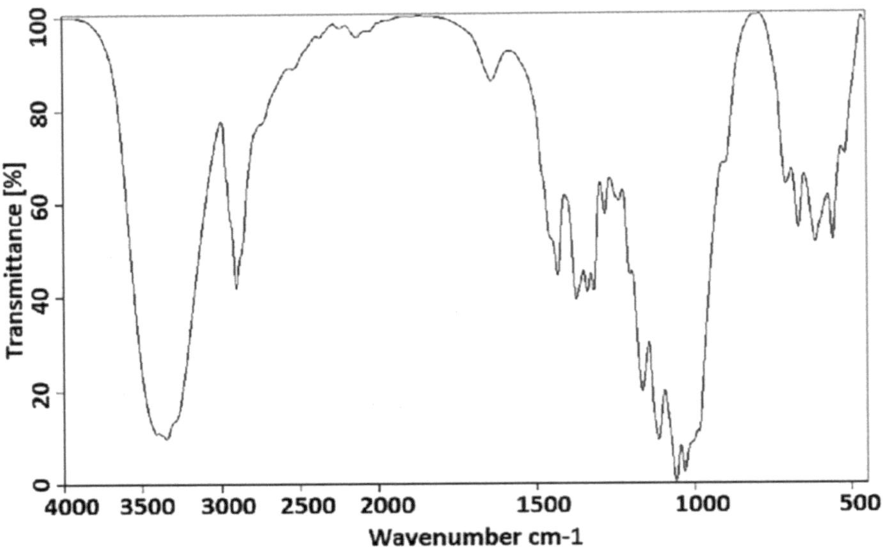

Fig. 3 IR spectrum of cellulose adapted from Szpiłyk et al. [19]

the absorption. The exact profile of this O–H band (sometimes deconvoluted into multiple sub-bands) provides insight into the hydrogen-bonding network [21]. For instance, cellulose I and II have slightly different O–H stretch band shapes due to different interchain H-bond arrangements. Around 2900 cm^{-1}, a moderate band appears from C–H stretching vibrations (the C–H stretches of the pyranose ring CH and CH_2 groups) [22]. This C–H stretch band is often used as an internal reference for normalization, since it tends to remain relatively constant regardless of crystallinity. In the fingerprint region (1800–800 cm^{-1}), several important cellulose IR bands are observed:

- 1420–1430 cm^{-1} (the "crystallinity band")—Arises mainly from CH_2 scissoring (bending) vibrations that are sensitive to chain ordering. In cellulose I, a distinct band near 1430 cm^{-1} is observed and is often used as an indicator of crystalline content. Cellulose II tends to show this band at a slightly lower frequency (1420 cm^{-1}) and with different intensity, due to its different packing arrangement. This band, when compared to an amorphous reference band (such as the 893 cm^{-1} C–O stretch), forms the basis of the *Lateral Order Index (LOI)* for crystallinity [23, 24].
- 1365–1375 cm^{-1}—A band due to C–H bending (wagging/twisting) modes, also sometimes related to crystallinity. A ratio of A_{1375}/A_{2900} has been defined in some studies as the *Total Crystallinity Index (TCI)*, assuming the 1375 cm^{-1} band height correlates with overall crystallinity. However, this ratio can also be influenced by other constituents (e.g., lignin or hemicellulose in plant fibers), so its interpretation needs care [22, 25].

- 1315–1330 cm^{-1}—Various CH_2 wagging and O–H in-plane bending modes. In cellulose II, a notable band around 1317 cm^{-1} often appears, associated with that polymorph's structure. This band is weaker or shifted in cellulose I.
- 1160 cm^{-1}—An asymmetric bridge C–O–C stretching vibration of the glycosidic linkage, coupled with ring vibrations. This band is strong in cellulose and, along with the next bands, constitutes a prominent feature in the 1200–900 cm^{-1} region [24].
- 1110–1120 cm^{-1}—C–O stretching and C–C skeletal vibrations. Often a shoulder or secondary peak near 1105 cm^{-1} is seen (assigned to C–O stretch of secondary alcohol groups), and another near 1050 cm^{-1} (C–O stretch of primary alcohol at C_6, plus C–C stretch). These overlapping contributions result in a broad absorption envelope. In cellulose II, sometimes a distinct peak around 1110–1115 cm^{-1} is noted, reflecting its altered C–O–C and C–C linkage environment.
- 1020–1050 cm^{-1}—Coupled C–O/C–C stretching and ring breathing modes. A peak around 1030 cm^{-1} is common in cellulose I and II [25].
- 895 cm^{-1} (approximately 893–898 cm^{-1})—Assigned to the C–O–C stretching (and/or O–C–H bending) in the β-glycosidic link. This band is prominent in cellulose and is often referred to as the "amorphous" band in semi-quantitative analyses because it tends to be more pronounced (relative to the 1420 cm^{-1} band) in less ordered cellulose. The intensity ratio A_{1420}/A_{893} (or A_{1430}/A_{893}) is widely used as a measure of relative crystallinity: higher values indicate a greater crystalline fraction. For example, in one study the A1430/A893 ratio was about 0.91 for a semicrystalline cellulose and higher for highly crystalline cotton cellulose [23].

Additionally, cellulose IR spectra often show a small band near 1640 cm^{-1}, which is not due to cellulose's structure but rather to the bending mode of absorbed water (H–O–H bend) in the sample. The presence and intensity of this water band can indicate the sample's moisture content. Drying the sample tends to reduce this band's intensity, whereas more humid samples show it more strongly.

4.2 Crystallinity Analysis by IR

As noted above, IR has long been used to estimate cellulose crystallinity via empirical band intensity ratios. Two commonly used metrics are

- *Lateral Order Index (LOI)* = A_{1420}/A_{893}, which compares the absorption at 1420 cm^{-1} (crystalline CH_2 bending) to that at around 893 cm^{-1} (amorphous/reference C–O stretch). A higher LOI implies a more ordered (crystalline) structure. Pure cellulose I (highly crystalline) yields a high LOI, whereas amorphous cellulose or cellulose II (which has a lower 1420 and higher 893 band) yields a lower LOI [25].

- *Total Crystallinity Index (TCI)* = A_{1375}/A_{2900}, which compares a CH bending band to the C–H stretching band. This ratio tends to increase with crystallinity as well (CH bending modes become better defined and relatively stronger in crystalline domains) [22].

While these indices provide a convenient gauge of relative crystallinity, they are semi-quantitative. Factors like lignin or hemicellulose overlaps, baseline corrections, and instrumental parameters can affect absolute values. Nevertheless, such indices are useful for comparing similar samples (e.g., before vs. after a treatment). For instance, mercerization (alkaline treatment) that converts cellulose I to cellulose II usually leads to a lower LOI and TCI, reflecting the introduction of more disorder and the structural differences of cellulose II. FT-IR is often used alongside X-ray diffraction (XRD) for crystallinity analysis, with generally good correlation reported between the techniques.

4.3 Hydrogen Bonding Information

The broad O–H stretching band in the IR spectrum contains information about the hydrogen-bond network in cellulose. Cellulose I (especially Iβ) typically shows resolved maxima or shoulders within the O–H stretch envelope, corresponding to different types of OH groups (for example, O_3H-O_5 intrachain hydrogen bonds vs. O_2H-O_6 interchain bonds). IR spectroscopy has been used to distinguish these by fitting sub-peaks, although the extensive overlap makes this challenging. Still, changes in this region can indicate changes in hydrogen bonding. For example, a shift of the overall O–H band to higher frequency (and a narrowing of its width) might suggest disruption or weakening of some hydrogen bonds (as might occur upon heating or derivatization). Conversely, when cellulose forms stronger hydrogen bonds (e.g., upon drying or annealing to increase order), the O–H stretch band can shift to lower frequencies and broaden. Another proposed metric is the *Hydrogen Bond Intensity (HBI)*, defined by some as the ratio A_{3340}/A_{1330} (absorbance at 3340 cm^{-1} versus at 1330 cm^{-1}) [22]. This ratio is thought to relate to the average hydrogen bond strength in the sample (higher HBI when more O–H groups are involved in strong H-bonds that shift O–H absorption to lower frequency). Correlations have been observed, HBI tends to increase with increasing crystalline order, but HBI is not an absolute measure of any single hydrogen bond; rather, it is a holistic indicator of hydrogen-bonding status.

4.4 Functional Group Identification and Chemical Changes

One of IR spectroscopy's greatest strengths is the clear identification of functional groups via their signature vibrations. In cellulose chemistry, many derivatives and modifications are conveniently tracked by IR spectra.

- Cellulose esters (e.g., cellulose acetate): Introduction of acetyl groups yields a strong C=O stretching band near 1740–1750 cm^{-1} (acetyl carbonyl) and new C–O stretching bands near 1235 cm^{-1} and 1040 cm^{-1} (ester C–O stretches). The intensity of the carbonyl band is directly related to the degree of substitution (DS). In fact, a recent study used ATR-FTIR combined with machine learning to predict the DS of cellulose acetate with high accuracy from the IR spectrum. As DS increases, the O–H bands in the spectrum decrease and eventually disappear (fully substituted cellulose triacetate has essentially no O–H band and shows only the ester carbonyl and C–O bands) [26, 27].
- Cellulose ethers (e.g., carboxymethyl cellulose, CMC): Carboxymethylation replaces some hydroxyls with $–CH_2-COO^-$ groups. IR readily identifies this substitution by the appearance of a strong *asymmetric* carboxylate stretch near 1600 cm^{-1} and a *symmetric* COO^- stretch around 1420 cm^{-1} (the latter can overlap with the CH_2 bending band at 1420). Additionally, new C–O stretching bands from the ether linkage ($–CH_2-O–$) and the carboxylate ($–C–O^-$) appear near 1320 cm^{-1} and 1060 cm^{-1}. The broad O–H band may become somewhat less intense if many –OH are substituted, but since CMC is often analysed in its sodium salt form ($–COO^-$ Na^+), the O–H band still largely represents remaining cellulose OH plus absorbed water. (*Note*: Raman spectroscopy can also detect carboxymethylation; for example, a new Raman band around 1380 cm^{-1} can appear for the symmetric COO^- stretch, but IR is more sensitive to these highly polar functional groups) [28].
- Oxidation and carbonyl formation: Aside from deliberate functionalization, oxidative degradation of cellulose (e.g., during aging or bleaching of pulp) introduces carbonyl groups (aldehydes, ketones, or acids). IR is the go-to method to detect these changes. A carbonyl band at 1730 cm^{-1} (unconjugated C=O) or in the 1650–1700 cm^{-1} range (conjugated or hydrogen-bonded C=O) indicates the presence of oxidized groups. For example, periodate oxidation (which cleaves vicinal diols to dialdehydes) yields an IR band 1730 cm^{-1} (aldehyde C=O), whereas TEMPO oxidation (which selectively converts C_6 primary alcohols to carboxylates) yields a carboxylate band at 1600 cm^{-1} (as discussed above for CMC). Tracking the growth of these carbonyl bands relative to the cellulose backbone bands allows quantification of oxidation extent. In studies of paper aging, the *Carbonyl Index* (C=O absorbance normalized to an internal reference band, e.g., at 1320 cm^{-1}) is a key metric of degradation [27].
- Hydrogen-bonding changes and additives: IR can monitor interactions of cellulose with other substances. For instance, when cellulose interacts with water or plasticizers, the O–H stretching profile changes (indicating new hydrogen bonds being formed). If cellulose is blended in a composite with a polymer that has carbonyl groups (e.g., poly(vinyl acetate) or certain polyurethanes), one may observe shifts in the polymer's carbonyl IR band due to hydrogen bonding with cellulose, a sign of good interfacial interaction. The O–H band of cellulose might also shift or change shape if it strongly hydrogen-bonds to another polymer or penetrant molecule. In essence, IR can be used to probe compatibility and interactions in cellulose-containing mixtures [28].

4.5 ATR-FTIR Advantages

It is worth noting the advantages of ATR-FTIR for cellulose analysis. ATR-FTIR allows direct analysis of cellulose fibers, fabrics, or films without complex preparation. It is widely employed for quick identification of cellulose in mixed materials (for example, confirming cellulose content in composites or verifying chemical treatments on fibers). The shallow penetration depth of ATR (on the order of 1–2 μm at mid-IR wavelengths) means it primarily probes the surface layers of a sample. This is ideal for studying modifications that occur on the fiber surface (such as surface derivatization, adsorption of molecules, or weathering effects). For example, ATR-FTIR has been used to examine in real time how water or solvents interact with cellulose surfaces or how the surface composition of cellulose changes during chemical reactions. The speed and convenience of ATR-FTIR make it a powerful tool for routine cellulose analysis and *in situ* monitoring of cellulose modification processes.

In summary, IR spectroscopy provides a comprehensive view of cellulose's functional groups and bonding. It excels at identifying chemical modifications (through diagnostic bands like C=O, C–O, and COO^-) and at quantifying bulk structural indices like crystallinity and hydrogen-bond order. In the context of cellulose polymorphs, IR can distinguish different forms by subtle spectral differences (for example, cellulose II often shows a stronger 898 cm^{-1} band relative to 1420 cm^{-1} and a unique band near 1260 cm^{-1} attributed to its CH_2 rocking mode). However, some IR bands of the various cellulose allomorphs overlap significantly, making Raman, with its generally sharper skeletal-mode bands, sometimes more discriminating for polymorph identification. Thus, both IR and Raman play important roles in the spectroscopic characterization of cellulose's many allomorphs and derivatives. We will now delve into specific categories of cellulose materials and discuss how these spectroscopic features manifest and are interpreted in each case.

5 Different Forms of Cellulose and their Spectroscopic Characterization

5.1 Native Cellulose (Cellulose Iα and Iβ)

5.1.1 Structure and Properties

Native cellulose (Cellulose I) refers to the form produced biosynthetically by plants, algae, and certain bacteria. It consists of parallel-chain cellulose in a crystalline arrangement. Native cellulose actually occurs as a mixture of two closely related allomorphs: Cellulose Iα and Iβ. Cellulose Iα has a triclinic unit cell (dominant in algal and bacterial cellulose), and Cellulose Iβ has a monoclinic unit cell (dominant in higher plants). These forms differ in how adjacent chains are layered and hydrogen-bonded, but both are forms of cellulose I and are not easily distinguished by bulk techniques except by specialized methods (like solid-state NMR or certain diffraction techniques). Both Iα and Iβ consist of sheets of hydrogen-bonded chains, with three primary intrachain H-bonds (e.g., O_3H-O_5 and O_2H-O_6) and additional interchain bonds that give substantial stability. Native cellulose typically has a high degree of polymerization (DP can range from a few hundred to 10,000 glucose units in wood cellulose) and forms microfibrils a few nanometers in width. Impurities such as residual hemicellulose or lignin can be present in native cellulose from biomass.

5.1.2 Spectroscopic Characteristics

The IR and Raman spectra of native cellulose reflect its well-ordered structure. In IR, native cellulose shows the characteristic bands of cellulose I: for instance, a prominent 1430 cm^{-1} CH_2 bending band and a relatively weaker 893 cm^{-1} band, yielding a high LOI indicative of high crystallinity. The O–H stretching band in IR often shows multiple peaks or shoulders (for example, two discernible maxima around 3340 and 3270 cm^{-1} have been reported for highly crystalline bacterial cellulose), assigned to different hydrogen-bonded OH groups in the cellulose I structure. In Raman spectra, cellulose I is recognizable by a distinctive doublet in the 1095–1120 cm^{-1} region (stemming from two C–O stretches) and by the presence of the 380 cm^{-1} band (actually a doublet around 355/380 cm^{-1} for Iβ). If Iα and Iβ are present in different proportions, very subtle differences can occur. For instance, some studies noted slight shifts or intensity differences in the 380 cm^{-1} Raman band between Iα-rich vs. Iβ-rich samples. However, generally both Iα and Iβ give very similar spectra; distinguishing them solely by IR or Raman is difficult. (Sum-frequency generation (SFG) spectroscopy, a nonlinear optical technique, is more sensitive to Iα vs. Iβ content due to symmetry differences, but that is beyond our scope here.)

Importantly, native cellulose in plant fibers usually contains both crystalline and amorphous regions. Spectroscopically, this means that in addition to peaks from well-ordered cellulose, features from less-ordered regions appear. For example, a completely amorphous cellulose would show a much weaker 1430 cm^{-1} IR band and a stronger 893 cm^{-1} band (as well as a broader, slightly higher-frequency O–H stretch) compared to a crystalline sample. Native cellulose's spectrum can be viewed as a combination of crystalline and amorphous contributions. In Raman, amorphous regions lead to broadened peaks and a lower I_{380}/I_{1095} intensity ratio. Thus, spectra of native cellulose isolated from plants (like cotton, jute, and wood pulp) often have slightly broader bands than those of highly crystalline reference celluloses (for example, algal cellulose or tunicate cellulose, which can be extremely crystalline).

Vibrational spectroscopy is frequently used to monitor treatments of native cellulose and their effect on structure. For instance, during mercerization (alkaline treatment of cellulose fibers), one can observe the conversion of cellulose I to cellulose II via IR: the 1430 cm^{-1} band diminishes and shifts toward 1415 cm^{-1}, and the 893 cm^{-1} band grows more prominent. Likewise, acid hydrolysis that preferentially removes amorphous regions causes the IR crystallinity indices (LOI, TCI) to rise, since the remaining material is richer in crystalline segments. Raman spectra of partially hydrolysed cellulose (e.g., to make microcrystalline cellulose) show sharpening of the 380 and 1095 cm^{-1} bands and an increase in their intensity ratio, consistent with loss of disordered domains.

In sum, native cellulose (cellulose I) provides the baseline spectra for cellulose. It is characterized by spectroscopic indicators of high crystallinity and strong hydrogen bonding. Variations in these spectra can reveal information about the cellulose source and purity. For example, the presence of residual lignin or hemicellulose in "raw" biomass adds extra bands: lignin contributes aromatic ring stretches near 1510 cm^{-1}, and hemicellulose (if acetylated) can contribute a carbonyl band around 1730 cm^{-1}. These extraneous bands disappear upon purification of cellulose. Detecting such features is useful; for instance, a decrease in the 1730 cm^{-1} band (acetyl C=O) and 1510 cm^{-1} band after chemical treatment would indicate removal of hemicellulose and lignin, respectively.

5.2 *Cellulose II (Regenerated or Mercerized Cellulose)*

5.2.1 Structure and Formation

Cellulose II is the thermodynamically more stable polymorph of cellulose, obtained either by regenerating cellulose from solution or by mercerization (treatment of cellulose I with strong alkali, causing the chains to rearrange). In cellulose II, the molecular chains adopt an antiparallel arrangement (in contrast to the parallel alignment in cellulose I), and the unit cell is monoclinic [29]. The hydrogen-bonding pattern in cellulose II differs from that in cellulose I: notably, the O_3H-O_5 and O_2H-O_6 hydrogen bonds are arranged differently, and an additional O_6H-O_3' interchain hydrogen bond is present, leading to a more extensively hydrogen-bonded network per pair of chains. Cellulose II typically has slightly lower crystallinity (depending on processing conditions) and a density comparable to cellulose I.

5.2.2 Spectroscopic Signature

Cellulose II's vibrational spectra have distinct differences from cellulose I, making spectroscopy a convenient way to confirm conversion of one form to the other. In IR, conversion to cellulose II is evidenced by broadening and shifting of the O–H stretch band: cellulose II often shows a single broad O–H stretching peak around 3350 cm^{-1} (lacking the multiple sub-peaks seen in highly crystalline cellulose I) due to its different hydrogen bonding arrangement. The 1420–1430 cm^{-1} band (crystalline CH_2 bend) is generally weaker in cellulose II and shifted toward 1415–1420 cm^{-1} [30]. Concurrently, the 893 cm^{-1} "amorphous" band tends to increase in relative intensity (or the 1430 cm^{-1} band decreases), reflecting the higher proportion of disordered structure in cellulose II or simply the polymorphic change in the CH_2 bending mode intensity ratio. As a result, the LOI (I_{1420}/I_{893}) and similar IR crystallinity indices drop when cellulose I is converted to II. Another IR indicator of cellulose II is a band near 1260 cm^{-1}, sometimes observed as being enhanced in cellulose II (attributed to a CH_2 rocking mode unique to this form). Additionally, the CH_2 bending region of cellulose II often appears as two resolved bands (around 1420 and 1380 cm^{-1}) because the symmetric and asymmetric CH_2 bends become inequivalent in the cellulose II crystal structure.

Raman spectroscopy provides especially clear markers for cellulose II. A well-known diagnostic is the appearance of a Raman band at 577 cm^{-1}, which is absent in cellulose I. This band is a unique lattice mode or ring deformation associated with the cellulose II allomorph and serves as a fingerprint for even small amounts of cellulose II in a sample. Additionally, the Raman band associated with the CH_2OH conformation shifts: cellulose II shows a strong band at 1460 cm^{-1} (attributed to the gg conformer of CH_2OH) and a weaker one at 1480 cm^{-1}, essentially the inverse of the cellulose I pattern. The C–O–C stretching region (near 1095–1120 cm^{-1}) also differs; cellulose II tends to have a single dominant peak around 1096 cm^{-1} with a shoulder near 1115 cm^{-1}, whereas cellulose I showed a more pronounced doublet (approximately 1095 and 1120 cm^{-1}). These differences arise from the antiparallel chain arrangement and the resulting symmetry changes when cellulose I is converted to II [8].

Because of these clear spectral differences, one can use Raman or IR to monitor the progress of mercerization or regeneration in real time. For example, as alkali mercerization proceeds on cellulose fibers, the 577 cm^{-1} Raman peak of cellulose II grows in intensity, indicating formation of the cellulose II polymorph. Simultaneously, in IR one would observe the decrease of the 1430 cm^{-1} band and growth of the 893 cm^{-1} band. In regenerated cellulose fibers (like rayon, which is almost pure cellulose II), the spectra match those of mercerized cellulose. Any residual cellulose I in such fibers can be sensitively detected by checking for the telltale cellulose I features (e.g., a Raman band at 380 cm^{-1} or an IR band at 1430 cm^{-1}) that indicate incomplete conversion [13, 31].

In summary, cellulose II's spectroscopic profile is well characterized by its unique Raman band at 577 cm^{-1} and subtle shifts in other bands, as well as an IR pattern with lower crystallinity indices and slightly shifted band positions compared to cellulose I. These features allow analysts to confirm a material's polymorphic form after processing. For instance, if a treated biomass exhibits the 577 cm^{-1} Raman signal and a diminished 1430 cm^{-1} IR band, one can conclude that cellulose I has been largely converted to cellulose II, a change that can affect properties like digestibility or reactivity.

5.3 *Cellulose III and IV Polymorphs*

5.3.1 Structure and Formation

Cellulose III and IV are less common polymorphs derived from cellulose I or II by specific treatments. Cellulose III can exist in two forms: cellulose III_I (produced from cellulose I by treatment with liquid ammonia or certain amines) and cellulose III_{II} (produced from cellulose II by similar treatment). These involve an *ammoniation* step: swelling cellulose in liquid ammonia, then removing the ammonia, which yields a metastable cellulose III structure characterized by expanded lattice spacings. Cellulose IV (IV_I and IV_{II}) is obtained by heating cellulose III_I or III_{II}, respectively, at high temperatures (e.g., 200 °C in glycerol). This treatment partially reverts the structure to a more ordered form without returning exactly to cellulose I or II. The precise structures of cellulose IV are less well understood, but they are considered highly ordered modifications [6, 31].

5.3.2 Spectroscopic Features

Both cellulose III and IV have unique spectroscopic characteristics that facilitate their identification, although the differences from cellulose I/II can be subtle. For cellulose III_I (from cellulose I), the IR spectrum in many respects resembles that of cellulose II (both lack the strong 1430 cm^{-1} band of cellulose I). However, distinguishing bands are reported. Literature indicates cellulose III shows new or shifted IR bands near 1470 cm^{-1} and 893 cm^{-1}, and differences in the O–H stretch region possibly due to residual ammonia or altered H-bonding. In Raman spectra, cellulose III_I often resembles cellulose II in having the CH_2OH-related band near 1460 cm^{-1} (gg conformer dominant). Indeed, in Raman studies the spectrum of cellulose III can look similar to that of cellulose II in certain regions. For example, cellulose III lacks the 1480 cm^{-1} band that is prominent in cellulose I and instead shows the 1460 cm^{-1} band like cellulose II. However, cellulose III can be distinguished by other spectral features. Notably, cellulose III_I *does not* show the 577 cm^{-1} Raman band of cellulose II (since cellulose III has its own lattice modes). Instead, cellulose III_I exhibits a few low-frequency Raman bands at different positions (some reports mention a band near 350 cm^{-1} and another around 380 cm^{-1}) that differ from those of cellulose I or II. Additionally, cellulose III's high-frequency C–H stretch region may shift slightly compared to cellulose I/II, reflecting small differences in conformation [8].

Cellulose IV_I and IV_{II}, being heat-treated forms of cellulose III_I and III_{II}, respectively, have spectra closer to cellulose I or II but still distinct. They are rarely encountered outside of laboratory preparations. Spectroscopically, heating cellulose III tends to partially restore some features of cellulose I or II, but not completely. For example, an annealed cellulose III (cellulose IV) might show the 1430 cm^{-1} IR band reappearing to some extent, indicating some return of order, yet other bands remain shifted, indicating it hasn't fully reverted [31].

In practice, identification of cellulose III or IV is often done primarily by X-ray diffraction, with IR/Raman as supporting evidence. Still, a trained eye on the IR spectrum can note subtle clues; for instance, cellulose III sometimes shows an IR absorption around 1645 cm^{-1} if ammonia is trapped (an N-H bending mode) or minor shifts in the 1200–1000 cm^{-1} region relative to cellulose II. In Raman, the absence of the 577 cm^{-1} band (to rule out cellulose II) coupled with the presence of a strong 1460 cm^{-1} band (suggesting gg CH_2OH, like cellulose II and III) and perhaps a slightly altered low-frequency profile might signal cellulose III [31].

From a review standpoint, the key point is that cellulose III and IV, while uncommon, can be recognized by subtle but unique IR/Raman features, and their creation via specific chemical treatments can be confirmed spectroscopically. For example, treating cellulose I with liquid ammonia is expected to yield cellulose III_I; correspondingly, one would observe the loss of the original 1430 cm^{-1} IR band and the disappearance of the 380 cm^{-1} Raman band (from cellulose I) without the emergence of the 577 cm^{-1} band (cellulose II), confirming the structural change to cellulose III.

5.4 Cellulose Nanocrystals (CNCs)

5.4.1 Structure and Properties

Cellulose nanocrystals (CNCs) are rod-like crystalline nanoparticles obtained by acid hydrolysis of native cellulose. The acid (commonly sulfuric acid) preferentially hydrolyses amorphous regions, releasing crystalline rod fragments typically a few hundred nanometers long and 5–20 nm in cross-section. CNCs maintain the cellulose I crystal structure of their parent source (often predominantly cellulose Iβ for plant-based CNCs), though sulfuric-acid-prepared CNCs carry negatively charged sulphate half-ester groups ($-OSO_3^-$) on some surface hydroxyls, which introduce slight structural distortions and confer colloidal stability in water. CNCs are essentially highly crystalline cellulose I, with crystallinity often >80–90%, and they can form chiral nematic liquid crystal suspensions at high concentrations. They represent nearly "perfect" crystalline cellulose nanoparticles [8, 32].

5.4.2 Spectroscopic Characteristics

Because CNCs are essentially highly crystalline cellulose I domains, their spectra closely resemble those of native cellulose I but with some notable differences arising from their higher crystallinity and any surface chemistry:

- Sharpening of bands: CNCs often show sharper IR and Raman bands compared to their source cellulose due to the removal of disordered regions. For example, the Raman 380 cm^{-1} band becomes relatively stronger and narrower in CNCs than in the original pulp. Likewise, the IR crystallinity indices (LOI, TCI) usually increase for CNCs relative to the starting cellulose. One study noted that after acid hydrolysis, the Raman I_{380}/I_{1096} ratio increased, indicating enrichment in ordered domains.

- Surface chemistry effects: If sulfuric acid is used in preparation, CNCs bear sulphate ester groups (1 per 20–25 glucose units typically). These groups introduce new IR bands around 1220–1250 cm^{-1} (S=O stretching of sulphate ester) and sometimes a band near 805 cm^{-1} (C–O–S stretch). The presence of sulphate can be confirmed by these IR signals. (Neutralizing to the sodium salt form may cause slight shifts in the S=O band.) Raman spectroscopy is less sensitive to sulphate, but a weak band around 1050 cm^{-1} can sometimes be attributed to the sulphate symmetric stretch overlapping with cellulose C–O bands. If other acids are used (e.g., phosphoric acid), different surface groups would be introduced and would have their own signature bands [32].
- Crystallinity and polymorph: CNCs from plant cellulose are cellulose Iβ, so their spectra reflect that polymorph. CNCs can also be produced from algae or bacterial cellulose (rich in Iα) or from tunicate cellulose (highly crystalline Iβ). In general, since CNCs are highly crystalline, one sees very clear cellulose I "fingerprints," e.g., a strong 1430 cm^{-1} IR band and the characteristic 1095/1120 cm^{-1} Raman doublet. The overall spectral features correspond to those of native cellulose I, with minimal contributions from amorphous content.
- Comparison to native cellulose: The absence of non-cellulosic components in purified CNCs simplifies interpretation. For example, raw biomass might show an IR band at 1730 cm^{-1} (due to hemicellulose acetyl groups or pectin), which disappears in CNCs after those components are removed. Similarly, lignin's aromatic bands (around 1510 and 1600 cm^{-1} in Raman or IR) vanish upon isolation of CNCs. Thus, CNC spectra are "cleaner." If one compares the IR spectra of a plant fiber versus the CNCs extracted from it, one notes the disappearance of those non-cellulose bands and an increase in the relative intensity of the cellulose bands. This purification effect is documented in studies converting agricultural waste into CNCs.

Overall, CNCs being essentially crystallite fragments of Cellulose, do not introduce new cellulose peaks (except those associated with surface sulphate groups, if present). Instead, their significance in spectroscopy is that they allow one to study nearly perfect cellulose crystals. They are often used as a reference for the spectra of fully crystalline cellulose in research. For instance, in evaluating the disorder in a pulp, one might compare its spectrum to that of CNCs derived from that pulp to gauge what fraction of certain band intensities correspond to the ordered phase. One should note that CNCs prepared with sulfuric acid have a lower thermal stability, and their IR spectra upon heating might show early signs of sulphate decomposition (changes in the 1200 cm^{-1} region). Also, if CNCs self-assemble into films (nanopaper), some reorganization can occur; for example, hydrogen bonds might rearrange upon drying, subtly affecting the broad O–H IR band shape [32].

In summary, cellulose nanocrystals present the spectra of highly crystalline cellulose I, modified only by any surface substituents from their preparation. Spectroscopically, they serve as a benchmark for crystalline cellulose and exhibit only minor differences (primarily due to surface chemistry) from the native cellulose I spectrum. Their well-defined spectra make them useful for fundamental studies of cellulose vibrations and for calibration of spectral methods for crystallinity.

5.5 *Microcrystalline Cellulose (MCC) (Fig. 4)*

5.5.1 Structure and Properties

Microcrystalline cellulose (MCC) refers to partially depolymerized, relatively crystalline cellulose obtained by controlled acid hydrolysis of purified pulp, followed by neutralization and drying. Commercial MCC (e.g., Avicel) is a powder composed of

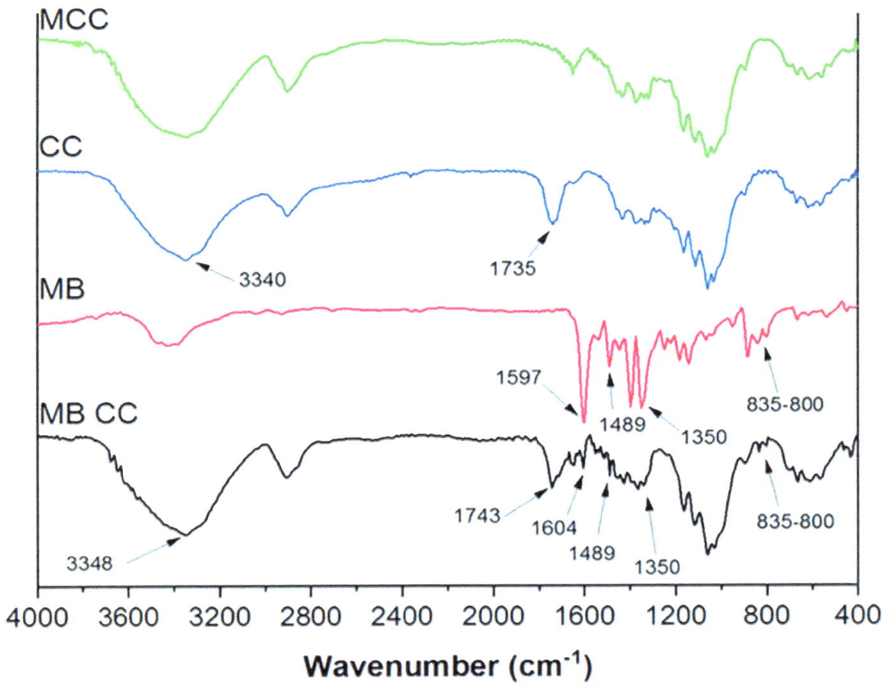

Fig. 4 FT-IR spectra of microcrystalline cellulose (MCC, green), cellulose citrate (CC, blue), methylene blue (MB, red), methylene blue adsorbed on cellulose citrate (MB-CC, black) adapted from Olivito et al. [33]

aggregated cellulose crystallites on the order of tens of microns in size. MCC preparation is similar to that of CNCs but usually involves shorter hydrolysis (not to the point of yielding individually dispersed nanocrystals) and results in a dry powder rather than a colloidal suspension. MCC retains a significant crystalline core (with DP perhaps 150–300) and includes some residual disordered regions on particle surfaces. In essence, MCC is Cellulose I that has been enriched in crystallinity by the removal of some amorphous regions, but it still contains both crystalline and amorphous fractions [34].

5.5.2 Spectroscopic Characteristics

MCC's spectra are very similar to those of native cellulose I, with strong signals indicative of crystallinity. Because MCC still has some fraction of amorphous cellulose (usually more than CNCs do), its crystallinity indices are intermediate between those of the original pulp and a fully nanocrystalline product. For example, Avicel PH-101 (a common MCC) has a crystallinity index in the range of 70–80%. Spectroscopically:

- IR: MCC shows a strong 1420 cm^{-1} band and a relatively lower 893 cm^{-1} band, reflecting its substantial crystallinity (often LOI 1.0 or higher for MCC). The O–H stretch band remains broad (MCC still contains accessible OH and often is measured in a conditioned state with some moisture). However, MCC being typically dry and of lower DP means it may have slightly different O–H band characteristics (more chain ends and less bound moisture can slightly alter the O–H band shape).
- Raman: MCC likewise shows the strong 380 cm^{-1} band and well-defined 1095 cm^{-1} peak typical of cellulose I. Raman crystallinity measures (like the I_{380}/I_{1095} ratio) for MCC are high, though not as high as for CNCs. Indeed, Agarwal et al. have used MCC and ball-milled MCC mixtures to calibrate Raman crystallinity methods. They demonstrated MCC's Raman spectrum as a "control" crystalline spectrum in a calibration set.
- MCC is often used as a substrate or excipient in formulations (e.g., tablets), and its identification by IR or Raman is straightforward. The presence of cellulose's characteristic peaks (as described in the overview) confirms MCC. Since MCC is essentially purified cellulose, it does not have unique bands beyond those of cellulose itself. (Notably, MCC prepared with hydrochloric acid leaves no persistent substituents, in contrast to CNCs prepared with sulfuric acid that have sulphate groups.) Thus, MCC generally lacks the 1200 cm^{-1} sulphate band that sulfuric-acid CNCs show, making MCC's IR spectrum "pure" cellulose in that sense.

In effect, MCC can be considered a bulk form of highly crystalline cellulose I. Any differences between MCC's spectrum and that of the original pulp can help identify which spectral features originate from disordered segments. For instance, progressive ball-milling of MCC reduces its crystallinity; IR will show the 1430 cm^{-1} band dropping and the 893 cm^{-1} band rising, and Raman will show the 380 cm^{-1} band weakening relative to others. These trends have been documented, confirming how MCC's spectra shift toward those of amorphous cellulose upon mechanical degradation [8, 35].

5.6 Cellulose Derivatives (Ethers and Esters)

5.6.1 Structure and Properties

Cellulose can be chemically modified to produce a wide range of derivatives. Common examples include cellulose ethers (such as carboxymethyl cellulose (CMC), hydroxypropyl cellulose, and methylcellulose) and cellulose esters (such as cellulose acetate, cellulose nitrate, and cellulose xanthate, the latter being an intermediate in viscose rayon production). In these derivatives, some of the hydroxyl groups on the glucose rings are replaced by other substituents (ether or ester groups). The degree of substitution (DS, 0 to 3 per anhydroglucose unit) and the distribution of substituents affect the derivative's solubility and crystallinity heavily substituted cellulose derivatives are often amorphous or only loosely ordered. (In fact, many cellulose derivatives dissolve in water or organic solvents when DS is above a certain threshold, resulting in loss of the native crystalline structure.)

Cellulose derivatization can be performed in various media, including novel solvents like ionic liquids that dissolve cellulose. The resulting materials range from slightly modified, still-crystalline cellulose (for low DS) to completely amorphous polymers (for high DS, such as fully substituted cellulose triacetate).

5.6.2 Spectroscopic Characteristics

Introduction of new functional groups yields characteristic new bands in the IR (and sometimes Raman) spectrum of cellulose derivatives.

- *Carboxymethyl cellulose (CMC)*: In its sodium salt form, CMC shows strong IR bands at 1600 cm^{-1} (asymmetric COO^- stretch) and 1420 cm^{-1} (symmetric COO^- stretch, often overlapping the cellulose CH_2 bend). The O–H stretch band in IR may become slightly less intense or shifted due to the reduced number of free -OH groups and increased hydrophilicity (CMC is water-soluble at DS 0.7). In Raman, CMC exhibits new bands for the carboxylate: often a band around 1380 cm^{-1} (the symmetric COO^- stretch) can appear, and some cellulose backbone bands (e.g., around 1330 cm^{-1}) may shift or weaken if those OH positions are substituted. The cellulose fingerprint is still present, but alterations in intensity (for example, the 1330 cm^{-1} OH bend may diminish and merge with others) indicate the substitution. (*Note*: CMC is often analysed in the Na^+ salt form, so in IR the broad O–H band still largely reflects residual OH plus absorbed water, as mentioned above.) [36, 37].

- *Cellulose acetate*: This ester shows the strong ester carbonyl C=O band at 1740 cm^{-1} in IR, as noted earlier, along with C–O stretching bands at 1230 cm^{-1} and 1040 cm^{-1} (from the acetate groups). As the DS increases, the O–H stretching band (around 3400 cm^{-1}) decreases and vanishes for fully substituted cellulose triacetate. Raman spectra of cellulose acetate (especially triacetate) can show the carbonyl band around 1730 cm^{-1} (Raman is less sensitive to C=O than IR, but with a strong laser line one can observe it) and changes in the fingerprint region; for example, the 1095 cm^{-1} cellulose band may shift or split because the acetylated glucose rings have altered vibrational modes. The disappearance (or great reduction) of the O–H IR band is a clear sign of a high DS; a partially acetylated sample would still show an O–H band [38].
- *Cellulose nitrates (nitrocellulose)*: Nitration introduces -NO_2 groups. IR spectra show the -NO_2 asymmetric and symmetric stretches typically around 1650–1630 cm^{-1} and 1280–1270 cm^{-1}, respectively (exact positions depend on the degree of nitration). These are very distinctive bands. Nitrocellulose (especially if highly nitrated) also shows no O–H band. Raman spectra of nitrocellulose can show an -NO_2-related band near 1300 cm^{-1}, but analysis is mainly by IR.
- *Cellulose xanthate*: This is a transient derivative (an intermediate in the viscose process) where cellulose is treated with CS_2 to form cellulose xanthate (with –O–$CSS^{-}Na^{+}$ groups). Xanthate groups have strong IR bands near 1170–1200 cm^{-1} and 950–1000 cm^{-1} (C=S and C–S stretches). Though not typically characterized in detail via IR/Raman in literature (because of its instability), these bands would clearly indicate xanthate formation.

In general, IR spectroscopy is a primary tool to confirm substitution on cellulose. For instance, in synthesizing CMC, one monitors the rise of the 1600 cm^{-1} carboxylate band and the drop of the 3340 cm^{-1} O–H band. In producing cellulose acetate, IR is used to measure DS by the ratio of the C=O band to the O–H band or by correlations like the ratio of 1375/2900 to 1740/2900 (though now machine learning models can do this more accurately). Raman is a useful complementary technique; it can sometimes differentiate certain derivatives even when IR bands overlap with those of solvents or water. For example, Raman can detect an aromatic ring stretch if cellulose is derivatized with an aromatic group (such as in some cellulose carbamates for chromatography applications) where IR might be less clear. Derivatization often disrupts cellulose's native hydrogen-bonded crystalline structure, especially at high DS. The cellulose backbone conformation might change, and the material often becomes more amorphous. Spectroscopically, heavily derivatized cellulose shows the loss of "crystalline cellulose" features. For example, methylcellulose or sodium CMC in solid form often shows an IR spectrum lacking the 1430 cm^{-1} crystallinity band (indicative of an amorphous structure). In Raman, such derivatives might show a much weaker 380 cm^{-1} band and generally broader peaks, reflecting loss of long-range order. This is useful analytically: one can tell by Raman if a sample is still largely native crystalline cellulose or a substantially amorphous derivative by the presence/absence of those sharp crystalline marker bands.

In summary, cellulose derivatives are readily identified by vibrational spectroscopy through the new functional group signals they introduce. Both IR and Raman have roles: IR is excellent for detecting groups like $-COO^-$, $-C(O)CH_3$, $-NO_2$, etc., while Raman can help identify others (like $-CH_3$ or aromatic substituents) that have distinctive polarizability signatures. Vibrational spectroscopy is indispensable for confirming chemical modifications of cellulose and assessing the extent of substitution.

5.7 Cellulose Grafts (Fig. 5)

5.7.1 Structure and Properties

In cellulose graft copolymers, polymer chains are covalently attached ("grafted") onto the cellulose backbone. This is often achieved by initiating a polymerization from the cellulose –OH groups or by pre-functionalizing cellulose with initiator

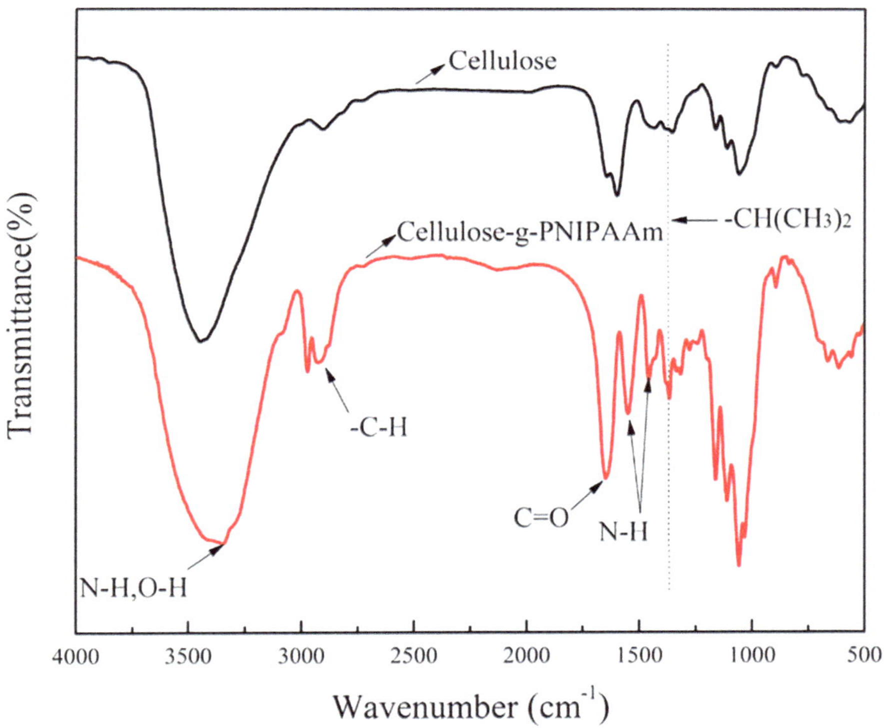

Fig. 5 The FTIR spectra of the cellulose and cellulose-g-PNIPAAm adapted from Chen et al. [39] demonstrates modification effects via grafting PNIPAAm onto cellulose, ideal for thermoresponsive polymer sections

moieties and then attaching polymers. Common grafted polymers include poly(ethylene glycol) (PEG), poly(acrylic acid) (PAA), poly(vinyl alcohol) (PVA), poly(N-vinylpyrrolidone) (PVP), Poly(N-Isopropylacrylamide) (PNIPAAm), and others. Grafting can dramatically alter cellulose's properties: for example, grafting PEG can improve cellulose's flexibility and water solubility; grafting PAA can introduce pH-responsive behavior and high charge (carboxylate groups). Typically, the degree of grafting (often expressed as weight percent gain or grafting ratio) is controlled to tune properties.

5.7.2 Spectroscopic Characterization

Raman and IR spectroscopy are key to confirming grafting and analyzing the structure of cellulose graft copolymers.

- Presence of graft polymer bands: The clearest evidence of grafting is the appearance of the characteristic bands of the grafted polymer superimposed on the cellulose spectrum. For instance, if PEG is grafted, one would see the ether C–O–C stretch of PEG around 1100 cm^{-1} in the IR spectrum (this overlaps with cellulose's C–O stretch, but an increase in intensity in that region or the appearance of a distinct shoulder can indicate PEG). In Raman, PEG has a strong band near 845 cm^{-1} (C–C–O stretching/twisting mode) and CH_2 twisting modes around 1295 cm^{-1}, which could appear in the grafted material's Raman spectrum [40]. As another example, if poly(acrylic acid) (PAA) is grafted, one would observe the PAA's carbonyl stretch near 1700–1720 cm^{-1} in IR (broad, if COOH, or around 1560 cm^{-1} if deprotonated COO^-), and perhaps the 1450–1550 cm^{-1} region reflecting –COO^- in Raman or IR. Grafted PVA (which is essentially –CH_2-CH(OH)– chains) might be harder to detect since PVA's spectrum is similar to cellulose (both have O–H and C–H stretches, etc.), but an increase in intensity of broad O–H and aliphatic C–H vibrations might be noted. Grafted PVP would introduce a lactam C=O around 1650 cm^{-1} and C–N vibrations that could be picked up in IR or Raman.
- Changes in cellulose's own bands: Successful grafting means some of cellulose's -OH groups are consumed or their environment is changed. IR might show a decrease in the overall O–H stretching band intensity (if a significant fraction of OH groups are esterified/etherified by grafts). Some cellulose backbone bands could shift slightly due to the covalently attached bulky side chains altering the vibrational coupling. For instance, the 1160 cm^{-1} C–O–C stretch or the 1050 cm^{-1} band might shift or change shape if the ring is now bonded to a large polymer chain. In Raman, one might observe reduced intensity in certain cellulose bands if the graft polymer coating partially obscures or dampens the Raman response of the cellulose (for example, if the graft polymer forms a shell around cellulose crystallites, the effective Raman signal from cellulose could diminish proportional to graft coverage).
- Interfacial interactions: IR and Raman can also probe interactions between cellulose and the grafted polymer beyond the covalent bond. If the graft polymer

forms hydrogen bonds or other interactions with the cellulose backbone, this can cause spectral shifts. For example, grafting PAA onto cellulose: the PAA –COOH groups may hydrogen-bond with remaining cellulose -OH groups. This could broaden or shift the O–H stretching band in IR and shift the PAA C=O band to lower frequency (if hydrogen-bonded). Similarly, grafted polymers might adopt different conformations on the cellulose surface than in their free state, which could be investigated by Raman (though such analysis can be complex).

- Degree of grafting: The relative intensity of graft polymer bands to cellulose bands in the spectra can provide a rough measure of the degree of grafting. For example, a stronger PEG C–H stretch (2870 cm^{-1}) relative to the cellulose 2900 cm^{-1} band suggests more PEG content. Often, one calibrates such measurements with known standards or uses elemental analysis for accuracy, but spectroscopy provides a quick check. In the literature, one sometimes sees a grafting ratio confirmed by the ratio of an IR band of the graft polymer to one of cellulose across a series of samples.

To illustrate, in cellulose-g-PEG (often made to improve biocompatibility or flexibility), IR will show additional $–CH_2–$ stretching bands around 2870 cm^{-1} from PEG and an increase around 1100 cm^{-1} for the PEG C–O stretch. Raman would show PEG's 843–845 cm^{-1} band and perhaps a CH_2 bending around 1460 cm^{-1} (which can overlap with cellulose's CH_2 bend). Meanwhile, the cellulose's 897 cm^{-1} band might be partially obscured by PEG's nearby vibrations, but the 380 cm^{-1} band of cellulose would still be visible unless the grafting is extremely dense [40].

In Cellulose-g-PAA, the IR spectrum would show PAA's broad O–H and C=O features (a broad band around 1700 cm^{-1} for COOH, which might actually appear as a very broad envelope from 1720 down to 1500 cm^{-1} if partially ionized or H-bonded) and still the cellulose bands. Raman might detect some of PAA's modes (though poly(acrylic acid) is not strongly Raman-active except perhaps C–H bends around 1450 cm^{-1}). The main cellulose Raman peaks would remain, but their relative intensities might change if PAA coats the surface.

The main point is that both components' spectral features are observable in a grafted system, confirming the hybrid nature of the material. The challenge is often deconvoluting overlapping bands, which can be aided by comparing spectra of the pure components to that of the graft copolymer. Multivariate analysis can also help separate contributions.

5.8 Cellulose Composites

5.8.1 Structure and Properties

Cellulose composites refer to materials where cellulose (in some form) is combined with other constituents (polymers, resins, inorganic fillers, etc.) to form an integrated material. Examples include cellulose fiber–reinforced plastics (where cellulose fibers, micro- or nano-fibrillated cellulose, or CNCs are dispersed in a polymer matrix like polypropylene, polyethylene, epoxy, or polyurethane), as well as cellulose–inorganic hybrids (e.g., paper or aerogels loaded with silica or clay). The goals are typically to enhance mechanical properties, reduce cost, or impart functionality (for instance, using nanocellulose to reinforce biodegradable plastics).

5.8.2 Spectroscopic Characterization

In composites, IR and Raman spectroscopy can be used to identify each component and to study interfacial interactions.

- Identifying components: The IR spectrum of a composite is approximately a superposition of the spectra of cellulose and the matrix material. For instance, in a cellulose-polypropylene composite, one would see cellulose's O–H and C–O bands along with polypropylene's CH stretching (2950 and 2870 cm^{-1}) and CH bending (1455 and 1375 cm^{-1}) bands [41]. In Raman, polypropylene has characteristic CH_2 stretching and twisting modes (e.g., a prominent Raman band 809 cm^{-1} for CH_2 rocking in PP) that would appear alongside cellulose's bands. By monitoring known bands (e.g., the 1095 cm^{-1} band for cellulose and the 809 cm^{-1} band for polypropylene), one can confirm the presence of both components. Moreover, Raman or IR mapping can show the spatial distribution of each [42]. For example, one could use Raman imaging to map cellulose vs. polymer domains by integrating the intensity of a cellulose band vs. a polymer band across a sample.
- Interfacial interactions: If the cellulose surface interacts strongly with the matrix (via hydrogen bonding, ester linkages, etc.), subtle spectral shifts can occur. For example, in a cellulose–polyurethane composite, the urethane C=O IR band might shift or broaden if it hydrogen-bonds with cellulose, indicating good interfacial adhesion. In a cellulose-epoxy composite, one might look for evidence of chemical bonding; for instance, if an epoxide group reacts with cellulose -OH to form an ether, the IR spectrum might show a reduced O–H band and possibly the disappearance of the epoxide ring band (915 cm^{-1}). In general, the cellulose O–H stretching band is a sensitive indicator: its position and shape in a composite can reveal if cellulose's -OH groups are engaged in hydrogen bonds with the matrix (often manifested as a slightly shifted or narrowed O–H band). Another clue is the H–O–H bending band at 1640 cm^{-1} (if present from residual water): changes in its intensity might indicate differences in how much water cellulose holds when embedded in the matrix vs. alone [43, 44].
- Phase separation and dispersion: Spectroscopy can sometimes indicate the dispersion state of cellulose in a composite. If cellulose is well-dispersed at a molecular level, the spectrum will be consistent throughout the sample. If cellulose aggregates in certain regions, microspectroscopy (FTIR microscopy or micro-Raman) might show spectra with variable cellulose band intensities in different spots. Polarized IR or Raman can also be used in fiber composites to

assess fiber orientation: for example, in a unidirectional cellulose fiber composite, one can use polarized Raman along vs. across the fiber direction to gauge the alignment of cellulose and possibly the polymer matrix molecules [42].

A noteworthy advanced method is Raman imaging of composites, which has been used to visualize the dispersion of cellulose nanocrystals or fibers in matrices. By tuning the analysis to a specific Raman band of cellulose (say the 1095 cm^{-1} band) and mapping its intensity, one can "see" the distribution of cellulose within the composite. Similarly, FTIR micro-spectroscopy or imaging can map concentrations by looking at the intensity of a cellulose-specific band vs. a matrix-specific band. These techniques help assess whether cellulose forms a percolating network or is evenly integrated.

6 Future Trends and Challenges

Looking forward, several trends and challenges are likely to shape the field of cellulose vibrational spectroscopy:

- In-situ and real-time monitoring: A major trend is moving spectroscopic analysis from the lab to the production line. This involves developing robust IR and Raman probes for real-time monitoring of cellulose processing. For example, inline ATR-FTIR sensors could track the degree of substitution in a cellulose acetylation reactor by measuring the increase of the C=O band in real time. Portable Raman devices (with stable solid-state lasers and fiber optics) might be used to monitor cellulose fiber treatments in textiles or to detect cellulose degradation in historical papers on-site. In fact, there are already examples of handheld Raman spectrometers used to detect cellulose nitrate in old film reels (to prevent fire hazards in archives) and portable NIR devices used by recycling centers to identify cellulose-based fibers vs. synthetics on sorting lines. Future improvements in miniaturization and stability will expand such uses. The challenges to address include background fluorescence in real-world samples (which can swamp Raman signals) and calibration for varying sample forms (moisture content, thickness, etc.) [45]. However, progress in algorithms (for baseline correction and signal extraction) and hardware (e.g., shifted-excitation Raman to mitigate fluorescence) is reducing these issues.
- Advanced data analysis and machine learning: As noted earlier, machine learning is making inroads into spectral analysis, and this will likely accelerate. Deep learning models (e.g., neural networks) could be trained on large libraries of cellulose IR and Raman spectra from different sources and treatments to automatically classify samples and quantify features. For instance, a convolutional neural network might predict crystallinity, allomorph content, and even crystal size from a Raman spectrum that is too complex for a human analyst to fully interpret. Early work in adjacent fields (like using neural networks on polymer spectra for composition prediction) shows great promise. Applying these meth-

ods to cellulose could lead to, for example, an instrument that takes a Raman spectrum of pulp and instantly outputs metrics like alpha-cellulose content, crystallinity, or residual lignin content based on a trained model. The challenge here is the need for comprehensive training data covering the many variations (different wood species, different instruments, etc.), but collaborative efforts and shared spectral databases can overcome that. Moreover, AI-driven spectral interpretation (where the algorithm suggests chemical origins for subtle new peaks) could assist in discovering new markers for cellulose polymorphs or early signs of degradation [46, 47].

- Integration with other analytical techniques: Future cellulose characterization will benefit from *hyphenated* techniques that combine vibrational spectroscopy with other methods. One example is simultaneous Raman or IR with X-ray diffraction (XRD). Some synchrotron facilities now allow IR spectra and XRD patterns to be collected on a sample under the same conditions (temperature, humidity, mechanical strain), enabling direct correlation of molecular and crystalline structural changes. Similarly, combining nuclear magnetic resonance (NMR) with IR/Raman can marry insights about chemical structure (NMR) with those about order and bonding (vibrational spectra). We may also see more integration in imaging: for instance, correlative Raman and electron microscopy (SEM or TEM) to connect chemical information with ultrastructure. In plant biology and materials science, coupling vibrational spectroscopy with mechanical testing (performing *operando* spectroscopy during the deformation or compression of a cellulose material) could elucidate how molecular structures respond to stress in real time. The challenge for such integrations is complexity and instrument availability. Fully integrated systems are expensive and not widespread. But as interdisciplinary research grows, more labs are devising creative combinations to gain a multi-faceted understanding of cellulose.
- New vibrational techniques: We anticipate further adoption of emerging vibrational methods in cellulose studies. Terahertz (far-IR) spectroscopy, which probes very low-frequency collective modes, could provide additional fingerprints of different hydrogen-bond networks in cellulose polymorphs beyond what current far-IR or low-frequency Raman does. It might, for instance, differentiate cellulose Iα vs. Iβ better or assess nanoscale crystal size via phonon confinement effects. Two-dimensional IR (2D-IR) spectroscopy, a nonlinear technique that provides correlation maps of vibrational couplings, is another frontier. 2D-IR could be applied to cellulose to study couplings between O–H stretches or between O–H and C–O modes, mapping out hydrogen bond networks in a way that linear IR cannot. These techniques are currently specialized and available in only a few labs, but as they become more accessible, they could unveil new details about cellulose's structure (for instance, 2D-IR might be able to distinguish overlapping OH bands and assign them to specific intra- vs. inter-chain hydrogen bonds) [8].
- Standardization and reproducibility: With the expanding use of vibrational spectroscopy in both research and industry, there is a growing push to standardize methods for cellulose analysis. This includes establishing reference spectra and calibration protocols so that results are comparable across laboratories. For

example, agreeing on how to baseline-correct and normalize a cellulose IR spectrum for crystallinity measurement or which bands to use for a particular index would make it easier to compare results from different studies. Interlaboratory studies might be conducted to ensure that an IR crystallinity index measured in one lab can be translated to another's results. Standard reference materials (SRMs) of cellulose with known properties (crystallinity, DP, etc.) could be used to calibrate instruments (analogous to how a polystyrene film is used to calibrate IR frequency and intensity scales). This is both a trend and a challenge: cellulose's variability (moisture content, polymorphic mix, etc.) makes standardization non-trivial. However, as techniques mature, establishing such protocols (possibly via organizations like TAPPI or ISO) will be important for making vibrational spectroscopy a truly reliable quality control tool in cellulose product manufacturing.

- Extending to modified and complex systems: Finally, future work will increasingly involve complex cellulose-based systems such as cellulose-reinforced nanocomposites, partially carbonized cellulose (for carbon materials), and cellulose dissolved in novel solvents (ionic liquids, deep eutectic solvents). Vibrational spectroscopy will play a key role in studying these but may require adaptations. For instance, Raman spectroscopy of cellulose in carbon composites (e.g., a composite where cellulose is partially converted to carbon) must contend with strong fluorescence or absorption from carbonized components. Overcoming this might involve shifted-excitation Raman, resonance Raman at different wavelengths, or UV-Raman to avoid visible fluorescence. Studying cellulose in *situ* in new solvent systems (like ionic liquids) via IR/Raman will provide insights into solvent interactions and dissolution mechanisms, guiding better processing methods. Each new application poses its own challenges; for example, overlapping signals from a solvent or matrix that obscure the cellulose bands, but lessons learned from traditional cellulose spectroscopy will guide how to adjust techniques (such as by subtracting solvent spectra or using polarization to isolate certain bands).

In summary, the future of cellulose characterization by vibrational spectroscopy is bright, with technology pushing the boundaries of *when*, *where*, and *how* we can analyse cellulose. The major challenges revolve around dealing with complex, real-world samples and ensuring consistency and accuracy of interpretation. Through interdisciplinary innovation combining spectroscopy with new analysis methods and other techniques, these challenges are being met, enabling deeper insight into cellulose structure and transformations.

7 Conclusion

Vibrational spectroscopy has established itself as an indispensable tool for the characterization of cellulose and cellulose-based materials. In this chapter, we have reviewed how Raman and infrared spectroscopy provide complementary perspectives on cellulose's structure, from its molecular conformation and hydrogen-bonding network to its crystalline polymorphs and interactions in composites. The introduction highlighted cellulose's fundamental structure and the motivation for spectroscopic analysis, setting the stage for understanding how vibrational techniques unravel cellulose's complexities. The theoretical background section outlined why IR and Raman respond to different aspects of molecular vibrations, explaining their complementarity in analysing a polysaccharide like cellulose.

In the core sections, we examined the spectroscopic signatures of cellulose in its many forms. The overview of Raman characterization detailed how Raman bands (e.g., the 1095 cm^{-1} C–O stretch and the 380 cm^{-1} ring deformation) serve as fingerprints for cellulose and indicators of order, while polarized Raman and intensity ratios yield information on orientation and crystallinity. The overview of IR characterization similarly described IR's sensitivity to functional groups and hydrogen bonds, with features like the I_{1420}/I_{893} cm^{-1} LOI correlating with crystalline content and new bands revealing chemical modifications. We then delved into different forms of cellulose: we saw that native cellulose I's spectra differ subtly from regenerated cellulose II (for example, the emergence of a 577 cm^{-1} Raman band and a shift of the CH_2 bending IR band to 1420 cm^{-1} in cellulose II), and how the rarer cellulose III/IV polymorphs can be recognized by unique spectral shifts like the absence of the 577 cm^{-1} band and presence of a split CH_2 bend. We discussed cellulose derivatives, grafts, and composites, emphasizing how vibrational spectroscopy identifies introduced functional groups ($-COO^-$, $-C{=}O$, $-NO^-_2$, polymer backbones) and monitors interfacial interactions (such as hydrogen bonding in composites). Throughout these sections, references to recent studies and figures illustrated practical spectral differences and their structural implications.

Finally, we looked ahead at future trends and challenges. As cellulose finds new applications in materials, energy, and biotechnology, vibrational spectroscopy will be at the forefront of monitoring and understanding these developments. In-situ spectroscopy and portable devices will bring cellulose analysis out of the lab; advanced algorithms will extract more information from spectra than ever before; and integration with other methods will solve multifaceted problems. Ongoing efforts to standardize methods and to address complex sample matrices are paving the way for more reliable and universally applicable spectroscopic techniques. In conclusion, Raman and IR spectroscopy, used in synergy, have dramatically advanced our understanding of cellulose, from the arrangement of atoms in its polymer chains to the performance of cellulose products in everyday use. They provide a non-destructive, information-rich window into this vital biopolymer. Ongoing innovations promise to further deepen our insight, helping to develop new cellulose-based materials and improve processes in a sustainable bioeconomy. As this review demonstrates, the vibrational spectroscopic characterization of cellulose is a rich and evolving field, mirroring the versatility and importance of cellulose itself.

References

1. Heinze T (2015) Cellulose: Structure and properties. Adv Polym Sci 271:1–52. https://doi.org/10.1007/12_2015_319
2. Chaka KT (2022) Extraction of cellulose nanocrystals from agricultural by-products: a review. Green Chem Lett Rev 15:582–597
3. Makarem M, Lee CM, Kafle K, Huang S, Chae I, Yang H, Kubicki JD, Kim SH (2019) Probing cellulose structures with vibrational spectroscopy. Cellulose 26:35–79
4. Mbisana M, Keroletswe N, Nareetsile F, Mogopodi D, Chibua I (2024) Nanocellulose composites: synthesis, properties, and applications to wastewater treatment. Cellulose 31:10651–10678. https://doi.org/10.1007/s10570-024-06268-y
5. Seddiqi H, Oliaei E, Honarkar H, Jin J, Geonzon LC, Bacabac RG, Klein-Nulend J (2021) Cellulose and its derivatives: towards biomedical applications. Cellulose 28:1893–1931
6. Lee CM, Mittal A, Barnette AL, Kafle K, Park YB, Shin H, Johnson DK, Park S, Kim SH (2013) Cellulose polymorphism study with sum-frequency-generation (SFG) vibration spectroscopy: identification of exocyclic CH2OH conformation and chain orientation. Cellulose 20:991–1000. https://doi.org/10.1007/s10570-013-9917-3
7. Rongpipi S, Ye D, Gomez ED, Gomez EW (2019) Progress and opportunities in the characterization of cellulose—an important regulator of cell wall growth and mechanics. Front Plant Sci 9:1894
8. Agarwal UP (2019) Analysis of cellulose and lignocellulose materials by Raman spectroscopy: a review of the current status. Molecules 24:1659
9. Dassanayake RS, Dissanayake N, Fierro JS, Abidi N, Quitevis EL, Boggavarappu K, Thalangamaarachchige VD (2023) Characterization of cellulose nanocrystals by current spectroscopic techniques. Appl Spectrosc Rev 58:180–205
10. O'Sullivan AC (1997) Cellulose: the structure slowly unravels. Cellulose 4:173–207. https://doi.org/10.1023/A:1018431705579
11. Sugiyama J, Persson J, Chanzy H (1991) Combined infrared and electron diffraction study of the polymorphism of native celluloses. Macromolecules 24(9):2461–2466
12. Vanderhart DL, Atalla RH (1984) Studies of microstructure in native celluloses using solid-state 13C NMR. Macromolecules 17(8):1465–1472
13. Agarwal UP, Ralph SA, Baez C, Reiner RS (2021) Detection and quantitation of cellulose II by Raman spectroscopy. Cellulose 28:9069–9079. https://doi.org/10.1007/s10570-021-04124-x
14. Cao Y, Shen D, Lu Y, Huang Y (2006) A Raman-scattering study on the net orientation of biomacromolecules in the outer epidermal walls of mature wheat stems (Triticum aestivum). Ann Bot 97:1091–1094. https://doi.org/10.1093/aob/mcl059
15. Rosso DF, Negrão DR, Driemeier C (2022) Unveiling the variability and multiscale structure of soybean hulls for biotechnological valorization. Waste Biomass Valoriz 13:2095–2108. https://doi.org/10.1007/s12649-021-01655-z
16. Librando V, Lorusso S, Minniti Z (2011) Conservation science in cultural. Heritage. https://doi.org/10.6092/issn.1973-9494/2700
17. Tang Z, Li W, Lin X, Xiao H, Miao Q, Huang L, Chen L, Wu H (2017) TEMPO-oxidized cellulose with high degree of oxidation. Polymers 9:90421. https://doi.org/10.3390/polym9090421
18. Gieroba B, Kalisz G, Krysa M, Khalavka M, Przekora A (2023) Application of vibrational spectroscopic techniques in the study of the natural polysaccharides and their cross-linking process. Int J Mol Sci 24:2630
19. Szpiłyk M, Lubczak R, Lubczak J (2021) The biodegradable cellulose-derived polyol and polyurethane foam. Polym Test 100:107250. https://doi.org/10.1016/j.polymertesting.2021.107250
20. Cichosz S, Masek A (2020) IR study on cellulose with the varied moisture contents: insight into the supramolecular structure. Materials 13:1–22. https://doi.org/10.3390/ma13204573
21. Lee CM, Kubicki JD, Fan B, Zhong L, Jarvis MC, Kim SH (2015) Hydrogen-bonding network and OH stretch vibration of cellulose: comparison of computational modeling with

polarized IR and SFG spectra. J Phys Chem B 119:15138–15149. https://doi.org/10.1021/acs.jpcb.5b08015
22. Khai DM, Nhan PD, Hoanh TD (2017) An investigation of the structural characteristics of modified cellulose from acacia pulp. Vietnam J Sci Technol 55:452. https://doi.org/10.15625/2525-2518/55/4/9216
23. Vârban R, Crişan I, Vârban D, Ona A, Olar L, Stoie A, Ştefan R (2021) Comparative Ft-IR prospecting for cellulose in stems of some fiber plants: flax, velvet leaf, hemp and jute. Appl Sci 11:8570. https://doi.org/10.3390/app11188570
24. Oh SY, Yoo DI, Shin Y, Seo G (2005) FTIR analysis of cellulose treated with sodium hydroxide and carbon dioxide. Carbohydr Res 340:417–428. https://doi.org/10.1016/j.carres.2004.11.027
25. Szymańska-Chargot M, Cieśla J, Pękala P, Pieczywek PM, Oleszek W, Żyła M, Szkopek Z, Zdunek A (2022) The influence of high-intensity ultrasonication on properties of cellulose produced from the hop stems, the byproduct of the hop cones production. Molecules 27:92624. https://doi.org/10.3390/molecules27092624
26. Rhein F, Sehn T, Meier MAR (2025) Efficient and accurate determination of the degree of substitution of cellulose acetate using ATR-FTIR spectroscopy and machine learning. Sci Rep 15:2904. https://doi.org/10.1038/s41598-025-86378-0
27. Nunes S, Ramacciotti F, Neves A, Angelin EM, Ramos AM, Roldão É, Wallaszkovits N, Armijo AA, Melo MJ (2020) A diagnostic tool for assessing the conservation condition of cellulose nitrate and acetate in heritage collections: quantifying the degree of substitution by infrared spectroscopy. Herit Sci 8:33. https://doi.org/10.1186/s40494-020-00373-4
28. Consumi M, Leone G, Tamasi G, Magnani A (2021) Water content quantification by FTIR in carboxymethyl cellulose food additive. Food Addit Contam Part A Chem Anal Control Expo Risk Assess 38:1629–1635. https://doi.org/10.1080/19440049.2021.1948619
29. Wang T, Hong M (2016) Solid-state NMR investigations of cellulose structure and interactions with matrix polysaccharides in plant primary cell walls. J Exp Bot 67:503–514
30. Lee CM, Kafle K, Huang S, Yang H, Kubicki JD, Kim SH (2019) Probing cellulose structures with vibrational spectroscopy. Cellulose 26:35–79
31. Agarwal UP (2022) Beyond crystallinity: using Raman spectroscopic methods to further define aggregated/supramolecular structure of cellulose. Front Energy Res 10:857621
32. Agarwal UP (2017) Raman spectroscopy in the analysis of cellulose nanomaterials. In: Nanocelluloses: their preparation, properties, and applications. American Chemical Society, Washington, pp 75–90
33. Olivito F, Algieri V, Jiritano A, Tallarida MA, Tursi A, Costanzo P, Maiuolo L, De Nino A (2021) Cellulose citrate: a convenient and reusable bio-adsorbent for effective removal of methylene blue dye from artificially contaminated water. RSC Adv 11:34309–34318. https://doi.org/10.1039/D1RA05464C
34. Queiroz ALP, Kerins BM, Yadav J, Farag F, Faisal W, Crowley ME, Lawrence SE, Moynihan HA, Healy A-M, Vucen S, Crean AM (2021) Investigating microcrystalline cellulose crystallinity using Raman spectroscopy. Cellulose 28:8971–8985. https://doi.org/10.1007/s10570-021-04093-1
35. Fathy M, Abdel Moghny T, Mousa MA, El-Bellihi A-HA-A, Awadallah AE (2016) Absorption of calcium ions on oxidized graphene sheets and study its dynamic behavior by kinetic and isothermal models. Appl Nanosci 6:1105–1117. https://doi.org/10.1007/s13204-016-0537-8
36. Tasaso P (2015) Optimization of reaction conditions for synthesis of carboxymethyl cellulose from oil palm fronds. Int J Chem Eng Appl 6:101–104. https://doi.org/10.7763/ijcea.2015.v6.460
37. Onojake L, Apuyor KE, Apuyor SE (2023) Production of carboxymethyl cellulose (CMC) from rubber seed (Hevea brasiliensis) shells. Afr Sci 24:283–289. https://doi.org/10.26538/africanscientist.2.202306017
38. Marlina D, Novita M, Anwar MT, Kusumo H, Sato H (2021) Raman spectra of polyethylene glycol/cellulose acetate butyrate biopolymer blend. J Phys Conf Ser 1869:012006

39. Chen W, He H, Zhu H, Cheng M, Li Y, Wang S (2018) Thermo-responsive cellulose-based material with switchable wettability for controllable oil/water separation. Polymers 10:592. https://doi.org/10.3390/polym10060592
40. Kumar A, Kebaili I, Boukhris I, Vaish R, Kumar A, Park HKB, Joo YH, Sung TH (2023) Cotton functionalized with polyethylene glycol and graphene oxide for dual thermoregulating and UV-protection applications. Sci Rep 13:5923. https://doi.org/10.1038/s41598-023-31415-z
41. Mustapha S, Lease J, Eksiler K, Sim ST, Ariffin H, Andou Y (2022) Facile preparation of cellulose fiber reinforced polypropylene using hybrid filler method. Polymers 14:1630. https://doi.org/10.3390/polym14081630
42. Lewandowska AE, Eichhorn SJ (2016) Raman imaging as a tool for assessing the degree of mixing and the interface between polyethylene and cellulose nanocrystals. IOP Conf Ser Mater Sci Eng 139:012030
43. Salih AM, Vakili-Tahami F, Biglari H (2025) Cellulose nanocrystals effect to improve mechanical properties of epoxy-based glass fiber-reinforced composites. Iran Polym J. https://doi.org/10.1007/s13726-025-01468-0
44. Zhu Q, Zhou R, Liu J, Sun J, Wang Q (2021) Recent progress on the characterization of cellulose nanomaterials by nanoscale infrared spectroscopy. Nano 11:1353
45. Povolotckaia A, Dynnikova M, Lipovskaya V, Grigorieva I, Kaputkina S, Borisov E, Vasileva A, Pankin D (2025) Investigation of cellulose nitrate film-based photographic materials (1902–1917) using a portable Raman spectrometer and chemometric methods. J Cult Herit 71:217–224. https://doi.org/10.1016/j.culher.2024.12.003
46. Gao W, Zhou L, Liu S, Guan Y, Gao H, Hui B (2022) Machine learning prediction of lignin content in poplar with Raman spectroscopy. Bioresour Technol 348:126812. https://doi.org/10.1016/j.biortech.2022.126812
47. Luo R, Popp J, Bocklitz T (2022) Deep learning for Raman spectroscopy: a review. Analytica 3:287–301

Vibrational Spectroscopic Characterization of Starch

Mohomad Niyas Sherin Thaj, Thisal Induwara Vitharana Denagama Vitharanage, and Hemanth Noothalapati

Abstract Starch, an essential plant biopolymer, plays a central role in nutrition, industry, and materials science. This review explores how vibrational spectroscopy, particularly Raman and infrared (IR) techniques, has transformed our understanding of starch's structural complexity. By harnessing their complementary strengths, we can now analyze everything from granule morphology and crystalline polymorphs (A-, B-, C-, and V-types) to processing phenomena like gelatinization, retrogradation, and chemical modification. We begin by grounding readers in the fundamentals: Raman excels at probing glucose-ring and glycosidic backbone vibrations (notably the 480 cm^{-1} band), while IR reveals polar functional groups and hydrogen-bond networks (e.g., the 1047/1022 cm^{-1} doublet). Across polymorphs, spectroscopy uncovers distinctive spectral fingerprints such as O–H band broadening in hydrous B-type starch, mixed features in C-type, and carbonyl peaks signaling V-type inclusion complexes. We then chronicle recent technological leaps: confocal Raman micro-mapping, handheld devices coupled with machine learning (e.g., single-granule amylose quantification with $R^2 > 0.98$, <5% error), mid-IR quantum-cascade laser imaging, and multimodal approaches that fuse IR/Raman (or Raman + NIR) for enhanced structural and compositional insight. We further discuss emerging frontiers, standardized calibration protocols, deep-learning spectral interpretation, nanoscopic techniques (TERS, nano-IR), and in vivo starch monitoring, highlighting their promise and technical challenges. Finally, we envision a

M. N. S. Thaj · T. I. V. D. Vitharanage
Faculty of Applied Science, Department of Polymer Science, University of Sri Jayewardenepura, Nugegoda, Sri Lanka

H. Noothalapati (✉)
Department of Biomedical Engineering, Chennai Institute of Technology, Chennai, Tamil Nadu, India

Department of Chemical Engineering, Indian Institute of Technology Hyderabad, Kandi, Sangareddy, Telangana, India

Faculty of Life and Environmental Sciences, Shimane University, Matsue, Japan
e-mail: nvhnag@life.shimane-u.ac.jp

N. Mazumder et al. (eds.), *Vibrational Spectroscopy for Biopolymer Characterization*, Biological and Medical Physics, Biomedical Engineering,
https://doi.org/10.1007/978-3-032-15306-7_3

future of fully integrative vibrational platforms: non-destructive, in situ probes that deliver rapid, reagent-free analysis spanning molecular design to digestibility, nutrition, and industrial use. Collectively, this chapter demonstrates that vibrational spectroscopy is not just a tool, it's a powerful lens into starch science, bridging molecular details to real-world applications.

Keywords Starch · Spectroscopy · Chemical bonds · Polymer

1 Introduction

Starch is a major biopolymer produced by plants as an energy reserve and is a primary carbohydrate in the human diet [1]. It typically consists of two glucose polymers: amylose, a mostly linear chain of α (1 → 4)-linked D-glucopyranose units, and amylopectin, a highly branched polymer with α (1 → 4) backbones and α (1 → 6) branch points. These polymers coalesce into semi-crystalline starch granules, wherein clusters of amylopectin double helices form ordered lamellar domains interspersed with amorphous regions. The double-helical crystallites of amylopectin give starch granules a characteristic short-range molecular order (organization of the helices within lamellae) and long-range order (overall crystalline packing), analogous to the hierarchical structure of cellulose fibers (Fig. 1).

Most native starches contain 20–30% amylose by weight (with "waxy" starches <5% and high-amylose variants >50%), [2, 3] which influences granule structure and properties. Starch granules are typically 1–100 μm in size and exhibit diverse morphologies (spherical, oval, polygonal) depending on botanical origin; for example, cereal starches (maize, wheat, rice) have small polygonal or lenticular granules, whereas tuber starches (potato) have larger oval granules. Starch's abundant availability and biodegradability make it important not only in foods but also in biofuel production, papermaking, pharmaceuticals, and biodegradable polymer materials. For instance, starch from corn kernels is enzymatically converted to ethanol at an

(a) (b)

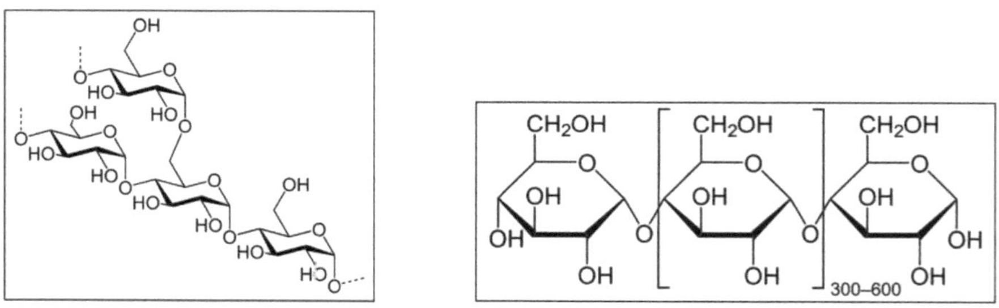

Fig. 1 Diagrammatic representation of the starch molecule in the two distinct forms, (**a**) branched amylopectin form and (**b**) linear amylose form

industrial scale, and starch-based plastics are developed as sustainable packaging. Understanding starch's structure-property relationships is therefore crucial for food science and biomaterials engineering. Vibrational spectroscopy has emerged as a powerful, non-destructive tool to characterize starch at the molecular level, probing its glycosidic linkages, helical order, hydrogen bonding, and chemical modifications. In this chapter, we review the use of vibrational spectroscopy, predominantly focusing on Raman and IR spectroscopy, in elucidating starch structure [1]. We begin with a brief theoretical background on these spectroscopic techniques, then discuss their application to starch, including typical spectral features and what they reveal about starch's different physical forms (native polymorphs A, B, and C; gelatinized and retrograded starch) and chemical modifications. We highlight recent advances such as micro-Raman imaging, portable spectroscopy, and combined IR/Raman analytical approaches and consider future trends and challenges in starch analysis. Throughout, we emphasize a critical, comparative perspective, showing how vibrational spectroscopy provides unique insights into starch chemistry and structure [4].

2 Theoretical Background

Vibrational spectroscopy encompasses techniques that measure molecular vibrations. Infrared (IR) spectroscopy detects absorption of IR radiation by vibrational transitions, whereas Raman spectroscopy detects inelastically scattered light (the Raman effect) arising from those vibrations. The two techniques obey different selection rules and thus offer complementary information.

- Raman scattering occurs when incident monochromatic light (usually a laser) interacts with molecular vibrations and scatters with a shift in energy corresponding to a vibrational mode. Most photons scatter elastically (Rayleigh scattering), but a tiny fraction undergo Raman scattering: if the molecule gains vibrational energy, a lower-energy Stokes-shifted photon is emitted; if the molecule loses vibrational energy from an excited state, a higher-energy anti-Stokes photon is emitted. For a vibrational mode to be Raman-active, it must involve a change in the molecule's polarizability (the ease of distorting the electron cloud) during the vibration. Starch molecules, being polysaccharides, have many C–C and C–O bond vibrations that significantly change polarizability, making Raman spectroscopy well-suited to detecting backbone and ring modes. Symmetric vibrations (e.g., skeletal stretches) are often strong in Raman even if IR-inactive. An important practical advantage is that water produces only weak Raman signals, so one can analyze starch in aqueous environments or gels with minimal interference from water. However, Raman signals are inherently weak, and fluorescence (from impurities or chromophores in unrefined starch) can overwhelm the spectrum. Using lasers in the near-infrared (e.g., 785 nm or 1064 nm) or applying spectrum processing (such as shifted-excitation Raman difference spectroscopy) are strategies to mitigate fluorescence in starch samples [5].

- Infrared absorption occurs when a vibrational mode causes a change in the molecule's dipole moment, allowing it to absorb IR light quanta of matching energy. Thus, modes involving polar bonds or asymmetric deformations (e.g., O–H, C=O, C–O stretches) are usually IR-active [6]. In an IR spectrum (often collected by Fourier-transform IR, FTIR), each peak corresponds to a vibrational frequency of a functional group, with peak intensity related to the transition dipole moment. Starch shows characteristic IR bands for O–H, C–H, and C–O vibrations of its glucosidic units. Because IR absorption of water (especially O–H stretch around 3400 cm^{-1}) is very strong, starch IR spectra are usually measured on dry samples or in attenuated total reflectance mode to minimize interference. Attenuated Total Reflectance (ATR) FTIR is particularly convenient for starch powders or films, as it probes only the sample's surface by an evanescent wave, requiring little sample preparation. ATR-FTIR spectra of starch are comparable to transmission spectra aside from minor intensity differences due to penetration depth.

Owing to their selection rules, vibrations that are intense in IR may be weak in Raman and vice versa [7]. For example, the broad O–H stretching band (3000–3600 cm^{-1}) arising from hydroxyl groups and hydrogen bonds in starch is very strong in the IR spectrum but hardly visible in Raman. Conversely, vibrations of the glycosidic C–O–C linkages and the glucose ring (800–1200 cm^{-1} region) give sharp, strong Raman peaks but relatively weaker IR absorption [7]. This complementarity is a central theme: IR spectroscopy excels at identifying functional groups, hydrogen bonding, and helical/order parameters via dipole-active modes, while Raman spectroscopy is especially sensitive to the carbon backbone conformation, ring vibrations, and crystalline vs. amorphous domain structure via polarizability-active modes. By combining both techniques, one can obtain a more complete vibrational "fingerprint" of starch than either alone. Modern instrumentation has improved both IR and Raman capabilities. Stable lasers (e.g., 532 nm, 785 nm) and CCD detectors now allow rapid acquisition of Raman spectra with low noise and minimal fluorescence. FTIR spectrometers with fast, sensitive detectors and ATR accessories enable quick, reproducible IR measurements on starch samples. Additionally, advanced methods like polarization measurements and two-dimensional correlation IR spectroscopy can probe starch chain orientation and interaction changes, while microscopic techniques (confocal Raman microscopy, IR micro-spectroscopy) can analyze starch at the micron scale. These innovations set the stage for detailed spectroscopic characterization of starch structure and its changes under various treatments.

3 Raman Spectroscopic Characterization of Starch

Starch has a rich Raman spectrum arising from its glucose ring vibrations, C–C/C–O skeletal stretches, and C–H and O–H deformations. Because starch shares the basic glucopyranose structure with cellulose, many band assignments are analogous, though differences in the α (1 → 4) linkage and helix packing of starch yield distinct spectral features. A prominent feature in starch Raman spectra is a strong band near 480 cm^{-1} [8], attributed to ring skeletal vibrations (primarily C–C–C and C–O–C bending modes of the glucose ring). This 480 cm^{-1} band is a diagnostic marker for starch: it is typically intense in organized (crystalline or helical) starch and has been used to identify starch in complex samples and to distinguish starches by botanic origin. For example, Raman analysis of corn vs. cassava starch found the 480 cm^{-1} band to be a major contributor to differentiating the two, correlating with their differing amylose content. Generally, higher amylose starches show lower relative intensity at 480 cm^{-1} (amylose being mostly amorphous in granules), whereas amylopectin-rich starches (with abundant double helices) give a stronger 480 cm^{-1} band, consistent with it being an indicator of ordered helical content [8].

In the high-frequency region, starch shows C–H stretching bands around 2900 cm^{-1} (C–H and CH_2 symmetric/asymmetric stretches) and O–H stretching around 3000–3600 cm^{-1} (a very broad, weak feature in Raman). The fingerprint region (approximately 200–1500 cm^{-1}) contains numerous overlapping bands from C–O, C–C, C–H, and ring deformations. Notable Raman bands for starch often reported include a band near 935–940 cm^{-1} (assigned to C–C stretching in the glucose ring or possibly C–O stretch of the glycosidic bond), bands around 860–880 cm^{-1} (C–H bending and glycosidic ring modes, sometimes associated with α-1,4 linkages), and a band near 1300 cm^{-1} (CH_2 twisting or bending mode). The precise positions and intensities of these bands can vary with starch structure. For instance, the band around 940 cm^{-1} may shift or split depending on helical conformation (amylose single helix vs. amylopectin double helix), and changes in the 860–880 cm^{-1} region have been noted upon amylose-lipid complex formation (V-type starch) [5].

Raman spectroscopy is particularly sensitive to starch crystallinity and polymorphic form. In native starch granules, semi-crystalline growth rings yield characteristic Raman signatures: the crystalline lamellae (amylopectin double helices) produce sharp bands (like the 480 cm^{-1} mode), whereas amorphous or less ordered regions contribute broader features (e.g., a broad band around 510 cm^{-1} may appear with more amorphous content). During starch gelatinization (disordering by heating in water), Raman spectra show a decrease in intensity and broadening of the 480 cm^{-1} band and other skeletal modes, reflecting the loss of helical order [9]. Conversely, when starch retrogrades (re-forms helices upon cooling), these Raman features regain intensity. Polarized Raman measurements can further probe molecular orientation in starch; although starch granules are polycrystalline, stretching or aligning starch (as in drawn films) could produce orientation-dependent Raman changes, analogous to what is observed in oriented cellulose fibers [10]. In practical applications, Raman spectroscopy has been used to quantify starch composition and structure. One notable success is the determination of amylose content via FT-Raman combined with multivariate analysis. Chemometric models based on Raman spectra (using regions like the 480 cm^{-1} band) allow rapid prediction of apparent amylose in starch samples with accuracy comparable to the standard iodine-binding method. This exemplifies how Raman spectral features correlate with starch's molecular makeup. Overall, the Raman spectrum of starch provides a fingerprint of its structure, with certain key bands serving as indicators of crystallinity, helical order, and even botanical or compositional differences, making Raman a valuable technique for starch characterization.

4 IR Spectroscopic Characterization of Starch

Starch's infrared spectrum is characterized by distinct absorption bands corresponding to its functional groups and the vibrational modes of the glucan chains. In the O–H stretching region (3000–3600 cm^{-1}), [11] starch shows a very broad, intense band due to the hydrogen-bonded hydroxyl groups on glucose units. The breadth of this IR band reflects a distribution of hydrogen bond strengths in starch (from both inter- and intramolecular O–H interactions), and its exact shape can be sensitive to moisture content and molecular order. For example, a fully gelatinized (amorphous) starch might exhibit a somewhat broader O–H band (and shifted peak maximum) compared to a native crystalline starch due to the disruption of organized hydrogen-bond networks. In the C–H stretching region (2800–3000 cm^{-1}), starch shows medium-intensity peaks around 2930 and 2880 cm^{-1} from –CH_2 and –CH stretches of the glucopyranose ring (Fig. 2). These C–H bands are usually similar across

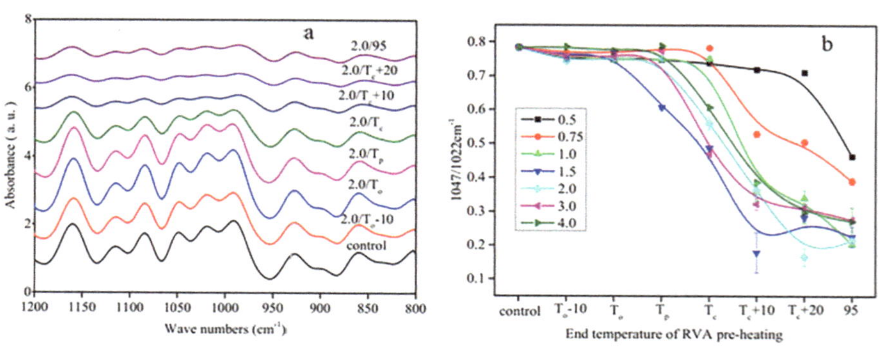

Fig. 2 The FTIR spectra (**a**) and ratios of absorbances at 1047/1022 cm^{-1} (**b**) of starch samples adapted from Wang et al. [12]

starch types, since all starches have the same glucose monomer (differences in these bands mostly indicate impurities or other components) [12].

The most information-rich part of the starch IR spectrum is the "fingerprint" region (1800–800 cm^{-1}). Here, complex overlapping bands arise from C=O (if present), C–O, C–C, and O–H bending vibrations in the starch molecule. Native starch (with no carbonyl groups) has no significant absorption in the 1800–1700 cm^{-1} region, a useful baseline; any peaks in this region usually indicate chemical modifications (e.g., acetyl or carbonyl groups, discussed later). One important set of IR bands in starch is around 1155–1080 cm^{-1}, attributed to C–O stretching vibrations in the C–O–C glycosidic linkages and ring C–O bonds. Within this region, two particular band positions have been extensively studied: approximately 1047 cm^{-1} and 1022 cm^{-1} [11]. These two IR bands (sometimes described as a doublet) are sensitive to the short-range order of starch; the 1047 cm^{-1} band is associated with the vibration of glucose rings in ordered helical environments, while the 1022 cm^{-1} band is associated with less ordered or amorphous structures. The ratio of absorbances at 1047 and 1022 cm^{-1} has been widely used as an index of starch crystallinity or "helix order" [12]. A higher 1047/1022 ratio signifies a greater proportion of double-helical order in the starch (as in native crystalline lamellae), whereas a lower ratio indicates more disordered chains (as in gelatinized or amorphous starch). In practice, this IR order parameter decreases upon gelatinization and increases upon retrogradation, correlating with X-ray diffraction crystallinity. For example, native cereal starch might show a relatively strong 1047 cm^{-1} band and a weaker 1022 cm^{-1} band, but after complete gelatinization, the 1047 cm^{-1} band diminishes significantly while the 1022 cm^{-1} band (and an even lower-frequency band, 995 cm^{-1}, associated with completely amorphous structure) becomes prominent [11].

Other notable IR bands for starch include 1155–1145 cm^{-1} (C–O stretch coupled with C–C stretch, often a shoulder in spectra), 1080 cm^{-1} (C–O stretch, sometimes overlapping with 1047 cm^{-1}), and 995–930 cm^{-1} (C–H bending and ring breathing modes; bands in this region can indicate different helical forms, e.g., V-type single helices can give a characteristic band near 930–935 cm^{-1}). Additionally, starch exhibits an IR band near 1645 cm^{-1} when moisture is present, due to the bending mode of absorbed water. A small peak around 1640–1650 cm^{-1} in a "dry" starch spectrum often actually indicates residual water (this band is usually much reduced after thorough drying). In the bending region below 800 cm^{-1}, complex modes from ring deformation occur, but these are less commonly analyzed in IR (since the ATR crystal cutoff often limits measurement below 600 cm^{-1}) [12].

Attenuated Total Reflectance IR has made it easy to rapidly scan starch samples for these key features. For instance, ATR-FTIR can be used to monitor starch during processing: as starch is heated in excess water, one can observe in real time the decrease of the 1047 cm^{-1} peak and increase of the 1022 cm^{-1} band, alongside broadening of the O–H stretch band, signaling the loss of crystalline order (gelatinization). Upon cooling and storage, partial recovery of the 1047 cm^{-1} band indicates retrogradation (recrystallization), which is often incomplete or different in polymorphic form from the native starch. Indeed, IR spectroscopy, together with techniques like DSC and XRD, has been applied to study starch retrogradation kinetics. Lu et al. used FTIR to analyze retrograded glutinous rice starch gels and observed characteristic changes in the O–H and C–O stretching regions as the starch recrystallized [13]. In summary, the IR spectrum of starch provides a wealth of information: the O–H stretch band reflects hydrogen bonding and moisture, the 1047/1022 cm^{-1} doublet reflects molecular order, and new peaks indicate chemical modifications. By tracking these spectral features, one can deduce structural changes in starch during processing and modification with high sensitivity.

5 Different Forms of Starch and Their Spectroscopic Characteristics

Starch can exist in several distinct structural forms, either due to different botanical origins or as a result of processing. Native starch granules exhibit different crystalline polymorphs (notably A-, B-, and C-type), and starch can be transformed into other physical states such as gelatinized (amorphous) or retrograded (recrystallized) forms. Furthermore, starch can be chemically or physically modified to alter its properties. Each of these forms leaves a "fingerprint" in the vibrational spectra (IR/Raman), which can be used to identify and characterize the form. In this section, we examine the major forms of starch A, B, and C polymorphs, gelatinized and retrograded starch; and chemically modified starches, describing their structure and the corresponding spectroscopic signatures reported in the literature (Fig. 3).

5.1 *A-Type Starch (Cereal Starch)*

5.1.1 Structure and Properties

A-type starch is the polymorph typically found in cereal grains (e.g., maize, wheat, rice) [14]. At the molecular level, A-type starch is characterized by amylopectin double helices packed in a monoclinic crystalline lattice with relatively low water content. Each unit cell of the A-type crystal allomorph contains 8 glucose residues (from two double helices) and a small number of water molecules locked in voids between helices. The double helices in A-starch are relatively tightly packed in parallel fashion, resulting in a high-density arrangement. This yields strong diffraction peaks at around 15°, 17°, 18°, and 23° 2θ in X-ray patterns (defining the A-type). A-type starch granules tend to have higher gelatinization temperatures and lower swelling power in water compared to B-type, owing to this compact structure and stronger internal hydrogen bonding (Fig. 4). Botanically, A-type starches are often polygonal under the microscope and smaller in size (e.g., cornstarch granules are 5–20 μm). They usually have moderate amylose content (20–30%) dispersed in the amorphous growth rings [14].

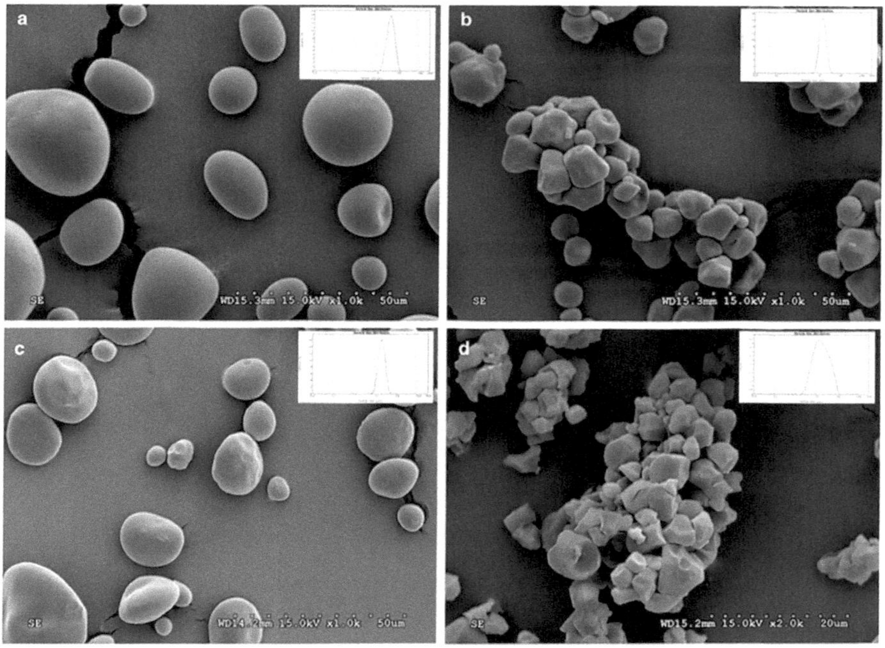

Fig. 3 SEM micrographs of (**a**) potato starch, (**b**) corn starch, (**c**) wheat starch, and (**d**) rice starch revealing the difference in granular structures from different starches. Reprinted from Domene-Lopez et al. (2019) under a Creative Commons Attribution license (https://creativecommons.org/licenses/by/4.0/)

5.1.2 Spectroscopic Characteristics

The vibrational spectra of A-type starch reflect its well-ordered double-helical structure. In IR spectroscopy, A-type starch typically shows a relatively high 1047/1022 cm^{-1} absorbance ratio, consistent with its substantial short-range order (abundant double helices) [1]. The O–H stretching band in A-starch IR spectra may appear as a broad band with one or more resolved maxima, indicating somewhat heterogeneous hydrogen bonding; however, because A-type has less inter-helical water, the O–H band can be slightly narrower or at a higher frequency compared to B-type (where more water broadens and shifts it). In the Raman spectrum, A-type starch exhibits a strong 480 cm^{-1} band (ring mode) as a hallmark of its crystallinity and other sharp features in the 800–1100 cm^{-1} region corresponding to its organized backbone vibrations. Some studies have noted that cereal starches (A-type) can be distinguished from tuber starches (B-type) by the intensity ratio of certain Raman bands; for instance, the 480 cm^{-1} Raman band is relatively more intense in A-type, whereas B-type shows some shifting and intensity redistribution in the 400–500 cm^{-1} region. Overall, A-type starch shows the "crystalline" spectral signatures most strongly: high IR order indices and sharp Raman peaks. Upon gelatinization, these

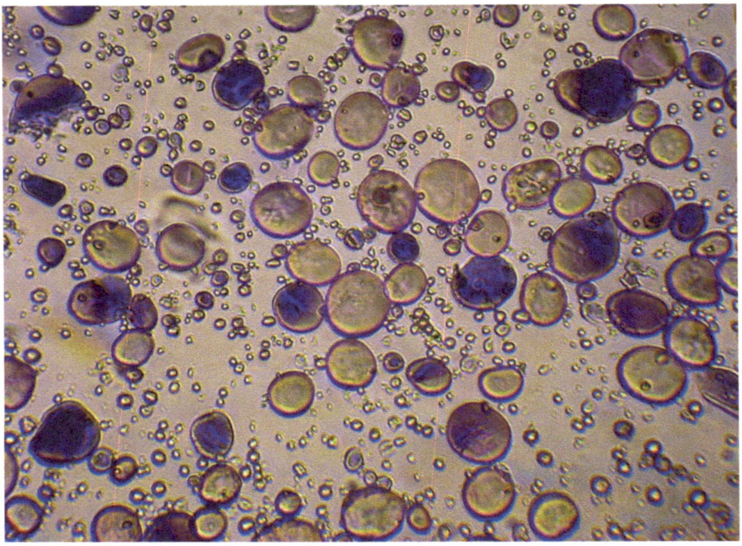

Fig. 4 Optical microscopic image of wheat starch granules stained with iodine

features diminish markedly. A-type starch can be converted to B-type by certain treatments (like long-term annealing or high-moisture heat treatment), and such a polymorphic transition would be evidenced by the corresponding changes in IR/Raman (discussed below) [1].

5.2 B-Type Starch (Tuber Starch)

5.2.1 Structure and Properties

B-type starch is the polymorph found in tuber and high-amylose starches (e.g., potato, green banana, high-amylose maize) and is characterized by a hexagonal crystalline lattice with a higher degree of hydration [15] amylopectin double helices in B-type starch arrange in a hexagonal pattern, and the unit cell contains significant water (around 36 water molecules per unit cell in the ideal B-form) lodged in channels between helices. This results in a more "open" structure. B-type starch granules (like potato starch) are typically larger (20–100 μm) and have a characteristic Maltese-cross birefringence pattern under polarized light. They also have distinct X-ray diffraction peaks at 5.6°, 17°, 22°, and 24° 2θ that identify the B-form. B-type starches often have higher overall crystallinity but consist of larger crystallite domains separated by more water; they swell more in water (due to that water content) and gelatinize at slightly lower temperatures than A-type (for normal amylose content) (Fig. 5).

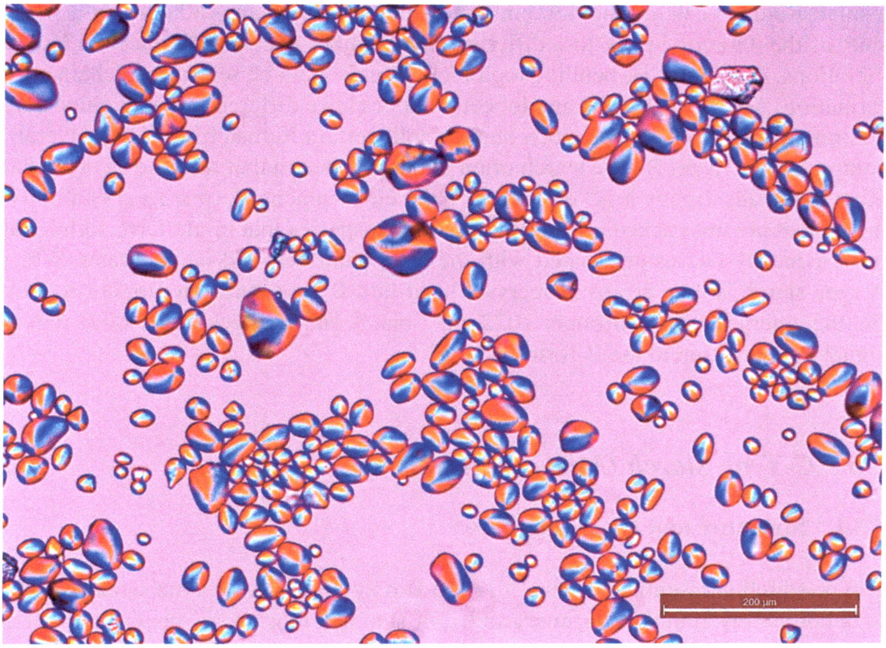

Fig. 5 Potato starch under an optical microscope (polarized light, magnification 100×). Examined under a microscope using a mixture of equal volumes of glycerol and distilled water, potato starch presents transparent, colorless granules, either irregularly shaped, ovoid, or pear-shaped, usually 30 µm to 100 µm in size but occasionally exceeding 100 µm, or rounded, 10 µm to 35 µm in size. Starch granules exhibit characteristic dark crosses in polarized light. (Adapted by Photon 400 750 licensed under the Creative Commons Attribution-Share Alike 4.0 International license)

5.2.2 Spectroscopic Characteristics

The vibrational spectra of B-type starch show subtle but important differences from A-type. In IR spectra, B-type starch tends to have a slightly lower 1047/1022 cm^{-1} ratio (indicating a somewhat lower fraction of double-helical order detectable by IR). This might seem counterintuitive since B-type is highly crystalline; however, the IR ratio is also sensitive to how tightly packed or uniformly oriented the helices are. The broader distribution of hydrogen bond environments in B-type (due to more water) can lead to a bit more intensity at 1022 cm^{-1} (amorphous-like component) relative to 1047 cm^{-1} [16]. The O–H stretching band in B-type IR spectra is typically very broad and may be centered at a slightly lower frequency compared to A-type, reflecting the extensive hydration (water interacting with hydroxyls). In fact, an IR broad peak around 3400 cm^{-1} with a shoulder near 3220 cm^{-1} is often noted, which can be deconvolved into contributions from inter-helical O–H bonds and water O–H bonds. In Raman spectra, B-type starch still exhibits the signature 480 cm^{-1} band [4] (since double helices are present), but its intensity and exact position can differ from A-type. Studies have found that B-type polymorphs may show a small Raman peak around 450 cm^{-1} in addition to the main 480 cm^{-1} or a slight shift of the 480 cm^{-1} peak to a different wavenumber due to differences in lattice vibrations. Also, the CH_2 bending region (1450 cm^{-1}) can be sensitive to chain conformation: A-type vs. B-type amylopectin helices have different skew in glucosidic torsion angles, which might subtly shift or split certain Raman bands [17]. Overall, compared to A-type, B-type starch often yields IR spectra that indicate a somewhat more hydrated, slightly less "organized" helix environment (despite its crystallinity, the helices are not packed as tightly) and Raman spectra that might have additional low-frequency modes associated with the larger unit cell and lattice water. When A-type starch is mercerized or recrystallized into B-type, one can monitor this by IR: the emergence of a stronger 1022 cm^{-1} bands and changes in the O–H region signal the transition to the B-form.

5.3 C-Type Starch (Mixed Polymorph)

5.3.1 Structure and Properties

C-type starch is essentially a mix of A and B polymorphs within the same granule. It is commonly found in legumes and root starches (e.g., peas, beans, and some sweet potatoes) [18]. The X-ray pattern of C-type starch is a combination of A and B reflections, typically showing peaks of both types (often a dominant A-type pattern with a small B-type peak at a low angle, or vice versa). On the molecular level, C-type starch granules contain some regions of A-type packing and some of B-type, possibly in different granular growth rings or as intergrown crystallites. The overall crystallinity of C-type starch is intermediate. Morphologically, C-type starch granules can resemble either parent type (pea starch granules look more A-like, while others more B-like). C-type starch may occur naturally in plants that store starch under intermediate moisture conditions or have specific branching patterns that favor mixed packing.

5.3.2 Spectroscopic Characteristics

As one might expect, the IR and Raman spectra of C-type starch often appear as an amalgam of A and B features. In IR, the 1047/1022 cm^{-1} ratio in C-type starch is typically intermediate between that of pure A and pure B. For example, pea starch (C-type) might show a moderate IR doublet ratio and an O–H band that is broad but not as broad as potato (B-type) [19]. The presence of both A- and B-type domains could cause multiple sub-peaks in the O–H stretching envelope (some hydroxyls in environments like A, some like B). In Raman spectra, C-type starch usually still has a prominent 480 cm^{-1} band (since both A and B have that ring mode). If the mixture is on a microscopic scale, the Raman spectrum might just be a weighted average. In some cases, careful analysis can detect the coexistence of both polymorphs: for instance, a small Raman band around 450 cm^{-1} might hint at B-type presence alongside the main 480 cm^{-1} of A-type. Similarly, minor differences in the 800–900 cm^{-1} region might indicate contributions from both helix forms. In practice, C-type starch is often identified by XRD rather than spectroscopy. However, FTIR with multivariate analysis has been used to classify starches by type: combinations of spectral features have successfully clustered starch samples into A, B, or C categories. For example, a portable ATR-FTIR or Raman device coupled with chemometric models was able to discriminate corn (A-type), wheat (A-type), cassava (A-type), and potato (B-type) starches, effectively recognizing that legume starches fell between the two groups. Thus, vibrational spectroscopy, especially with statistical pattern recognition, can identify C-type starches by their composite spectral signatures [20, 21].

5.4 V-Type Starch (Amylose Complexes)

5.4.1 Structure and Properties

In addition to the native polymorphs above, starch can form a V-type structure when amylose chains complex with small molecules such as lipids or iodine. V-type starch consists of single amylose helices (not double helices) wrapped around guest molecules (fatty acids, monoglycerides, iodine, etc.), typically crystallizing in a helical inclusion complex. The XRD pattern of V-type starch (e.g., amylose-lipid complexes) shows a strong peak around 20° 2θ and some lesser peaks, distinct from A or B. These complexes are of interest because they are a form of resistant starch (amylose-lipid complexes resist digestion).

5.4.2 Spectroscopic Characteristics

Vibrational spectroscopy can identify V-type starch by the presence of spectral features unique to single helices and the guest molecules. In IR, V-type amylose-lipid complexes will show the hallmark absorptions of the lipid (for example, a sharp carbonyl band around 1740 cm^{-1} if a fatty acid is present, which would be absent in normal starch) as well as changes in the starch fingerprint region. The formation of single helices often shifts the IR bands associated with helices: literature notes that V-type single helices lack the strong 1047 cm^{-1} band of double helices and instead might show an IR band around 1020–1022 cm^{-1} as the major peak (since the helices are isolated, behaving more like amorphous from the perspective of those IR modes) [22]. Also, a new IR band near 994 cm^{-1} has been reported for V_6 amylose complexes with iodine or butanol, corresponding to the helical conformation different from native double helices. In Raman spectra, amylose single helices can be identified by specific peaks of the guest and subtle differences in the backbone modes. For example, complexes with iodine give a distinctive iodine-starch Raman band (the classic blue iodine-starch complex has a band near not directly relevant for unlabeled systems). Amylose-lipid complexes might be detected by a Raman band of the lipid (like a C–H bend or C–C stretch of the alkyl chain) appearing in the starch spectrum. Importantly, V-type amylose helices typically still show the 480 cm^{-1} Raman band (because the glucose rings still vibrate), but the environment of amylose single helices can cause slight frequency shifts. A study on V-type starch reported that the 480 cm^{-1} band might shift to 475 cm^{-1} and lose intensity, and a new Raman band near 520 cm^{-1} can appear, which was absent in the double-helical form. This new band may correspond to lattice modes of the single-helix crystal. Thus, by comparing spectra before and after complex formation, one can confirm V-type starch. In practical terms, ATR-FTIR has been used to confirm amylose complexation in foods; for instance, when starch is cooked with lipids, the appearance of the lipid carbonyl peak and changes in the 1047/1022 ratio signal the creation of amylose-lipid V-complexes, which contribute to resistant starch content [23].

5.5 *Gelatinized Starch (Amorphous)*

Gelatinization refers to the process of heating starch in water (or excess moisture) such that the semi-crystalline granule structure is disrupted. The result is gelatinized starch, in which the molecular order (double helices and crystalline lamellae) has been mostly destroyed, producing an amorphous starch gel or paste. Amylopectin double helices unravel, and amylose leaches out of granules. On cooling, gelatinized starch may form a gel network (Fig. 6). Immediately after gelatinization (while hot), starch is largely amorphous (no crystallinity by XRD), though some residual short-range order can persist if not fully gelatinized [24].

The loss of molecular order upon gelatinization is dramatically reflected in vibrational spectra. IR spectroscopy shows a marked decrease in the intensity of the "crystalline" band at 1047 cm^{-1} and a relative increase of the 1022 cm^{-1} band. In fully gelatinized starch, the 1047 cm^{-1} band may nearly disappear into a broad envelope, whereas the 1022 cm^{-1} (amorphous) band and a broad band around 995 cm^{-1} dominate. This corresponds to the well-known drop in the IR doublet ratio. The broad O–H stretching band may shift or widen further as hydrogen bonds become more disordered and starch binds water in a less organized way. Raman spectroscopy of gelatinized starch likewise shows broadening and intensity loss of key bands. The 480 cm^{-1} Raman band (helix ring vibration) significantly diminishes upon gelatinization, since the organized double helices are gone. Instead, a broad band around 460–470 cm^{-1} may remain, indicative of the residual ring vibrations in a disordered environment. The sharp Raman peaks in the 800–1100 cm^{-1} region (like near 940 cm^{-1}) also broaden and merge. Essentially, the Raman spectrum of gelatinized starch takes on the appearance of a broad, featureless profile compared to the crisp spectrum of native starch. These spectroscopic changes provide a convenient way to assess the degree of gelatinization. In an industrial or laboratory setting, one could use ATR-FTIR on a dried starch sample to quickly check if the

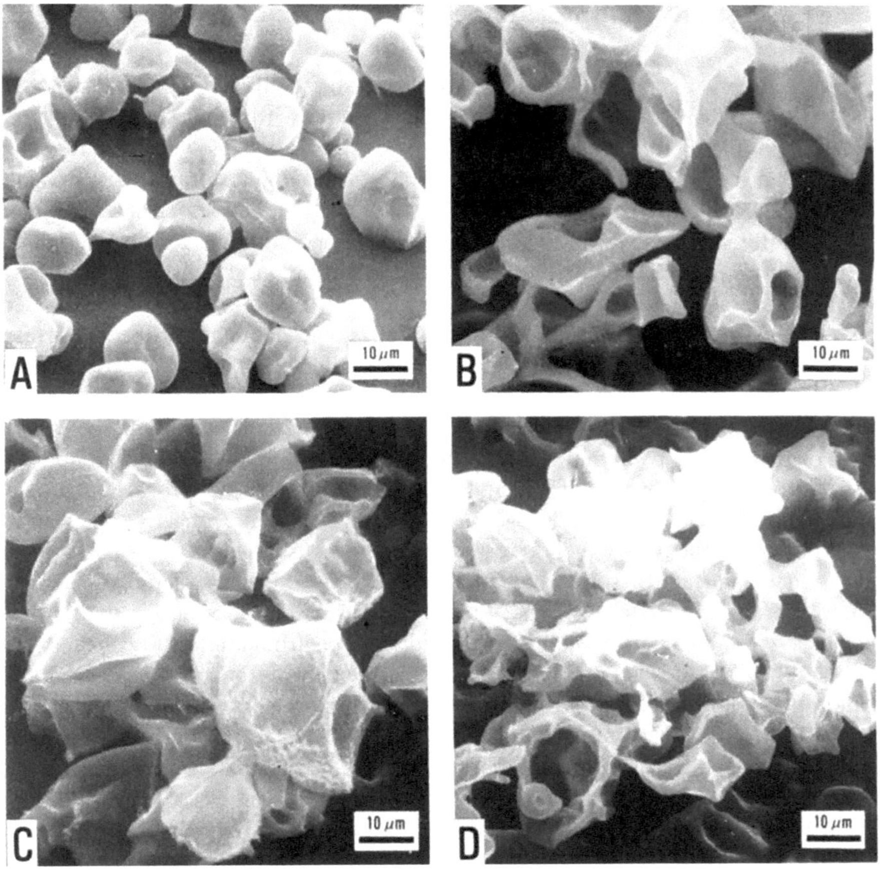

Fig. 6 SEM of corn starch swelling and gelatinization (Utah State Digital Commons https://digitalcommons.usu.edu/foodmicrostructure/vol1/iss1/3/ adapted by Christianson et al., 1982, CC BY 4.0)

starch is still largely crystalline (showing the 1047 cm^{-1} band) or fully gelatinized (absence of that band). Likewise, Raman spectroscopy can be used in-situ (since water doesn't interfere strongly) to monitor starch gelation in real time. One challenge, however, is that gelatinized starch often still contains some ungelatinized remnants unless heating is extreme; thus, spectra might show a mixture of features. Nonetheless, vibrational spectroscopy is an excellent indicator of the transition from the ordered to amorphous state in starch. For example, in the production of biomaterials where starch is thermally processed, Raman mapping can reveal heterogeneity spots of intact crystallites vs. gelatinized matrix by the presence or absence of the 480 cm^{-1} band. In summary, gelatinization causes a collapse of starch's spectral "fingerprint," with the disappearance of order-indicating bands, which is a clear spectroscopic signature of this form [11].

5.6 Retrograded Starch (Recrystallized)

Retrogradation is the process wherein gelatinized starch, upon cooling and storage, slowly recrystallizes. Amylose chains first reassociate (within hours or days), forming mostly disordered precipitates or V-type single helices, and amylopectin branches later rearrange over longer times (days to weeks) to form a more crystalline structure, often B-type polymorph crystals. Retrograded starch thus contains new crystallites distinct from the original granule crystallinity. This phenomenon contributes to the staling of bread and also to the formation of resistant starch (RS3 type). Retrograded starch is typically a mixture of double-helical regions (often B-type or a mix) embedded in an amorphous matrix of remaining unfrozen segments.

Retrograded starch's spectra reflect *partial* reordering. In IR, one observes the gradual return of the 1047 cm^{-1} band and increase of the 1047/1022 ratio as the starch retrogrades, though it usually does not reach the level of the native starch. The O–H stretching band may sharpen slightly or shift, indicating the formation of new hydrogen-bonded structures [25]. Notably, retrograded starch formed at typical ambient conditions often yields B-type crystallites; therefore, the IR spectrum of fully retrograded starch frequently resembles that of B-type starch: e.g., a prominent 1022 cm^{-1} band and a somewhat broadened 1045–1050 cm^{-1} shoulder, akin to potato starch. If amylose forms V-type complexes during retrogradation (for instance, with lipids present in the system), the IR spectrum might show the tell-tale carbonyl band of those complexes or other unique peaks. Raman spectroscopy can capture the reappearance of starch's organized structure as well. As retrogradation proceeds, the 480 cm^{-1} Raman band intensifies again, signifying re-formed helices. However, the shape of this band might differ from the original: retrograded amylopectin often crystallizes as B-type, so the 480 cm^{-1} band might be at a slightly different position or width (as described for B-type above) [25]. Furthermore, new Raman bands corresponding to V-type amylose complexes can appear; for instance, a complex with fatty acids could show the lipid's C–C stretch at 1060 cm^{-1} in Raman. In tracking retrogradation, time-resolved ATR-FTIR or periodic Raman measurements have been used. Hu et al. demonstrated that by using PLS regression on combined IR and Raman data, one could quantitatively predict the degree of retrogradation (amount of recrystallized starch) in stored gelatinized starch samples. Their models leveraged the gradual spectral changes, achieving good correlation ($R = 0.96$) with measured retrogradation levels. Spectroscopically, retrograded starch is essentially a "hybrid" of amorphous and re-crystalline features, and vibrational spectroscopy is sensitive to both. This makes IR/Raman ideal for studying factors that influence retrogradation (like temperature, moisture, or the presence of sugars or emulsifiers). For example, more extensive retrogradation (as from longer storage or cooling at 4 °C) yields spectra closer to native B-type starch, whereas inhibited or short-term retrogradation yields only small bumps in the formerly featureless gelatinized spectrum. In conclusion, retrograded starch can be identified by the partial re-emergence of order-indicative bands in IR/Raman spectra, and these techniques are frequently used to analyze starch retrogradation in food science.

5.7 *Chemically Modified Starches*

Structure and properties: Starch is often chemically modified to improve its functionality. Common modifications include etherification (e.g., hydroxypropyl starch), esterification (e.g., acetylated starch, octenyl-succinate starch), cross-linking (phosphate or adipate bridges between chains), and oxidation (introducing carbonyl/carboxyl groups). These modifications introduce new functional groups into the starch polymers or form chemical bonds between chains. For example, acetylated starch has acetyl (–C(O)CH_3) moieties substituting some hydroxyls, oxidized starch might have –CHO or –COOH groups on the glucose units, and phosphorylated starch contains phosphate monoester or diester groups. Such changes typically disrupt some of the native hydrogen bonding and can reduce crystallinity (many modified starches are more amorphous), but they impart properties like improved solubility, clarity, or stability. The degree of substitution (DS) average number of substituents per glucose unit, is usually low (e.g., DS 0.01–0.2 for food-grade modified starches), but even low levels are detectable by spectroscopic means.

5.7.1 Spectroscopic Characteristics

Introduction of new functional groups into starch gives rise to distinctive new bands in IR spectra and sometimes in Raman spectra, allowing one to readily confirm and characterize the modification. Some key examples include

Acetylated starch: The presence of acetyl groups (–O–C(O)CH_3) is evident in IR by the strong ester carbonyl (C=O) stretch near 1740 cm^{-1}. Native starch has no absorbance in this region, so even a small acetyl peak around 1735–1750 cm^{-1} indicates acetylation. The intensity of this band can be used to quantify the degree of substitution. Additionally, acetylation reduces the intensity of the O–H stretching band (3400 cm^{-1}) since many O–H groups are replaced. In fact, for fully acetylated starch (starch triacetate), the broad O–H band disappears entirely. Raman spectra of acetylated starch will also show the carbonyl band (though weaker, since C=O is a polar bond more IR-active). A Raman band around 1730 cm^{-1} can be observed with a strong laser. The substitution also shifts some backbone modes; for instance, the glycosidic C–O stretching band (around 1150–1080 cm^{-1}) may change shape due to the electron-withdrawing acetyl groups on nearby OHs.

Oxidized starch: Oxidation (e.g., with sodium hypochlorite) introduces carbonyl groups, commonly at C-2 or C-3 (forming dialdehyde starch if strong) or carboxylate groups (–COO^-) if further oxidized. IR spectroscopy is very sensitive to these changes: aldehyde C=O groups give a peak near 1730–1740 cm^{-1} (typically appearing if starch is oxidized to dialdehyde), and carboxylate (COO^-) groups show asymmetric and symmetric stretch bands at 1600 cm^{-1} and 1400 cm^{-1}, respectively. For example, a mildly oxidized starch with carboxylates will exhibit a new IR band around 1600 cm^{-1} (asymmetric -COO^-) and a shoulder near 1410–1420 cm^{-1} (symmetric -COO^-). These often overlap with the water-bending band (1645 cm^{-1}) or CH_2 bending (1415 cm^{-1}), but with careful analysis or after drying (to remove the water band), they are clearly distinguishable. Raman spectroscopy can also detect carboxylate groups: a –COO^- symmetric stretch can appear as a Raman band at 1380 cm^{-1} (which in IR appears at 1420 cm^{-1}, but Raman "sees" the symmetric mode around 1380). Thus, a new Raman peak near 1375–1385 cm^{-1} in an oxidized starch would indicate –COO^- introduction. Oxidation tends to break some glucosidic bonds and reduce molecular weight, often destroying crystallinity, so one will also notice in IR that the 1047 cm^{-1} band (order) diminishes and broad bands around 1022 and 1100 cm^{-1} dominate (amorphous character).

Phosphorylated (cross-linked) starch: Phosphate groups (mono- or di-esters) in starch introduce P=O and P–O vibrations. A diagnostic IR band for phosphate ester is around 1240–1260 cm^{-1} (P=O stretching). For example, sodium phosphate cross-linked starch often shows a band near 1250 cm^{-1} and another around 970–990 cm^{-1} (P–O stretch). These can overlap with starch's C–O stretches, but a bump in that region not present in the native starch spectrum reveals phosphorylation. Additionally, the presence of phosphate can broaden the O–H band (since phosphate groups can form strong hydrogen bonds and also bring in water) [26, 27].

Hydroxypropylated starch (etherified): The introduction of hydroxypropyl groups (–O–CH_2–CHOH–CH_3) adds new C–H and O–H features. IR will show additional C–H stretching (a small increase around 2970 and 2870 cm^{-1} from the –CH_3 of the propyl group) and possibly a weak C–O stretch around 1120 cm^{-1} (from the added secondary alcohol). These changes are subtle and often overlap with native starch bands, so quantification usually relies on noticing that the O–H band broadens (more hydroxyls overall) and the overall IR pattern indicates more amorphous character (hydroxypropylation heavily reduces crystallinity). Raman might detect the -CH_3 presence via a tiny band near 1450 cm^{-1} (CH_3 deformation) or 1380 cm^{-1} [9, 25, 28].

Cross-linking (e.g., with epichlorohydrin or adipates): Cross-links typically do not introduce a new IR band with a unique frequency (unless the cross-link has a distinct group like carbonyl in adipate). Instead, their spectroscopic signature is often indirect: cross-linking "locks" the starch structure, often preventing gelatinization, so after such treatment the starch IR spectrum might retain its 1047 cm^{-1} band even after heating (i.e., it resists becoming amorphous). If adipate or citrate is used to cross-link, their carbonyl peaks (1720–1740 cm^{-1} for ester linkages) may be observed.

In general, IR spectroscopy is the go-to method for confirming starch chemical modifications, given its clear identification of functional groups. For instance, in quality control, an ATR-FTIR scan can quickly verify if a starch has been acetylated (by the 1740 cm^{-1} band) or phosphorylated (by the 1250 cm^{-1} band). Raman spectroscopy provides complementary evidence, particularly useful if IR bands overlap or if one wants to verify certain substituents that are Raman-active. Moreover, vibrational spectroscopy can reveal how the modification affects the starch backbone: often one observes that modified starches have diminished crystallinity. An acetylated or hydroxypropylated starch typically has no significant 1047 cm^{-1} IR peak (the substitution disrupts helices, making the starch essentially amorphous). Correspondingly, the Raman spectrum will lack the sharp features of native starch (e.g., the 480 cm^{-1} band may be weak or broad). For example, a study noted that carboxymethyl starch's IR spectrum lost the characteristic "crystalline starch" 1045 cm^{-1} band, consistent with an amorphous structure. Thus, vibrational spectroscopy not only confirms the presence of new chemical groups but also indicates the secondary effect on starch's supramolecular structure. These techniques, combined with calibration models, have even been used to quantify the degree of substitution. For instance, the intensity of the 1740 cm^{-1} IR peak has been correlated to the acetyl content in starch, much like a recent report for cellulose acetate used ATR-FTIR with machine learning to accurately determine DS (analogous strategies can apply to starch acetates). In summary, chemically modified starches each carry a distinct spectral fingerprint, a boon for analysts verifying modified food starches or designing new starch derivatives for materials applications.

6 Recent Advances in Vibrational Spectroscopy of Starch

Advancements in spectroscopic technology and methodology over the past decade have significantly enhanced our ability to analyze starch structure. One notable development is the integration of imaging and micro-spectroscopy. Confocal Raman microscopy, for example, can map the composition of starch granules with submicron resolution. Researchers can collect thousands of Raman spectra across a granule and use techniques like principal component analysis (PCA) or k-means clustering to visualize internal heterogeneity (such as regions of differing amylose content or remnants of crystalline growth rings). Raman imaging has revealed, for instance, the non-uniform distribution of amylose and amylopectin in certain starch granules and has been used to observe how gelatinization starts at the granule hilum and progresses outward [29]. FTIR microscopy can be applied to starch: although starch granules are a bit small for conventional IR microspectroscopy without special optics, bulk mapping in starch-based foods (like mapping starch vs protein in a pasta cross-section) has been achieved. These imaging approaches extend vibrational spectroscopy from just providing average information to giving spatially resolved insight, which is crucial in understanding granular structure and interactions in composite systems [30].

Another area of progress is time-resolved and in situ spectroscopy. Modern spectrometers can acquire spectra very rapidly (on the order of one spectrum per second or faster). This enables monitoring processes like starch gelatinization or enzymatic hydrolysis in real time. For example, ATR-FTIR has been used to track gelatinization kinetics by observing the 1047/1022 cm^{-1} ratio decrease continuously as temperature ramps up. Starch hydrolysis by amylase can similarly be monitored via IR:

one sees the gradual loss of starch's characteristic bands and the appearance of bands for maltose or glucose. Such real-time measurements, when combined with two-dimensional correlation analysis, provide insight into the sequence of molecular events (which bonds break first, etc.). On the Raman side, a technique called stroboscopic Raman or time-gated Raman has been tested to capture transient states during starch gelatinization, though challenges remain in signal-to-noise. A very recent innovation is the use of soft mechanical nano-ablation (sMNA) combined with Raman spectroscopy to examine starch granules *in situ* within plant tissues. Farahi et al. demonstrated that gently abrading the surface layers of starch granules inside a plant cell wall and then performing Raman microspectroscopy on the exposed interior allowed chemical confirmation of starch without disrupting its native environment. This approach, linking nanomechanics with Raman, opens a new window to study starch granules in their natural context (e.g. in a leaf or trunk) where they are surrounded by cellulose and other polymers [31].

Enhanced spectroscopy techniques are also being explored. Coherent anti-Stokes Raman scattering (CARS) and stimulated Raman scattering (SRS) are nonlinear optical methods that can generate stronger signals and enable video-rate chemical imaging [32]. These have been applied to image starch granules in cells or to follow starch gelatinization at high speed, without the need for fluorescent dyes. While still mostly in the research domain, CARS microscopy could, for example, visualize starch in a living plant cell and monitor how it is mobilized during seed germination by tuning to starch's specific Raman frequencies. Surface-enhanced Raman spectroscopy (SERS), where a rough metal surface amplifies Raman signals, has seen limited but growing use for starch, mostly to detect starch-bound substances or very low-level components. For instance, SERS substrates have been used to detect iodine or fatty acids bound in amylose helices at much lower concentrations than normal Raman would allow. A recent study prepared cellulose-based Ag nanoparticle substrates and suggested they could be used for SERS detection of polysaccharides; similar concepts could apply to starch for sensitive detection of adulterants or minor constituents.

On the data analysis front, chemometrics and machine learning have greatly enhanced starch spectroscopy. We've mentioned how PLS regression can quantify amylose or retrogradation from spectra. More recently, machine learning models have been developed for starch source authentication. Hossain et al. introduced a handheld Raman spectroscopy method paired with machine learning to quantitatively determine amylose and amylopectin directly in single starch granules. By scanning hundreds of individual maize, cassava, and wheat granules, they employed principal component analysis and regression modeling to classify and predict composition in situ. Remarkably, their model achieved an R^2 greater than 0.98 and prediction errors below 5%, underscoring the method's precision and reliability. This work demonstrates that it's now possible to perform rapid, non-destructive, granule-level analysis in the field or on the factory floor, offering a practical and scalable tool for distinguishing starch from different botanical sources [30].

Finally, improvements in fundamental understanding have come from theoretical calculations and advanced analysis of starch spectra. Quantum chemical calculations (DFT) on small glucose and malto-oligosaccharide models have been used to assign vibrational modes in starch. While cellulose has seen extensive DFT studies to interpret its spectra, analogous work on maltose and amylose fragments is emerging. These calculations help deconvolute overlapping bands for example, clarifying which vibrations contribute to the 1047 vs 1022 cm^{-1} IR bands. They also aid in understanding the effect of substitution on vibrational frequencies, complementing experimental spectra. Additionally, 2D correlation spectroscopy (2D-COS) has been applied to starch IR data to tease apart changes during gelatinization or annealing. By correlating intensity fluctuations of bands as a function of temperature or time, researchers can determine the sequence of structure changes (e.g., which hydrogen bonds break first as temperature rises). This adds a dynamic, mechanistic dimension to starch spectroscopy that goes beyond static spectra [33].

In summary, recent advances have made vibrational spectroscopy of starch more powerful, quantitative, and insightful. Imaging techniques allow us to see starch's spatial composition, time-resolved measurements capture dynamic transformations, enhanced Raman methods and data fusion improve sensitivity and specificity, and sophisticated analyses deepen our understanding of starch at a molecular level. These developments collectively push vibrational spectroscopy to the forefront of starch research, turning it from a simple analytical tool into a comprehensive platform for starch characterization in both fundamental and applied contexts.

7 Combined Raman and IR Analysis of Starch

Given the complementary nature of Raman and IR spectroscopy for starch, combining the two techniques can provide a more holistic characterization than either alone. By jointly analyzing IR and Raman data, one can leverage the strengths of each: IR captures polar functional group information (excellent for detecting chemical modifications and hydrogen bonding), while Raman excels at fingerprinting the skeletal structure and crystalline order. In many cases, this combined approach has proven especially powerful.

One example is in measuring starch crystallinity or order. As discussed, IR provides the 1047/1022 cm^{-1} ratio as a measure of short-range order, and Raman provides the intensity of the 480 cm^{-1} band (or related features) as an indicator of ordered structure. Individually, each metric is useful but has limitations (IR might be affected by moisture, Raman by fluorescence). Using both, researchers can cross-validate results and get a more robust assessment. For instance, in starch retrogradation studies, FTIR might indicate reformation of helices via the IR doublet, while Raman concurrently tracks the return of the 480 cm^{-1} band; concordance of these two confirms true recrystallization. Discrepancies can be informative too: if IR indicates order but Raman does not, it might suggest the formation of structures that have dipole activity but low polarizability contrast (perhaps certain V-type complexes). Thus, the two techniques together give a complete vibrational fingerprint: essentially all vibrational modes of starch can be captured either in IR or Raman.

Modes that are IR-invisible (like symmetric C–C stretches) appear in Raman, and modes that are Raman-weak (like O–H stretches and C=O stretches) appear strongly in IR [25].

The combined approach also shines in multicomponent systems. In processed foods or bioplastics, starch often coexists with other ingredients (proteins, lipids, plasticizers). IR and Raman have different selectivity for these components. For example, proteins have amide bonds that give intense IR peaks (1650 and 1550 cm^{-1}) but relatively weak Raman signals, whereas starch has strong Raman signals but fewer distinctive IR peaks in that region. By combining IR and Raman, one can deconvolve the contributions of each component. A practical demonstration was by Hu et al., who fused MIR and Raman data to quantify starch retrogradation in the presence of other food ingredients [25]. The fusion model outperformed single-technique models, achieving higher prediction accuracy for retrogradation degree. The authors found that some information captured by Raman (related to starch backbone reordering) and some by IR (related to water and hydrogen bonds) were both necessary to accurately gauge retrogradation. Medium-level data fusion (i.e., combining processed features from each spectrum) gave the best results.

In starch authentication and classification, combined IR/Raman approaches have also emerged. For instance, one might record both an ATR-FTIR and a Raman spectrum of a starch sample and feed both into a machine learning algorithm for botanical origin classification. Zhang et al. reviewed such multimodal spectroscopy in food analysis and noted that dual-mode systems can compensate for the limitations of each method. In the case of starch, an IR spectrum might be very similar for two starches, but the Raman could be different (or vice versa) [33]. Using both ensures that subtle differences are captured. This is especially relevant for starches that are chemically modified: the IR will spot the chemical substituents clearly, while the Raman will reveal any changes in the polymer backbone and crystallinity as a result of modification. Combined, they fully characterize the modified starch.

Another aspect of combining IR and Raman is in structural interpretation. While X-ray diffraction gives the crystalline polymorph, IR/Raman together can provide insight into the hydrogen bonding pattern (from IR) and the conformation of glucosidic linkages (from Raman). For instance, when determining if a starch is truly in B-form or a mixture of forms, IR might confirm the presence of B-type O–H bonding (via a specific OH bending or stretching feature), while Raman confirms the presence of the unique lattice mode of B-type at 450 cm^{-1}. Together, this can strengthen conclusions about starch structure changes.

In practice, many researchers sequentially use both techniques on starch samples. The trend now is to integrate the datasets. Software and analytical pipelines are being developed to do synchronous analysis of IR and Raman spectra. One can even perform two-dimensional correlation analysis between IR and Raman, so-called hybrid 2D correlation, where variations in one spectrum as a function of some perturbation (temperature, time) are correlated with variations in the other spectrum. This can elucidate cross-information, such as "a decrease in the Raman 480 cm^{-1} band occurs simultaneously with an increase in the IR 1022 cm^{-1} band during process X," reinforcing interpretations about loss of helical order.

Ultimately, combined IR and Raman analysis aligns with the goal of obtaining a comprehensive molecular picture of starch. As one review put it, by considering both IR and Raman, one can capture “virtually all vibrational modes of cellulose” the same applies to starch. With both spectra, one has a complete set of fingerprints to use in multivariate models or spectral libraries. This is especially useful in forensic or quality control applications: e.g., verifying if a product contains starch and if so, what kind (IR picks up that there is starch via a 930 cm^{-1} C–H band and Raman confirms it via the 480 cm^{-1} band) giving high confidence identification [25].

In conclusion, the synergy of IR and Raman provides a more powerful analytical approach for starch than either technique on its own. By exploiting their complementary sensitivities, combined spectroscopic analysis enables more accurate quantification, better discrimination of starch types, and a deeper understanding of starch structure and its changes. As demonstrated in recent studies, merging IR and Raman data can improve predictive models, and as technology advances, we expect to see more integrated instruments and algorithms dedicated to dual-modal vibrational spectroscopy for starch and other biopolymers.

8 Future Trends and Challenges

As vibrational spectroscopy continues to evolve, several future directions and challenges can be anticipated in the context of starch characterization.

8.1 Standardization and Quantitative Calibration

One challenge is developing standardized protocols for starch analysis via IR/Raman. Starch spectra can be influenced by sample preparation (moisture content, particle size), instrument settings, and data processing. For the techniques to become routine analytical tools in industry (for example, to replace or supplement wet chemical methods for measuring starch amylose content or damage), robust standard methods are needed. This includes standardized procedures for collecting ATR-FTIR spectra of starch powders and perhaps reference spectral libraries. The inherent variability of starches (due to botanical source, moisture, etc.) makes standardization non-trivial. However, with the maturation of instrumentation and chemometric corrections, it is likely that organizations (like AOAC for food analysis) will move toward validating IR/Raman methods for certain starch parameters. For instance, a calibrated ATR-FTIR method for rapid amylose content determination could be developed, which would save time over traditional iodine colorimetry. Similarly, a Raman-based method for measuring starch crystallinity or gelatinization degree in flour could be standardized. Achieving this requires addressing sources of variability: pathlength differences in ATR, baseline fluorescence in Raman, etc., perhaps by using internal standards or consistent drying protocols. The reward would be rapid, reagent-free quality control tests for starch in various industries [25].

8.2 Dealing with Complex Matrices

In real-world samples, starch often exists in complex matrices (grains, foods, biocomposites). A future focus will be on improving the detection and analysis of starch in such complex systems. One challenge is spectral overlap. For instance, in a dough sample, starch IR bands overlap with protein and lipid bands. Advanced signal processing, like self-modeling mixture analysis or machine learning classification, will be necessary to untangle overlapping spectra. We foresee increased use of multivariate analysis and perhaps deep learning to recognize starch spectral patterns within mixtures. Another approach is coupling spectroscopy with separation or imaging; for example, using IR microscopy to target starch-rich regions of a sample or using polarized Raman to selectively observe aligned starch (like in extruded products). A specific challenge is fluorescence in plant tissues for Raman; progress in shifted-excitation Raman and time-gating may reduce this, enabling clearer starch signals even in fluorescent backgrounds. Also, integration with other techniques is a trend: Raman coupled with microscopy and X-ray could correlate spectroscopic data with crystallite structure in situ. We may see instruments that combine a Raman probe with a portable XRD or NIR device to do multi-modal analysis of grains on-site [25].

8.3 Nanoscale and in Vivo Analysis

Pushing the spatial resolution of starch analysis further is an ongoing pursuit. While confocal Raman can get to 1 μm, starch granule ultrastructure (growth rings, blocklets) is on the order of 100 nm to 1 μm. Techniques like Tip-Enhanced Raman Spectroscopy (TERS), which uses a metallic AFM tip to enhance and spatially resolve Raman signals, could in principle achieve nanoscale mapping of chemical features on a starch granule surface. If challenges of sample preparation and tip interference can be overcome, TERS might reveal, for example, the chemical composition differences between crystalline and amorphous lamellae on a 100 nm scale, something currently inferred indirectly. Similarly, advancements in nano-IR (AFM-based IR) could potentially map starch chemistry at the nanoscale, though the strong IR absorption of polymers and need for very thin samples are hurdles. There is also interest in in vivo or in planta analysis: using Raman microprobes to monitor starch metabolism inside living cells. A recent step in this direction was the noninvasive observation of starch granules in living plant tissues using nanomechanical opening and Raman confirmation. In the future, one might use minimally invasive fiber-optic Raman probes inserted into storage organs (like a potato tuber or cereal endosperm) to monitor starch conversion processes (e.g., real-time tracking of starch degradation during germination). Achieving this requires addressing interference from water and other cell components, but continuing improvements in probe technology and data analysis could make it feasible [25].

8.4 Advanced Data Analysis and Machine Learning

The flood of spectral data from imaging and high-throughput measurements calls for advanced analysis. Machine learning, beyond simple PCA/PLS, will likely play a growing role. Convolutional neural networks (CNNs) might be trained on Raman or IR spectral images of starch granules to classify regions into amylose-rich vs. amylopectin-rich or to identify defects. Likewise, generative models could be used to denoise or augment spectral data, improving the effective sensitivity of measurements. Another frontier is predictive modeling: combining vibrational spectra with models of reactivity or digestibility. For example, spectra might be used to predict the glycemic index of a starchy food, a complex function of its structure. Hossain et al. transformed handheld Raman spectroscopy into a powerful tool for non-destructive, single-granule quantification of amylose and amylopectin. In their innovative study, researchers scanned hundreds of maizes, cassava, and wheat starch granules using Raman micro-spectroscopy and applied a series of machine learning techniques PCA, LDA, logistic regression, and SVM to unravel complex spectral signatures. They pinpointed key marker bands at 856 cm^{-1} and 941 cm^{-1} for amylose and 871 cm^{-1} for amylopectin, which allowed them to build a *semi-supervised multivariate curve resolution model* capable of spatially mapping granule composition [30].

8.5 Expanding the Scope of Analysis

Future applications of starch vibrational spectroscopy may go beyond structure into monitoring reactions and interactions. For instance, enzyme-starch interactions: Raman spectroscopy could potentially follow how amylase enzymes bind and cleave starch in real time by tracking changes in specific Raman bands (perhaps with substrate analogs or isotope labeling to enhance contrast). Also, as starch finds its way into new materials (like starch-based hydrogels, 3D-printed starch scaffolds, and starch aerogels), IR/Raman will be crucial to characterize these novel forms. Starch combined with other biopolymers (such as starch-chitosan films or starch-PVA blends) presents complex hydrogen-bonding interactions; 2D IR correlation and Raman mapping can disentangle these interactions and guide formulation of biodegradable plastics.

8.6 Overcoming Instrumental and Fundamental Limits

On the instrumental side, improvements in detectors and lasers will continue to reduce noise and acquisition time. One challenge for Raman is to get better sensitivity without losing specificity; one trend here is resonance Raman using laser wavelengths that resonate with specific chromophores. Starch itself has no strong chromophore, but if labeled or complexed with iodine (which gives a blue color), resonance Raman could selectively boost the starch-iodine mode, which might be useful in analytical sensing of amylose (essentially a spectroscopic analog of the iodine test but with more information content). In IR, one limitation is water interference; developments in ATR-IR probes that can work directly in aqueous slurries (with absorbance subtraction techniques or probe coatings) might allow direct tracking of starch gelatinization in a boiling slurry by IR, which is currently challenging due to water absorption. Another promising development is mid-IR lasers and QCL (quantum cascade laser) imaging, which could allow video-rate IR imaging of starch-water systems at specific frequencies (like following the 1022 cm^{-1} band during gelatinization) [25, 30].

In facing these future challenges, the interdisciplinary nature of vibrational spectroscopy is evident. Collaboration between spectroscopists, data scientists, food scientists, and material scientists will drive innovation. For example, dealing with real-world samples may involve combining vibrational spectroscopy with other measurements (DSC, XRD, microscopy) in predictive models: a multi-technique approach that future chemometric models can incorporate (so called data fusion at a higher level). The ultimate vision is to fully exploit vibrational spectroscopy as a noninvasive, rapid diagnostic tool for starch structure in all its contexts, from a plant leaf to a factory production line to a piece of biodegradable plastic, providing insights that are currently obtainable only through slower or destructive methods.

In summary, the future is bright but demanding for starch vibrational spectroscopy. The technique is poised to become more quantitative, more integrative (with other techniques and with advanced analysis), and more in situ/in vivo. The challenges of variability, complexity, and sensitivity are being met by technological advances and smarter algorithms. By surmounting these challenges, vibrational spectroscopy will enable ever deeper insight into starch chemistry, ultimately enhancing our ability to utilize and manipulate this important biopolymer in food, materials, and biotechnology.

9 Conclusion

Vibrational spectroscopy has established itself as an indispensable tool for the characterization of starch structure in both research and industry. In this chapter, we have reviewed how Raman and infrared spectroscopy provide complementary perspectives on starch's molecular organization, from the conformation of its glucosidic linkages and the packing of helices to the detection of chemical modifications. The introduction outlined starch's fundamental composition (amylose and amylopectin) and the motivation for spectroscopic analysis, setting the stage for understanding how IR and Raman methods can unravel starch's complex semi-crystalline architecture. The theoretical background explained the different selection rules of IR vs. Raman and why these techniques yield distinct but complementary information for a polysaccharide like starch.

In the core sections, we examined the spectroscopic signatures of starch in its many forms. The overview of Raman characterization detailed how certain Raman bands (notably the 480 cm^{-1} ring-breathing mode) serve as fingerprints for starch and indicators of helical order, while chemometric analysis of Raman spectra can quantify properties like amylose content. The overview of IR characterization similarly described IR's sensitivity to functional groups and hydrogen bonding, with features like the 1047/1022 cm^{-1} doublet correlating with crystalline double-helix content and new bands revealing substitutions or interactions. We then explored different forms of starch: we saw that native A-type (cereal) and B-type (tuber) starches have subtle differences in their IR and Raman spectra reflecting their packing and hydration (e.g., B-type shows a broader O–H IR band and slight shifts in Raman lattice modes), while the less common C-type shows mixed features of both. Gelatinization causes the collapse of order-indicative bands (loss of the IR doublet and 480 cm^{-1} Raman peak), whereas retrogradation partially restores them, which can be tracked in situ. We discussed chemically modified starches, emphasizing how vibrational spectroscopy identifies introduced groups ($-COO^-$, $-C{=}O$, $-O-CH_3$, $-PO_4$) by their distinct bands and monitors the accompanying changes in starch's structural order (often a loss of crystallinity). Throughout these sections, references to recent studies and figures illustrated the practical spectral differences and what they mean in terms of molecular structure.

Recent advances have pushed the boundaries of what IR and Raman can do with starch. Improved instrumentation and imaging capabilities now allow us to map starch granule composition and monitor processes like gelatinization in real time. We highlighted, for example, the use of confocal Raman microscopy to visualize amylose and amylopectin distributions, portable Raman devices for on-site starch authentication, and data fusion approaches combining IR and Raman to enhance analytical performance. These advances not only increase the convenience and speed of starch analysis but also deepen its analytical power, extracting more information (spatial, kinetic, multicomponent) than was previously possible.

In considering future trends and challenges, it is clear that vibrational spectroscopy of starch will continue to grow in importance. Challenges such as standardizing methods for routine use, dealing with complex sample matrices, and pushing spatial/temporal resolution are actively being addressed by technological innovations and smarter data analysis techniques. The coupling of spectroscopy with machine learning and other analytical modalities is poised to unlock new frontiers, like predicting functional properties of starch-based materials from their spectra or probing starch within living plants noninvasively. Each new application, be it in food technology, bioenergy, or biodegradable materials, brings its own hurdles (fluorescence, overlap with other components, need for quantitation), but lessons learned from traditional starch spectroscopy guide these efforts.

In conclusion, IR and Raman spectroscopy offer a powerful, complementary pair of methods for starch characterization that is unparalleled in its combination of molecular specificity, speed, and non-destructive nature. From identifying starch polymorphs and monitoring gelatinization to detecting chemical modifications and enabling high-resolution starch imaging, vibrational spectroscopy provides deep insight into starch structure and transformations. As this chapter has shown, a critical, integrative approach using both IR and Raman and interpreting results in light of starch's unique structural chemistry allows us to fully characterize this fascinating biopolymer. Continued advances will no doubt further integrate vibrational spectroscopy into the toolkit of starch science, driving innovation in how we analyze and ultimately utilize starch in food, industry, and biotechnology. The story of starch, from granule to glucose and beyond, can be told through its vibrations, and we now have the tools to listen more keenly than ever.

References

1. Cozzolino D, Degner S, Eglinton J (2014) A review on the role of vibrational spectroscopy as an analytical method to measure starch biochemical and biophysical properties in cereals and starchy foods. Foods 3:605–621. https://doi.org/10.3390/foods3040605
2. Martens BMJ, Gerrits WJJ, Bruininx EMAM, Schols HA (2018) Amylopectin structure and crystallinity explains variation in digestion kinetics of starches across botanic sources in an in vitro pig model. J Anim Sci Biotechnol 9. https://doi.org/10.1186/s40104-018-0303-8
3. Castro-Campos FG, Morales-Sánchez E, Cabrera-Ramírez ÁH, Martinez MM, Rodríguez-García ME, Gaytán-Martínez M (2023) High amylose starch thermally processed by ohmic heating: electrical, thermal, and microstructural characterization. Innov Food Sci Emerg Technol 87:103417. https://doi.org/10.1016/j.ifset.2023.103417
4. Holder BH (2012) Characterization of starch by vibrational spectroscopy
5. Almeida MR, Alves RS, Nascimbem LBLR, Stephani R, Poppi RJ, de Oliveira LFC (2010) Determination of amylose content in starch using Raman spectroscopy and multivariate calibration analysis. Anal Bioanal Chem 397:2693–2701. https://doi.org/10.1007/s00216-010-3566-2
6. Man J-M, Cai C-H, Yan Q-X, Hu M-Z, Liu Q-Q, Wei C-X (2012) Applications of infrared spectroscopy in the analysis of ordered structure of starch grain. Acta Agron Sin 38:505. https://doi.org/10.3724/SP.J.1006.2012.00505
7. Wiercigroch E, Szafraniec E, Czamara K, Pacia MZ, Majzner K, Kochan K, Kaczor A, Baranska M, Malek K (2017) Raman and infrared spectroscopy of carbohydrates: a review. Spectrochim Acta A Mol Biomol Spectrosc 185:317–335
8. Almeida MR, Alves RS, Nascimbem LBLR, Stephani R, Poppi RJ, De Oliveira LFC (2010) Determination of amylose content in starch using Raman spectroscopy and multivariate calibration analysis. Anal Bioanal Chem, In, pp 2693–2701
9. Hakki, Boyaci I, Tümay Temiz H, Efe Geniş H, Acar E, Nur Yazgan N, Güven B, Selin Uysal R, Göktuğ Bozkurt A, Đlaslan K, Torun O, Dudak C, Şeker F, Hakkı Boyacı Đ (2015) Dispersive and FT-Raman spectroscopic methods. Food Anal. https://doi.org/10.1039/C4RA12463D
10. Nakajima S, Kuroki S, Ikehata A (2023) Selective detection of starch in banana fruit with Raman spectroscopy. Food Chem 401 134166. https://doi.org/10.1016/j.foodchem.2022.134166

11. Yu H, Cheng L, Yin J, Yan S, Liu K, Zhang F, Xu B, Li L (2013) Structure and physicochemical properties of starches in lotus (*Nelumbo nucifera* Gaertn.) rhizome. Food Sci Nutr 1:273–283. https://doi.org/10.1002/fsn3.37
12. Wang S, Zhang X, Wang S, Copeland L (2016) Changes of multi-scale structure during mimicked DSC heating reveal the nature of starch gelatinization. Sci Rep 6:28271. https://doi.org/10.1038/srep28271
13. Lu H, Ma R, Chang R, Tian Y (2021) Evaluation of starch retrogradation by infrared spectroscopy. Food Hydrocoll 120:106975. https://doi.org/10.1016/j.foodhyd.2021.106975
14. Kong L, Lee C, Kim SH, Ziegler GR (2014) Characterization of starch polymorphic structures using vibrational sum frequency generation spectroscopy. J Phys Chem B 118:1775–1783. https://doi.org/10.1021/jp411130n
15. kobe starch
16. Wang Y, Huang W, Zeng F, He J, Pan Z, Sahito ZA, Li X, Yang W (2025) A study focusing on the distinct fine structure of starch in mealy and waxy potatoes. Sci Rep 15. https://doi.org/10.1038/s41598-025-95977-w
17. Chen W, Xu W, Junejo SA, Feng Y, Gao T, Huang Q, Jiang Y, Zhang B, Jia X, Huang Q (2024) Analytical methods for starch retrogradation assessment. Food Biomacromol 1:107–126. https://doi.org/10.1002/fob2.12016
18. Guo Z, Jia X, Zhao B, Zeng S, Xiao J, Zheng B (2017) C-type starches and their derivatives: structure and function. Ann N Y Acad Sci 1398:47–61. https://doi.org/10.1111/nyas.13351
19. Yan Y, Peng B, Niu B, Ji X, He Y, Shi M (2022) Understanding the structure, thermal, pasting, and rheological properties of potato and pea starches affected by annealing using plasma-activated water. Front Nutr 9. https://doi.org/10.3389/fnut.2022.842662
20. Pozo C, Rodríguez-Llamazares S, Bouza R, Barral L, Castaño J, Müller N, Restrepo I (2018) Study of the structural order of native starch granules using combined FTIR and XRD analysis. J Polym Res 25. https://doi.org/10.1007/s10965-018-1651-y
21. Lu Z, Lu Y, Donner E, Liu Q, Bing D (2022) Physicochemical and nutritional properties of starches from nine Canadian-grown peas compared with six commercial starches. Legume Sci 4. https://doi.org/10.1002/leg3.133
22. Zhang W, Shi D, Dong W, Li H, Liu X (2025) Effects of glutathione on the physicochemical properties of high hydrostatically pressure gelatinized maize starch. Food Chem X 25:102158. https://doi.org/10.1016/j.fochx.2025.102158
23. Stawoska I, Staszak AM, Ciereszko I, Oliwa J, Skoczowski A (2020) Using isothermal calorimetry and FT-Raman spectroscopy for step-by-step monitoring of maize seed germination: case study. J Therm Anal Calorim 142:755–763. https://doi.org/10.1007/s10973-020-09525-x
24. Bernazzani P, Peyyavula VK, Agarwal S, Tatikonda RK (2008) Evaluation of the phase composition of amylose by FTIR and isothermal immersion heats. Polymer 49:4150–4158. https://doi.org/10.1016/j.polymer.2008.07.022
25. Hu X, Shi J, Zhang F, Zou X, Holmes M, Zhang W, Huang X, Cui X, Xue J (2017) Determination of Retrogradation degree in starch by mid-infrared and Raman spectroscopy during storage. Food Anal Methods 10:3694–3705. https://doi.org/10.1007/s12161-017-0932-0
26. Ram MS, Dowell FE, Seitz LM (2003) FT-Raman spectra of Unsoaked and NaOH-soaked wheat kernels, bran, and Ferulic acid. Cereal Chem 80:188–193
27. Huang P-H, Chen R-F, Chang W-C (2025) Effects of high-power ultrasonication combined with heat treatment on the 2 physicochemical and functional properties of wheat starch-long chain inulin 3 complexes. SSRN J. https://doi.org/10.2139/ssrn.5143360
28. Payne WZ, Kurouski D (2021) Raman spectroscopy enables phenotyping and assessment of nutrition values of plants: a review. Plant Methods 17:78
29. Dupuy N, Laureyns J (2002) Recognition of starches by Raman spectroscopy. Carbohydr Polym 49:83–90. https://doi.org/10.1016/S0144-8617(01)00304-6
30. Hossain IM, Pooja N, Kondeti SSC, Yamamoto T, Mazumder N, Noothalapati H (2025) Direct estimation of amylose and amylopectin in single starch granules by machine learning

assisted Raman spectroscopy. Carbohydr Polym 366:123929. https://doi.org/10.1016/j.carbpol.2025.123929

31. Farahi RH, Lereu AL, Charrier AM, Kalluri UC, Davison BH, Passian A (2020) Nanomechanics and Raman spectroscopy of in situ native carbohydrate storage granules for enhancing starch quality and lignocellulosic biomass production. ACS Omega 5:2594–2602. https://doi.org/10.1021/acsomega.9b02849
32. Slepkov AD, Ridsdale A, Pegoraro AF, Moffatt DJ, Stolow A (2010) Multimodal CARS microscopy of structured carbohydrate biopolymers. Biomed Opt Express 1:1347–1357
33. Zhang D, Kishimoto N (2024) Quantum chemical investigation into the structural analysis and calculated Raman spectra of amylose modeled with linked glucose molecules. Molecules 29. https://doi.org/10.3390/molecules29122842

Vibrational Spectroscopic Characterization of Deoxyribonucleic Acid (DNA)

Kankanige Nivuni Nihinsa Alwis and T. M. Sampath Udeni Gunathilake

Abstract The use of vibrational spectroscopy, especially the Fourier Transform Infrared (FTIR) and Raman methods, for the structural and functional characterisation of deoxyribonucleic acid (DNA) is covered in detail in this chapter. The chapter emphasizes the non-destructive, labelfree benefits of these methods while pointing out how sensitive they are to structural transitions (like helix-coil melting), conformational states (like A-, B-, and Z-DNA), and environmental variables like pH, temperature, and moisture. In order to accurately analyze structural integrity and molecular interactions, detailed vibrational band assignments are investigated for the phosphate backbone, base pairs, and sugar moieties of DNA. Along with reviewing the spectroscopic characteristics of single-versus double-stranded DNA, the chapter additionally examines at how external factors, such as medications, metal ions, proteins, and protective ligands, affect the stability and conformation of DNA. A special importance is given to the diagnostic potential of Raman and Surface-Enhanced Raman Scattering (SERS) in detecting epigenetic changes, including DNA methylation. The practical difficulties and limitations of existing approaches are critically assessed, along with concepts for incorporating vibrational spectroscopy into forensic and biological applications. All of these observations support the critical role that Raman and FTIR spectroscopy play in contemporary nucleic acid research and diagnostics.

Keywords DNA · Raman spectroscopy · FTIR spectroscopy

1 Introduction

DNA (deoxyribonucleic acid) is a biological macromolecule which acts as the primary carrier of genetic information. It also provides the information needed for the

K. N. N. Alwis · T. M. S. U. Gunathilake (✉)
Faculty of Applied Science, Department of Polymer Science, University of Sri Jayewardenepura, Nugegoda, Sri Lanka
e-mail: sgunathilake@sjp.ac.lk

N. Mazumder et al. (eds.), *Vibrational Spectroscopy for Biopolymer Characterization*, Biological and Medical Physics, Biomedical Engineering,
https://doi.org/10.1007/978-3-032-15306-7_4

growth, development, functioning, and reproduction of all living organisms [1]. The primary structure of DNA is a main valence chain of repeating units of phosphoric acid and sugar with purine and pyrimidine bases attached to the sugar residues [2]. DNA molecules are highly polymerized and unbranched [2] possesses an indistinguishable double helix structure, which is held together by hydrogen bonding between complementary base pairs in the two polynucleotide strands [3], with chains that are antiparallel, and the structure is stabilized by hydrogen bonding between complementary base pairs and van der Waals forces [2–4]. The standard bases of DNA are adenine (A), guanine (G), cytosine (C), and thymine (T). This highly complex structure provides DNA with the ability to store genetic information in its linear structure of nucleotides [3].

Dynamic organization of DNA is key for normal cell physiology, with conformational properties influencing stability, transcription regulation, and protein recognition [5]. DNA has ultimate importance in medicine and biology [6], and is not just the basis of heredity. Epigenetic modifications, such as DNA methylation [7], regulate gene expression without changing the inherent genetic code, while aberrant regulation of epigenetic marks is associated with severe diseases [8]. DNA methylation has been linked to a wide range of phenomena, from embryogenesis and environmental adaptation to various disease states, including cancer, lupus, muscular dystrophy, and birth defects [9]. Even the changes in DNA integrity, such as oxidative damage, also play an important role in factors like male fertility. Therefore, building an understanding of the structure and modification of DNA is essential for understanding normal biological processes and mechanisms causing disease [10].

Given DNA's central importance in biological systems, thorough molecular characterization is essential for a wide range of applications, including cancer diagnosis, drug treatment, and species identification. Traditional sequence-based methods for DNA analysis, while fundamental, are often time-consuming and require specialized skills or precise sequence information [1]. Moreover, the conformational flexibility of DNA and its susceptibility to subtle chemical modifications (like methylation, the addition of a methyl group to a DNA base) demand methods capable of detecting these minute changes. The spectroscopic behavior of DNA, including shifts in absorption bands, is influenced by its structural integrity, hydration state, and interactions with other molecules, necessitating techniques that can capture these complexities. There is a continuous demand for simpler, highly sensitive, and low-cost analytical protocols that can be translated into clinical applications for assessing DNA modifications.

FTIR and Raman spectroscopy offer complementary information about the energies of molecular vibrational modes, which is particularly important in characterizing DNA structures and conformational transitions, such as B-DNA, A-DNA, and Z-DNA forms, by detecting changes in peak intensities and shifts in spectral band positions. This includes monitoring "helix-coil" thermal transitions and denaturation. Vibrational spectroscopy is also useful in studying the interaction of DNA components with each other and with other substances, including hydrogen bonding between complementary bases, binding of metal ions, interactions with polypeptides, and investigating DNA methylation status. These techniques can distinguish methylated DNA from unmethylated DNA, identify specific methylated bases like 5-methylcytosine (5meC), and even pinpoint the position of methylation (e.g., N7 or O6 positions in guanine). While cytosine is the most commonly methylated DNA base, methylation can occur in all four DNA bases: cytosine, adenine, guanine, and thymine. This capability is crucial for understanding epigenetic mechanisms in health and disease.

Vibrational spectroscopic techniques, such as Fourier Transform Infrared (FTIR) and Raman spectroscopy, are vital for understanding DNA's molecular structure and changes by analyzing the unique vibrational patterns of its constituent molecules (Fig. 1).

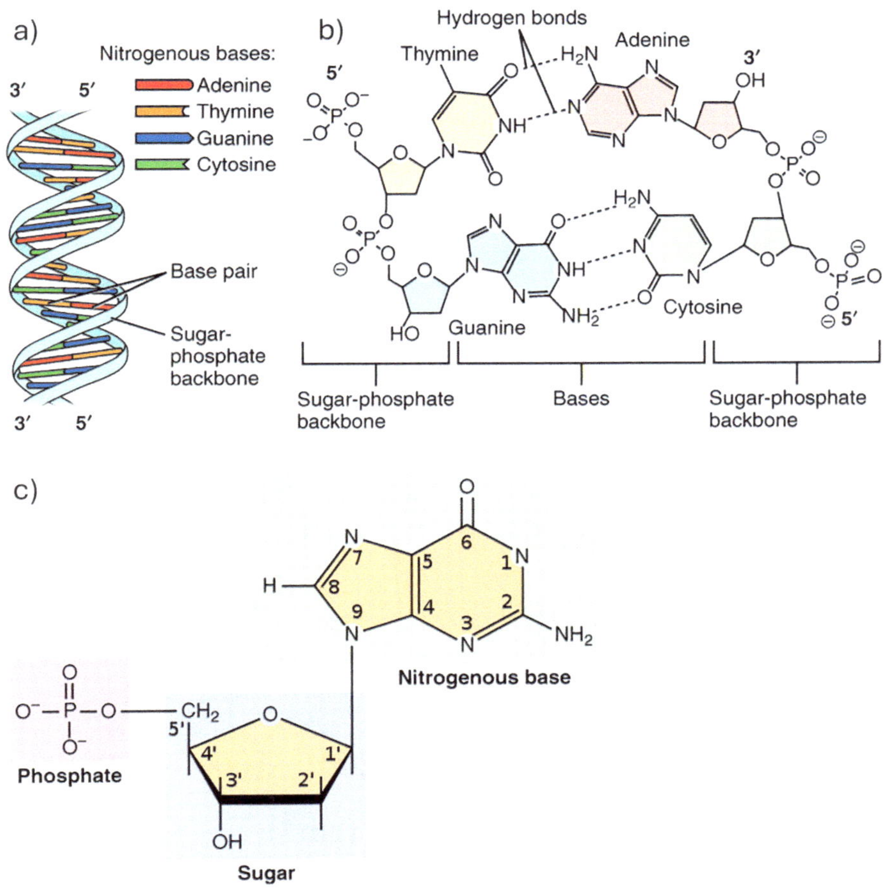

Fig. 1 (**a**) Diagram of the DNA double helix showing complementary base pairing and antiparallel strands. (**b**) Molecular structures of A–T and G–C base pairs, highlighting hydrogen bonding and the sugar-phosphate backbone. (**c**) Components of a DNA nucleotide: phosphate, deoxyribose sugar (with carbon numbering), and nitrogenous base

2 Common Vibrational Modes and Assignments in DNA (Fig. 2)

2.1 Phosphate Backbone

DNA's phosphodiester backbone produces characteristic vibrational bands, especially from the PO_2^- groups. The asymmetric stretching vibration $\nu AS(PO_2^-)$ appears around 1220–1245 cm^{-1} in FTIR spectra. This band is a well-known marker of B-form DNA, and its exact position is sensitive to environmental conditions. For instance, increased hydration causes a red-shift of $\nu AS(PO_2^-)$ from around 1240 to 1220 cm^{-1}. Complexation with cationic polymers can also enhance this band. The corresponding symmetric PO_2^- stretch $\nu AS(PO_2^-)$ appears near 1080–1095 cm^{-1}

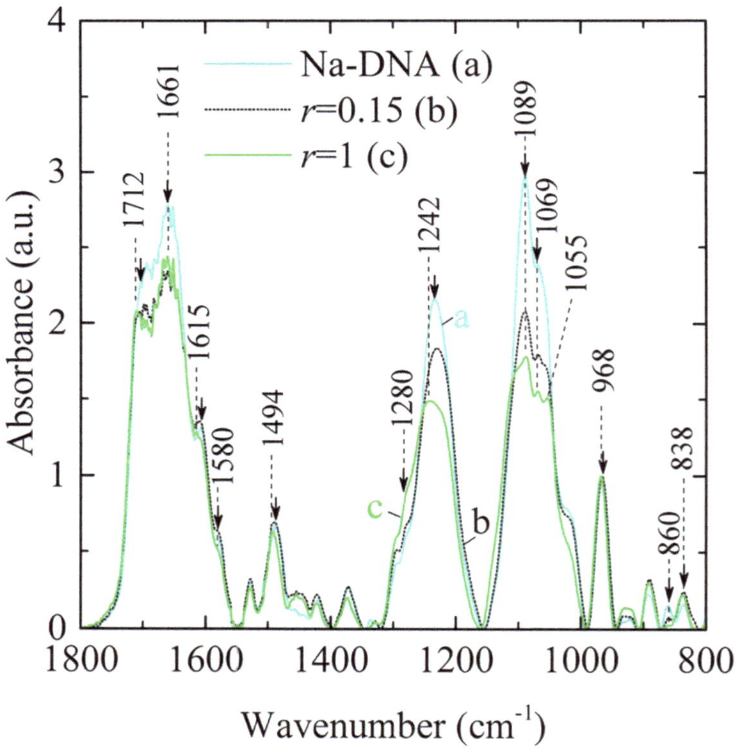

Fig. 2 FTIR spectra of DNA thin films in the region 800–1800 cm^{-1} measured at about 85% r.h. for representative Mg content. r = [Mg]/[P]: 0.15 (b—black, dotted) and 1 (c—green). Labels (dashed lines and numbers) identify bands undergoing prominent changes for r = 1. Reference spectrum of Na–DNA is shown for comparison (a—blue): vibrational bands, assigned to the base and backbone vibrations, are marked by arrows.

and is comparatively stable, showing little shift with DNA conformation or temperature. A lower-frequency backbone mode around 960–970 cm^{-1} is attributed to O–P–O bending vibrations of the phosphodiester linkages. In Raman spectra, the DNA backbone has prominent markers as well: B-form DNA shows intense Raman lines at 834 cm^{-1} and 787 cm^{-1}, assigned to deoxyribose-phosphate backbone vibrations (the 787 cm^{-1} line is a signature of the B-form helix). In contrast, A-form helices exhibit a Raman band near 806 cm^{-1} (characteristic of that conformation). Thus, backbone vibrations in both IR and Raman provide clear signatures of DNA's presence and conformational state [11–13].

2.2 *Base Vibrations (Purines and Pyrimidines)*

The heterocyclic bases produce numerous bands in the 1500–1700 cm^{-1} region. In FTIR, an important marker is the carbonyl stretching of guanine (G) and thymine (T) in base pairs, typically observed around 1705–1715 cm^{-1} for native double-helical DNA. This band (near 1705 cm^{-1} in non-deuterated DNA) is a signature of intact Watson-Crick base pairing and secondary structure. Upon DNA denaturation or loss of base pairing, it shifts to a lower frequency (around 1660 cm^{-1}), reflecting the change in hydrogen bonding (discussed further in the context of thermal melting). Thymine's C=O stretch also contributes around 1655–1660 cm^{-1}. Other base modes include a group of bands near 1575–1580 cm^{-1} (seen in both FTIR and Raman) primarily from guanine and adenine ring vibrations. In highly methylated DNA, the intensity of a band at 1578 cm^{-1} (guanine, adenine ring mode) is notably increased. Bands in the 1480–1500 cm^{-1} range are attributed to purine ring stretching (adenine and guanine). For example, Raman bands around 1484–1490 cm^{-1} arise from adenine and guanine ring vibrations. The cytosine and thymine bases show a distinctive vibration around 780 cm^{-1} (a ring breathing mode): in DNA, this IR/Raman band appears near 782–788 cm^{-1} and shifts slightly with base methylation (5-methylcytosine produces a band at 786 cm^{-1}). Methyl group vibrations in bases can be detected as well; for instance, a Raman band around 1330–1335 cm^{-1} can be assigned to CH_3 wagging/twisting modes, and a smaller peak near 1014 cm^{-1} is attributed to methyl rock. These become especially prominent in spectra of methylated DNA (discussed later in the context of epigenetics) [13, 14].

2.3 *Deoxyribose (Sugar) Vibrations*

The deoxyribose sugar rings contribute bands primarily in the 900–1100 cm^{-1} region. In FTIR, a notable sugar vibration is the C–O stretching of deoxyribose, observed around 1050–1070 cm^{-1}. This band tends to shift slightly with DNA conformation; for example, increasing hydration moves a deoxyribose C–O stretch from 1066 to 1053 cm^{-1} as DNA transitions from A to B-form. The sugar ring puckering modes appear in the high-frequency fingerprint region: B-form DNA typically shows a sugar-phosphate vibration near 968–970 cm^{-1} in IR spectra, whereas A-form and other conformations may split or shift this band. Another sugar marker is an IR band around 890–900 cm^{-1}, arising from deoxyribose ring breathing. Overall, sugar vibrations are somewhat weaker than base or phosphate bands, but they still provide insight into the DNA backbone conformation (such as distinguishing C3′-endo vs C2′-endo sugar puckers in different helix forms).

By analyzing this array of vibrational bands, one can assign spectral features to specific DNA structural components. For example, the presence of strong absorbances at 1220 and 1080 cm^{-1} confirms the phosphodiester backbone, while distinct bands near 1700, 1650, 1500, and 785 cm^{-1} indicate intact base pairs and overall helical structure. These assignments form the basis for using IR and Raman spectroscopy as a "molecular fingerprint" of DNA [15, 16].

3 Differences Between Double-Stranded and Single-Stranded DNA

Vibrational spectra sensitively reflect whether DNA is double-stranded (dsDNA) or single-stranded (ssDNA). Several characteristic changes occur upon helix unwinding into single strands. In the IR spectrum, the base pairing carbonyl band around 1705–1715 cm^{-1} (from guanine and thymine) diminishes and shifts when the DNA duplex is denatured. Specifically, a peak near 1600 cm^{-1} in dsDNA (arising partly from base in-plane vibrations) shifts to about 1650 cm^{-1} in ssDNA. This shift corresponds to the disruption of hydrogen-bonded base pairs. Another indicator is that the intensity of the band near 1480 cm^{-1} (a combined vibration of several bases) remains relatively unchanged between double- and single-stranded forms, whereas other bands undergo reductions. For instance, the absorbance around 1400 cm^{-1}, which includes contributions from all bases (primarily cytosine and adenine ring modes and backbone C–H deformation), is strong in dsDNA but significantly decreases in ssDNA. A band near 1300 cm^{-1} (associated largely with cytosine vibrations) also diminishes upon strand separation.

Importantly, the sugar-phosphate backbone signals in the 1200–800 cm^{-1} region generally become more pronounced in single strands. Single-stranded DNA shows higher overall absorbance in this region than duplex DNA, due to increased exposure of backbone groups to the solvent. Despite this intensity increase, the frequencies of key backbone modes remain nearly the same; for example, the $\nu AS(PO_2^-)$ and $\nu S(PO_2^-)$ still appear around 1220 cm^{-1} and 1080 cm^{-1}, respectively, in ssDNA as they do in dsDNA. The ratio of these two phosphate bands ($\nu AS/\nu S$) tends to be higher for ssDNA than for dsDNA, reflecting changes in the phosphate environment. Experimentally, it's observed that this ratio increases when the DNA helix unwinds, and it can vary with the base composition of single-stranded regions. In both forms, however, the asymmetric stretch at 1220 cm^{-1} is typically more intense than the symmetric stretch at 1080 cm^{-1}.

These spectroscopic differences align with the physical changes upon denaturation. The loss of Watson-Crick base pairing (dsDNA to ssDNA) removes the infrared band at 1710 cm^{-1} and causes the aforementioned shifts and intensity changes. In summary, one can diagnose single- vs. double-stranded DNA by noting a downshift of the 1600 cm^{-1} band to 1650 cm^{-1}, significant decreases in the 1400 and 1300 cm^{-1} bands, enhanced backbone signals from 1200–800 cm^{-1} without major peak shifts, and an increased $\nu AS(PO_2^-)/\nu S(PO_2^-)$ intensity ratio. These spectral diagnostics are particularly useful for monitoring DNA denaturation or strand separation in situ [17–19].

4 Identifying A-, B-, and Z-DNA Conformations Via Spectral Shifts

Different double-helical DNA conformations (B-DNA, A-DNA, Z-DNA, etc.) exhibit distinct vibrational fingerprints. B-form DNA (the typical form in aqueous, physiological conditions) has characteristic IR bands and Raman lines that distinguish it from other forms. In IR spectra, B-DNA shows a strong base carbonyl band above 1710 cm^{-1} (indicating standard Watson-Crick pairing), a marker band at 1223 cm^{-1} for the antisymmetric phosphate stretch, and a prominent band near 968–970 cm^{-1} (assigned to the backbone C–C stretch and/or sugar-phosphate vibration). B-form also yields an IR "sugar" band in the 830–842 cm^{-1} range, related to deoxyribose vibrations. In Raman spectra, B-DNA is identified by lines such as a robust 682 cm^{-1} band (purine ring breathing) and the previously mentioned 834 cm^{-1} and 787 cm^{-1} backbone bands. Additionally, Raman bands at 1374 cm^{-1} (adenine) and 1363 cm^{-1} (guanine) are often noted in B-form DNA. These bands correspond to specific base modes, and their intensities can vary with base sequence [20–22].

4.1 A-Form DNA

A-form DNA, which occurs in dehydrated samples or RNA/DNA hybrids, shows systematic shifts relative to B-form. IR spectra of A-DNA have the base carbonyl stretch slightly lower, around 1705 cm^{-1}. The sugar-phosphate vibrations split into a doublet around 1418–1401 cm^{-1}, reflecting A-form's C3′-endo sugar puckers. The thymidine (thymine deoxynucleotide) vibration that was at 1280 cm^{-1} in B-form shifts to 1274 cm^{-1} in A-form. The antisymmetric phosphate stretch moves to 1244 cm^{-1} (a bit higher than in B), and a distinct furanose (sugar) vibration appears near 1188 cm^{-1}. In the fingerprint region, A-DNA shows a pattern of multiple bands between 978 and 865 cm^{-1} instead of a single 968 cm^{-1} band and a characteristic IR band at 808 cm^{-1} (in contrast to B's 830–840 cm^{-1}). Raman spectroscopy of A-DNA also differs: the B-form Raman markers at 834 and 787 cm^{-1} are replaced by A-form backbone signals at 806 and 777 cm^{-1}. The purine ring breathing mode shifts slightly (a Raman line at 663 cm^{-1} is noted for A-form). Overall, these shifts in both IR and Raman correspond to the tighter, more compact helical geometry and altered glycosidic bond angle of A-DNA [23–25].

4.2 Z-Form DNA

Z-form DNA is the left-handed helix formed by alternating purine/pyrimidine sequences under high salt or certain chemical conditions and has dramatic spectral signatures. In IR spectra, Z-DNA is indicated by unique bands and shifts: a band at 1408–1413 cm^{-1} arises from C3′-endo deoxyribose vibrations (a hallmark of the Z form's sugar pucker). A cluster of IR bands around 1352–1357 cm^{-1} and 1320 cm^{-1} corresponds to purines in *syn* conformation (particularly guanine) with C3′-endo sugars. The antisymmetric phosphate stretch in Z-DNA appears near 1215 cm^{-1}, slightly lower than in B-DNA. Notably, the Z-form shows an IR band at 929–930 cm^{-1} that replaces the B-form's 890 cm^{-1} deoxyribose band. This new band is attributed to an O–P–O backbone vibration unique to Z-DNA. In Raman spectra, the conversion from B to Z can be tracked by shifts of base bands: the adenine ring mode around 1363 cm^{-1} in B shifts to 1362 cm^{-1}, and guanine's 1374 cm^{-1} (B-form) moves to 1354 cm^{-1} in Z. Z-DNA Raman also features an intense new band at 1314 cm^{-1} (related to guanine in *syn* and the altered backbone) and characteristic backbone peaks at 810 cm^{-1} and 744 cm^{-1}. A low-frequency purine vibration around 625 cm^{-1} appears in Z as well.

The spectral differences can be summarized by how certain marker bands shift when transitioning from B to A and A to Z forms. For example, the IR phosphate $\nu AS(PO_2^-)$ band shifts: B (1220–1225 cm^{-1}) to A (1244 cm^{-1}) and A to Z (1215 cm^{-1}). The deoxyribose ring mode near 890 cm^{-1} in B is absent in Z, replaced by a 930 cm^{-1} band. In the bases, B-form's guanine carbonyl at 1710 cm^{-1} drops below 1710 cm^{-1} in A, and in Z-form the environment causes multiple new peaks (e.g., 1630 and 1318 cm^{-1}) to appear. Experimentally, IR monitoring of B to Z transitions (for instance, in alternating CG polymers) shows new IR bands emerging at 1630, 1318, 1265, 1124, 1014, and 925 cm^{-1} for the Z conformation, along with shifts of existing bands (e.g., a band at 1409 cm^{-1} in B shifting to 1420 cm^{-1} in Z). Often a splitting of certain bands is observed (for example, the B-form 784 cm^{-1} may split into 778 and 784 cm^{-1} in Z). These spectral changes provide a "fingerprint" for identifying DNA's conformation (Fig. 3). In practical terms, one can deduce whether DNA in a sample is in A, B, or Z-form by comparing the positions of a few key peaks (phosphate stretch, base carbonyl region, low-frequency backbone region) with known reference values [20, 26–28].

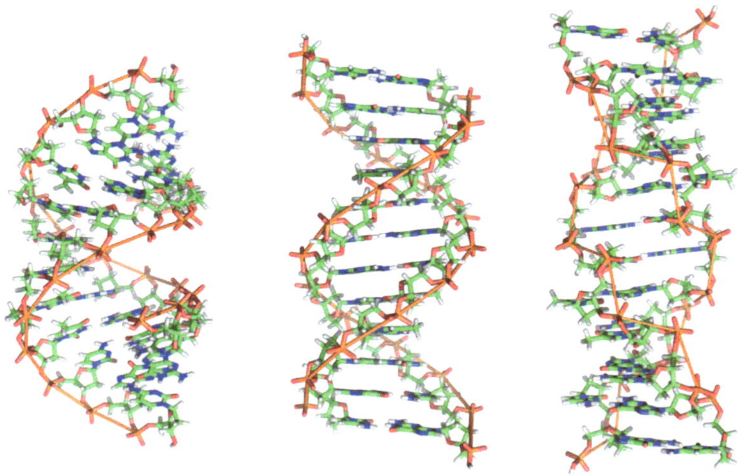

Fig. 3 Helical conformations of DNA: A-form, B-form, and Z-form, highlighting their structural differences. Image by Zephyris, licensed under Creative Commons Attribution-Share Alike 3.0 Unported (CC BY-SA 3.0)

5 Conformational Analysis: Helix–Coil Transitions and Environmental Effects

Vibrational spectroscopy is exceptionally useful for probing DNA's conformational transitions, such as helix-to-coil melting, and the influence of hydration, pH, and temperature on DNA structure. Changes in the intensity or frequency of specific IR bands and Raman lines act as reporters for these transitions.

5.1 Helix–Coil (Denaturation) Transitions

As DNA melts from a double helix to single strands (for example, with increasing temperature), the most prominent spectral change is in the base-pairing region. In the IR spectrum of native DNA (in solution or a hydrated film), an intense band around 1705 cm^{-1} (1685 cm^{-1} if the DNA is deuterated) is a signature of the helical secondary structure. This band is associated with the stretching vibrations of the base carbonyl groups (G and T) coupled with amino group in-plane vibrations, and it indicates intact hydrogen bonding between complementary bases. Upon denaturation, this band undergoes a clear shift to lower frequency for instance, from 1685 cm^{-1} down to about 1660 cm^{-1} in D_2O solution, as the base pairing is disrupted. Concomitantly, the intensity of the 1705/1685 cm^{-1} band drops, reflecting fewer paired bases. By monitoring this band's intensity as a function of temperature, one can generate a melting curve akin to that obtained by UV absorbance at 260 nm. Indeed, the relative IR absorbance at 1682 cm^{-1} vs. 1662 cm^{-1} tracks the DNA helix-coil transition, yielding a sigmoidal curve with a melting temperature (T_m) that matches the calorimetric or UV-determined T_m. This demonstrates that FTIR can accurately monitor DNA melting in real time, even in concentrated solutions or films where traditional UV melting studies are difficult. The advantage is that IR detects the loss of base pairing directly via the carbonyl vibration shift, providing a clear marker of the transition. Raman spectroscopy likewise can observe denaturation: certain Raman peaks (like those at 1485 and 1580 cm^{-1} from base rings) change intensity or position when DNA is denatured, although IR is generally more direct for this purpose [29, 30].

5.2 *Effects of Hydration*

DNA's secondary structure is highly dependent on hydration level. FTIR studies on oriented DNA fibers have confirmed classical X-ray findings that B-DNA is stable at high relative humidity (≥92% RH for sodium DNA), while A-DNA appears at lower humidity (75% RH for sodium DNA). Lithium salt DNA, for example, adopts B-form above 66% RH and a distinct "C-form" between 44% and 66% RH. As hydration decreases further, DNA eventually loses its regular structure. Infrared spectroscopy not only detects these transitions but also provides additional insights. For instance, as the relative humidity is raised from 0% up to 92%, specific band shifts occur: the base carbonyl band moves from 1705 cm^{-1} (dry) to 1715 cm^{-1} (fully hydrated), the antisymmetric phosphate stretch shifts from 1240 to 1220 cm^{-1}, and the deoxyribose C–O stretch shifts from ~1066 to 1053 cm^{-1}. These systematic red shifts with increasing water indicate that water molecules are progressively hydrogen-bonding to the phosphate and base carbonyl groups, lowering their vibrational frequencies. Notably, phosphate groups are the first to hydrate: between 0 and 66% RH, up to about six water molecules per phosphate are taken up, with minimal effect on base vibrations. Above 66% RH, additional water begins to interact with bases, causing the base carbonyl band to move significantly (1705 to 1715 cm^{-1}). These observations support a model where the DNA phosphate backbone forms a hydration shell earlier, and full B-form ordering (including base stacking geometry) occurs only when both the backbone and bases are well-hydrated. Furthermore, the quality of IR spectra improves with hydration as bands sharpen and the background diminishes as water content rises. This is partly because hydration reduces structural disorder, yielding more defined vibrational modes. Dichroic measurements (polarized IR on oriented fibers) show constant dichroic ratios within stability ranges of each DNA form, confirming that distinct hydration plateaus correspond to distinct conformations. In summary, IR spectroscopy verifies that B to A and A to C form transitions are driven by stepwise hydration changes, with identifiable spectral plateaus for each conformation. It also extends our understanding by quantifying hydration effects on specific bonds: e.g., $\nu as(PO_2^-)$ continuously red-shifts with increasing water content, reflecting how water binding softens the phosphate vibration [30, 31].

5.3 *Effects of pH (Base Protonation/Deprotonation)*

DNA bases have ionizable groups that can gain or lose protons, and vibrational spectroscopy can pinpoint these changes. Protonation of a base typically causes its C=O and C=N stretching frequencies to increase (due to stronger bonds when an adjacent site is protonated), whereas deprotonation or anion formation causes frequencies to decrease (due to conjugation or charge delocalization). Take cytosine as an example: at low pH, cytosine can accept a proton. IR spectra of protonated cytosine (or cytidine) show the C=O stretch shifting from 1655 cm^{-1} to 1680 cm^{-1}, and the $-NH_2$ scissoring vibration shifting from 1667 to 1725 cm^{-1}. Such large upward shifts indicate protonation at the ring nitrogen (N1) rather than at the exocyclic amino group. Adding a proton at N1 disrupts the cytosine ring's conjugation, strengthening the C=O bond and thus raising its vibrational frequency, which is analogous to protonating pyridine (where ring bond frequencies increase). Guanine, on the other hand, has a very low pK_a (1.6 for protonation), and when protonated (likely at the N_7 position), its IR bands from 1700–1400 cm^{-1} all shift to higher frequencies. For example, guanine's amino deformation mode moves from 1681 to 1718 cm^{-1} upon protonation, again consistent with proton attachment at a ring nitrogen (N_7) causing a generalized strengthening of double bonds. IR data combined with crystallography confirm N_7 as the protonation site in guanine. At high pH, bases lose protons. Guanine deprotonation (pK 9.4 for losing the N1 proton) leads to a decrease in its carbonyl stretch intensity and the emergence of new bands around 1638 and 1413 cm^{-1}. These new bands correspond to an enolate form and other tautomeric vibrations, indicating that the guanine C=O has partially converted to $C–O^-$ (lowering its frequency). The changes mimic the spectrum of 2-amino-6-methoxypurine (an analog), supporting that deprotonation occurs at N1 and results in charge redistribution. In summary, by observing where and how key base bands move, IR spectroscopy pinpoints sites of proton gain or loss: cytosine's response indicates N1 protonation, and guanine's changes indicate N_7 protonation and N1 deprotonation effects. This is valuable for studying DNA under varied pH (e.g., in acidic environments or alkali-induced strand separation). It also demonstrates that vibrational spectra can report on tautomeric shifts and protonation states of bases, information that complements UV or NMR techniques.

5.4 *Effects of Temperature*

Raising temperature influences DNA's vibrational spectrum beyond just causing denaturation at the melting point. Even below the melting temperature, heating produces gradual spectral shifts due to thermal perturbation of base stacking and backbone geometry. For instance, as temperature climbs above 40 °C, several base-related bands move to lower wavenumber, reflecting weakened hydrogen bonding and base unstacking. Thymine bands at 1699 cm^{-1} (carbonyl stretch) and 1661 cm^{-1} and an adenine band at 1491 cm^{-1} all shift downward with heating. This indicates that thermal motion lengthens certain bonds or changes the environment (e.g., breaking minor hydrogen bonds or π-stacking interactions). Concurrently, some backbone modes respond: the IR band at 1234 cm^{-1} (which includes the asymmetric phosphate stretch) actually shifts to higher frequency as temperature increases. This unusual up-shift suggests a transition toward A-form-like character, since dehydrating or heating DNA can promote A-like conformation. Indeed, by 70 °C, spectra often show features of A-DNA, implying that before complete strand separation, B-DNA may convert to a compact form at high temperature. One observation is that the deoxyribose ring vibration near 887 cm^{-1} does not shift much even beyond 70 °C, meaning the sugar puckering is less sensitive to temperature than base pairing is. Overall, temperature causes a continuous change in vibrational frequencies due to thermal expansion and dynamic averaging of structures. By tracking these, FTIR can detect premelting conformational shifts (B to A transition, base unstacking) that occur before the helix fully dissociates. The technique can precisely measure the DNA melting point and even the transition width (cooperativity) from the temperature-dependent intensity ratios of, e.g., 1662 vs 1682 cm^{-1} bands. Unlike UV absorbance, IR can do so in concentrated solutions or gels where optical density would hinder UV measurements. Thus, vibrational spectroscopy not only confirms well-known thermal melting behavior but also provides extra detail on gradual structural changes and is applicable in a broader range of sample conditions [32].

6 DNA Interactions with Drugs, Ligands, and Proteins

Vibrational spectroscopy is a powerful method for studying how various molecules bind to DNA. When a drug or ligand interacts with DNA, it often perturbs specific vibrational bands corresponding to the binding site (e.g., bases or backbone). By examining shifts in frequency or changes in intensity of these bands, one can deduce the mode of binding, affected functional groups, and even secondary structure changes in the DNA. Below we discuss several case studies illustrating this:

6.1 General Indicators of Binding

When a ligand binds in the DNA major groove or minor groove or intercalates between bases, it alters the local electronic environment, which in turn can shift the vibrational frequencies of nearby DNA bonds. For example, hydrogen bond formation between a drug and a base carbonyl will typically lower the C=O stretching frequency, whereas coordination of a metal to a phosphate will shift the PO_2^- stretching frequencies. Often, the phosphate asymmetric stretch (1220 cm^{-1}) is very sensitive to binding of cationic ligands, since they neutralize or coordinate the phosphate group. An increase in intensity or a slight low-frequency shift of $\nu AS(PO_2^-)$ can indicate direct interaction at the phosphate. Similarly, perturbations in base vibrational bands (e.g., shifting of the guanine C=O at 1717 cm^{-1} or the thymine band at 1663 cm^{-1}) signal that the ligand is contacting those base functional groups. By systematically analyzing the spectrum of a DNA-ligand complex and comparing it to free DNA, researchers identify which bands move and thus infer the ligand's binding sites and its effect on DNA's secondary structure [33, 34].

6.2 Aspirin–DNA Interaction

A concrete example is aspirin (acetylsalicylic acid), whose binding to DNA has been studied by FTIR and Raman difference spectroscopy. At low aspirin concentrations (drug/DNA phosphate molar ratio $r = 1/40$), the spectra show that aspirin primarily interacts with DNA's backbone phosphate groups and A·T base pairs. Specifically, upon adding aspirin, the DNA's $\nu as(PO_2^-)$ band around 1222 cm^{-1} is substantially perturbed (shifted and intensified), indicating direct binding to the phosphate groups. Concurrently, bands associated with A–T base pairing, such as thymine's carbonyl at 1663 cm^{-1} and adenine's vibration at 1609 cm^{-1}, are shifted and gain intensity. The fact that these A–T bands are affected (while G–C bands remain largely unchanged at this low ratio) suggests aspirin's salicylate anion is docking in the major groove at A–T rich regions, likely interacting with adenine N_7 and thymine O_2 atoms (which are not involved in Watson-Crick hydrogen bonds and thus accessible). Importantly, at $r = 1/40$ no major changes in the overall DNA helical structure are observed; the base pairing is still intact (no dramatic shift of the 1700 cm^{-1} band), indicating no significant helix destabilization yet.

As the aspirin concentration is increased ($r > 1/20$, for instance), new spectral changes emerge: now G·C base pair vibrations are strongly perturbed. The guanine carbonyl band at 1717 cm^{-1} and a cytosine/guanine band at 1494 cm^{-1} show large shifts and intensity changes, revealing that aspirin (or its deprotonated form) at higher loading also binds to G–C base sites, perhaps via guanine O_6 or cytosine O_2. At these higher ratios, the IR spectrum indicates partial helix destabilization; for example, a decrease in the intensity of the 1705 cm^{-1} helical band and slight shifts in phosphate and base bands consistent with a more A-like DNA conformation.

Indeed, evidence shows a B to A transition in DNA when heavily loaded with aspirin. The overall picture is that aspirin initially anchors to the DNA backbone and A–T edges, and in greater amounts it perturbs all base pairs and induces DNA to shift toward the A-form, possibly due to dehydration or steric strain. The binding mechanism involves aspirin's anionic carboxylate and carbonyl groups: these can form hydrogen bonds or electrostatic contacts with phosphate oxygens and base sites (directly, or mediated by bridging water molecules). This case illustrates how FTIR can delineate a stepwise binding: first identifying the specific groups involved, then showing the onset of structural distortion at higher drug concentrations [35].

6.3 *Metal Ion Interactions*

Metal cations are another important class of ligands for DNA. Vibrational spectroscopy distinguishes two modes of metal binding, ionic binding to the phosphate backbone, which is typical for small monovalent and divalent cations that stabilize DNA structure, and coordination to bases, which certain heavy metal ions prefer and which can destabilize the helix. Generally, *DNA-stabilizing ions* like Na^{+}, K^{+}, Mg^{2+}, Ca^{2+}, and Zn^{2+} (and even structural ions like Mn^{2+}, Co^{2+}, and Ni^{2+}) bind predominantly to the negatively charged phosphate groups. In IR spectra, such binding causes only minor shifts in base absorption bands; in fact, these cations usually *do not* displace the base vibrational frequencies. Their effect is mainly to neutralize charges: one observes slight adjustments in the phosphate νas and νs intensities or positions, but base-carbonyl bands around 1700 cm^{-1} remain at the same frequencies, indicating that base pairing and stacking remain undisturbed. For example, comparing IR spectra of native DNA vs. DNA with Mg^{2+} or Ca^{2+} shows negligible differences in the 1800–1500 cm^{-1} region (bases), implying the metal is bound to phosphates and does not interact with the hydrogen-bonded bases. This is consistent with conductometric titration studies, which found that alkaline earth metal ions bind equally well to native and denatured DNA (hence to phosphates rather than bases).

In contrast, destabilizing or heavy metal ions (Ag^{+}, Hg^{2+}, Au^{3+}, Cu^{2+}, Cd^{2+}, Pb^{2+}, Fe^{3+}, etc.) often target electron-rich sites on the bases (like N_7 of guanine or the keto oxygens) and can disrupt base pairing. IR spectra of DNA exposed to these heavy metals show significant shifts or intensity changes in the base region (1800–1500 cm^{-1}). For instance, Ag^{+} or Hg^{2+} binding can cause the guanine carbonyl band (1700 cm^{-1}) to shift or weaken, and the cytosine and adenine bands to alter, reflecting direct coordination (e.g., Ag^{+} bridging between N_7 of guanine and O6 of another guanine to form metal-mediated base pairs). Notably, Ag^{+} is known to prefer guanine N7 sites at low concentrations, forming linear Ag–N_7(G) bonds that can drive B-DNA toward Z-DNA under some conditions (since Z-DNA is favored by certain base coordination). The presence of Ag^{+}, Hg^{2+}, or Au^{3+} leads to "global" spectral shifts of all base peaks, essentially because these metals can bind to multiple base sites and even induce mismatched base pairing (Hg^{2+} can form

T–Hg–T pairs, and Ag^+ can form C–Ag–C pairs). As a specific example, when Ag^+ interacts with DNA, new IR bands or shifted bands have been noted around 1713, 1550, and 1490 cm^{-1}, corresponding to altered guanine and cytosine vibrations. Fe^{3+} (a strongly binding trivalent cation) provides an interesting case: Fe^{3+} binding to DNA causes clear shifts in the phosphate stretching bands (due to strong ion pairing with PO_2^- groups), but little to no change in base bands. This suggests Fe^{3+} primarily shields or bridges phosphate groups without penetrating the base stack, a behavior more like stabilizing ions, except that excess Fe^{3+} can precipitate DNA. Uranyl (UO_2^{2+}) is another heavy cation that binds DNA: IR studies show that uranyl binds strongly to phosphate oxygens, evidenced by shifts in the DNA phosphate νas and νs bands, while base-sensitive bands remain unchanged. This implies uranyl does not appreciably interact with base pairs (consistent with its hard-metal preference for oxygens in phosphates). In summary, vibrational spectroscopy can tell whether a metal's primary binding site is the backbone (phosphate vibrations shift) or the bases (base vibrations shift). It thereby helps in elucidating the mechanism of metal-induced DNA conformational changes or damage.

6.4 Cationic Polypeptides and Proteins (Histones)

DNA in cells is regularly bound by positively charged proteins (like histones) and peptides (like protamines), which neutralize the phosphate backbone and compact the DNA. FTIR spectroscopy has been extensively used to investigate such DNA–protein complexes. A common observation is that binding of lysine- or arginine-rich peptides shifts the DNA phosphate bands and changes their intensity. For example, when a lysine-rich histone (such as histone H1) or a poly-L-lysine peptide binds to DNA, the antisymmetric PO_2^- stretch band (around 1220–1225 cm^{-1} in free DNA) shifts to a slightly lower frequency (1218 cm^{-1}) and often increases in intensity. This shift to lower wavenumber indicates the phosphate groups are less tightly bonded (due to the cationic amines interacting with them), effectively reducing the force constant of the P=O stretch. The intensity increase comes from the enhanced polarizability of the phosphate group when "neutralized" by the cationic peptide. Additionally, lysine-rich proteins have their own amide vibrations; upon DNA binding, one often sees an amide I band around 1640–1650 cm^{-1} in the complex spectrum. In complexes with H1 or poly-L-lysine, an amide I band at 1645 cm^{-1} appears (from the protein's C=O stretch), overlapping with DNA's base and backbone signals. The presence of this band with increased intensity has been noted, for instance, in histone–DNA spectra. Meanwhile, arginine-rich peptides (like poly-L-arginine or protamine, which is rich in Arg) can produce slightly different effects. Studies show that protamine–DNA binding causes a notable enhancement of the IR band near 960 cm^{-1} (assigned to the DNA backbone/phosphodiester vibration). This suggests that arginine-phosphate interactions may induce a more ordered or constrained backbone structure, "optimizing" the DNA conformation in a way that intensifies that particular vibration. In general, different polycations cause differential perturbations: Lys-rich ones emphasize changes in the phosphate asymmetric stretch and introduce strong protein amide signals, whereas Arg-rich ones might sharpen or amplify certain backbone modes (possibly due to multidentate contacts with phosphates). Interestingly, even simple inorganic cations show a similar trend: ions that bind stronger to phosphate (and potentially form hydrogen bonds to it) tend to lower the $\nu as(PO_2^-)$ frequency, whereas those that bind weakly do not. Thus, by analogy, the spectral effect of a given protein can inform us about whether it primarily shields phosphates (shifting $\nu asPO_2^-$) or induces structural rearrangements in DNA (enhancing other bands). The net outcome is that DNA complexation with basic proteins invariably stabilizes the backbone (reducing repulsion, often leading to compact conformations), and this is manifested in the IR spectrum by subtle but diagnostic changes in phosphate and sugar vibrations [21, 36].

6.5 DNA–RNA Hybrid Duplexes

A DNA–RNA hybrid helix (one strand DNA, one strand RNA) is an interesting case where each strand has a different sugar (deoxyribose vs. ribose) and possibly different helical parameters (A/B hybrid geometry). FTIR spectroscopy of DNA–RNA hybrids reveals features of both types of nucleic acids. The IR spectrum contains intense absorption bands that are characteristic of double-helical DNA and others typical of RNA. One striking observation is about the phosphate group orientation in hybrids. Polarized IR (dichroic) studies showed that the absorption bands of the phosphate groups in a hybrid do not all align the same way. In pure DNA B-form, all phosphate groups have a certain orientation relative to the fiber axis, and similarly in pure RNA A-form, they have another. In the DNA–RNA hybrid, the dichroism data suggest that the DNA strand's phosphates behave like in B-DNA, while the RNA strand's phosphates behave like in A-RNA. In other words, each strand in the hybrid retains its "native" geometry to some extent: the DNA strand in the hybrid is not fully B-form (because it's constrained by the RNA partner), but its phosphate orientation is intermediate, and the RNA strand's phosphate orientation remains closer to that of an A-form helix. This finding, first suggested by IR dichroism, indicates that DNA–RNA hybrids have a chimeric structure with respect to backbone arrangement. Intense bands in the hybrid's spectrum, for example near 1240–1220 cm^{-1} (PO_2^- stretch) and 810–830 cm^{-1} (backbone/sugar modes), reflect contributions from both forms. This example underscores that vibrational spectra can even pick up subtleties in mixed nucleic acid structures, beyond the scope of simple DNA or RNA, by dissecting the orientation and environment of functional groups [2].

6.6 Protective Ligands (E.g., Myo-Inositol)

Certain small molecules protect DNA structure under stress. Myo-inositol is a polyol known to help cells survive dehydration and UV radiation. FTIR studies have shown that myo-inositol can bind to DNA and effectively replace water molecules in stabilizing the helix. Normally, as mentioned, drying a DNA film (reducing RH from 75% to 30%) causes a drop in the dichroic ratio of the 1670 cm^{-1} band (the base pairing band), indicating loss of helical order. However, if DNA is dried in the presence of myo-inositol, an entirely different outcome is seen. The polarization measurements show that the dichroic ratio of the 1670 cm^{-1} band remains constant from 84% down to 10% RH in the DNA–inositol complex. In other words, DNA retains the orientation of base-pairing even when dehydrated, as long as inositol is present. This directly supports the hypothesis that myo-inositol replaces water in the DNA structure. By hydrogen-bonding to phosphate and perhaps base functional groups, inositol fulfills the role of the hydration shell, thereby preventing the DNA from collapsing or transitioning to a different conformation upon drying. Spectrally, one would see that the shifts in bands that normally accompany dehydration (like the 1705 to 1715 cm^{-1} shift or the 1240 to 1220 shift) might be mitigated or altered in the presence of inositol, but the key observation was maintaining polarization (dichroism) of an IR band. This example illustrates how vibrational spectroscopy can evaluate co-solvent or additive effects on DNA and even confirm protective mechanisms (keeping DNA in B-form when it would otherwise go to A or denature).

In summary, IR and Raman spectra provide a direct, label-free window into DNA's interactions: whether it's a small drug intercalating or groove-binding, a metal coordinating, or a protein neutralizing the backbone, each scenario leaves a distinct spectral "fingerprint." By analyzing which DNA bands shift or change intensity, researchers deduce binding sites (phosphate vs. base), sequence preference (A–T vs. G–C), and any induced conformational changes (like B to A transitions or helix destabilization). This makes vibrational spectroscopy an indispensable tool for characterizing DNA complexes in structural biology and biophysics [37].

7 Monitoring DNA Damage and Structural Changes

Vibrational spectroscopic techniques can sensitively detect various forms of DNA damage and structural alterations, whether caused by oxidative stress, radiation, heat, or biological processes like apoptosis. Because such damage often involves breaking chemical bonds or forming new ones (e.g., strand breaks, base modifications), it produces telltale changes in IR and Raman spectra of DNA.

7.1 Oxidative DNA Damage

Reactive oxygen species can attack DNA, leading to strand breaks and base modifications. One practical example is oxidative damage in sperm nuclear DNA. Raman microspectroscopy and FTIR have been used to distinguish untreated sperm DNA from oxidatively fragmented DNA. In Raman spectra, damage to the DNA backbone often manifests as changes in the phosphate region (1000–1100 cm^{-1}). Specifically, an increase in the 1050 cm^{-1} Raman band (assigned to a PO_3 stretching mode or C–O stretch in a perturbed backbone) alongside a decrease in the 1095 cm^{-1} band (symmetric PO_2^- stretch) has been observed as DNA fragmentation increases. The ratio of these two intensities, $I_{(1050)/I(1095)}$, correlates linearly with the DNA fragmentation index (DFI) measured by independent methods. In FTIR, an analogous pair of bands is seen: a band near 1013 cm^{-1} (phosphate or deoxyribose vibration) grows, while the normal phosphate band at 1080 cm^{-1} diminishes with greater oxidative damage. The IR intensity ratio $I_{(1013)/I(1080)}$ likewise shows a linear correlation with DFI. These spectral changes make sense chemically: strand breaks or oxidized abasic sites likely introduce more free phosphate ends or altered sugar rings that enhance the 1050 (or 1013) cm^{-1} signal, while reducing the number of long, helical phosphate stretches that contribute to the 1095 (or 1080) cm^{-1} mode. In addition to backbone signals, oxidative DNA damage can indirectly affect other cellular components. A notable Raman marker around 1445–1447 cm^{-1} (methylene bending mode of lipids and proteins) has been reported to shift or change intensity when DNA is severely damaged. In oxidatively damaged sperm, a decrease in the 1446 cm^{-1} peak intensity was observed, suggesting perturbation of the lipid environment, possibly due to byproducts of DNA damage or co-localized oxidative hits on membranes. Thus, vibrational spectroscopy not only quantifies DNA backbone cleavage but can hint at collateral damage to surrounding biomolecules (e.g., lipid peroxidation indicated by CH_2 bend changes). Importantly, these spectral biomarkers are detectable even when conventional assays might require cell lysis or labeling, highlighting a non-invasive advantage [38].

7.2 Heat-Induced Damage in Bone DNA

In forensic or archaeological contexts, heating (e.g., in fires) can damage DNA within bone. FTIR spectroscopy has been applied to heat-treated bone to gauge the condition of DNA alongside the bone's organic (collagen) and mineral (hydroxyapatite) components. Up to moderate temperatures (150 °C), FTIR shows minimal change in bone's amide I (collagen) and phosphate (mineral) bands, correlating with intact collagen and retrievable DNA. Indeed, bone samples heated to 150 °C for short durations still yielded successful DNA profiling, and correspondingly the IR "amide I to phosphate" ratio (reflecting collagen content) remained steady. However, beyond 200 °C, significant changes occur: the collagen amide bands (around 1650 and 1550 cm^{-1}) weaken as collagen denatures and carbonizes, and the mineral fingerprint shows an increased crystallinity band around 1030–1100 cm^{-1}. By 250 °C, IR spectra reveal a marked loss of amide signals (collagen breakdown), and concurrently DNA becomes non-amplifiable. Essentially, both DNA and collagen degrade at similar thermal thresholds, likely because DNA depurination and strand scission occur in the same temperature range (>150–200 °C) as collagen gelatinization and charring. The parallel degradation patterns mean that FTIR can serve as a quick screening tool: if the bone's collagen IR signal is heavily diminished (relative to the mineral), it's a sign that the sample was exposed to high heat and any DNA is probably too damaged for analysis. In a study, it was noted that 150 °C for a few hours caused only minor bond breakage (water loss) and DNA remained sufficient, whereas above 200 °C the drop in amide IR signal correlated with failure to PCR-amplify DNA fragments. Thus, FTIR could be used to triage bones, identifying heat-compromised ones by their spectral signature before wasting effort on DNA extraction. This application shows vibrational spectroscopy's value in an interdisciplinary setting (materials/forensics meets genomics) by linking molecular structure changes to functional DNA preservation [39].

7.3 Apoptosis (Programmed Cell Death)

During apoptosis, internucleosomal DNA is cleaved into regular fragments (the classic "DNA ladder"), and chromatin undergoes condensation. FTIR spectroscopy has proven adept at detecting these changes in cells. As cells undergo apoptosis, the amount of intact high-molecular-weight DNA decreases, which FTIR registers as a reduction in certain DNA bands and the appearance of others related to fragmentation. For example, studies on leukemic cells have shown that IR absorbances around 1222 cm^{-1}, 1177 cm^{-1}, and 1045 cm^{-1}, all of which are associated with nucleic acids (particularly phosphodiester and sugar vibrations), change noticeably during apoptosis. In early apoptosis, increased enzymatic nicks and slight chromatin condensation might cause a small increase in the 1220 cm^{-1} asymmetric phosphate band (due to exposure of phosphates) and a shift in the 1177 cm^{-1} band (C–O stretch of ribose or vibrations of thymine/adenine). As apoptosis progresses to later stages with extensive fragmentation, these DNA bands decrease in intensity overall, reflecting the loss of DNA content in each cell (since DNA is being cleaved and in part degraded). One particularly useful marker is the IR band at 1713 cm^{-1}, attributed to the C=O stretching of base pairs in double-stranded DNA. In viable cells with fully intact nuclei, the 1713 cm^{-1} band is prominent (indicating plenty of base-paired DNA). During apoptosis, as DNA is cleaved into smaller fragments and eventually digested, the relative intensity of the 1713 cm^{-1} band diminishes. Researchers have quantified this to estimate apoptotic cell fractions. In fact, the strength of the 1713 cm^{-1} band has been proposed as an indicator of "DNA stability" in cells; a high value denotes a large amount of long, paired DNA (healthy cells), whereas a reduced value signals DNA fragmentation and loss of base pairing (apoptotic or otherwise DNA-compromised cells). Liu et al. demonstrated that IR spectroscopy could measure the percentage of apoptotic cells in a leukemia cell population by analyzing such spectral changes, with results correlating well to flow cytometry (TUNEL assay) and DNA ladder assays. Notably, IR could detect apoptotic changes (like protein secondary structure shifts and DNA loss) as early as 4 h after treatment with an anti-cancer drug, whereas flow cytometry needed 6 h to show clear changes. Besides the DNA 1713 cm^{-1} band reduction, other IR changes in apoptosis include an increase in overall lipid content (e.g., a stronger CH_2 stretch or bend band) and a shift in protein amide I indicating unfolding to random coil. All these are consistent with cells condensing and blebbing (hence more membrane per volume, thus higher lipid signal) and with caspase-mediated cleavage of proteins leading to unfolding. Therefore, vibrational spectroscopy offers a rapid, label-free way to diagnose cell death by simultaneously tracking DNA integrity, protein conformation, and lipid changes [36, 40].

7.4 DNA Methylation (Epigenetic Modifications)

Methylation of DNA (most commonly at the 5-carbon of cytosine in CpG dinucleotides, forming 5-methylcytosine) is a subtle chemical change, but it significantly alters the vibrational spectrum of DNA. Unlike damage or denaturation, methylation doesn't break the DNA backbone or disrupt base pairing directly; instead, it adds a $-CH_3$ group to the base. Vibrational spectroscopy, especially Raman-based methods, can detect these extra methyl groups and the consequent electronic changes in the base. All four bases can in principle be methylated (C, A, G, T), but 5-methylcytosine (5 mC) is the predominant epigenetic mark, and N_7- or O_6-methylguanine is important in mutagenesis and chemotherapy contexts (Fig. 4).

7.5 Base-Specific Marker Bands for Methylation

The addition of a methyl group introduces new vibrational modes (C–H stretches, bends, and methyl rock/twist) and also shifts existing base vibrations by changing electron distribution. In Raman spectra, a well-known signature of DNA containing 5-methylcytosine is a sharp band near 785–787 cm^{-1}, slightly upshifted from the 782 cm^{-1} cytosine ring breathing mode in unmethylated DNA. Heavily methylated DNA shows an increased intensity at 786 cm^{-1} compared to unmethylated DNA, making it a convenient marker for 5mC presence. Another diagnostic peak is around 1330–1335 cm^{-1}, which corresponds to CH_3 twisting and wagging motions coupled with backbone modes. Highly methylated cells or DNA samples exhibit a notable increase in this 1330 cm^{-1} band (sometimes specifically pinpointed at 1331 or 1335 cm^{-1}). Indeed, Raman spectra of methylated vs. non-methylated DNA

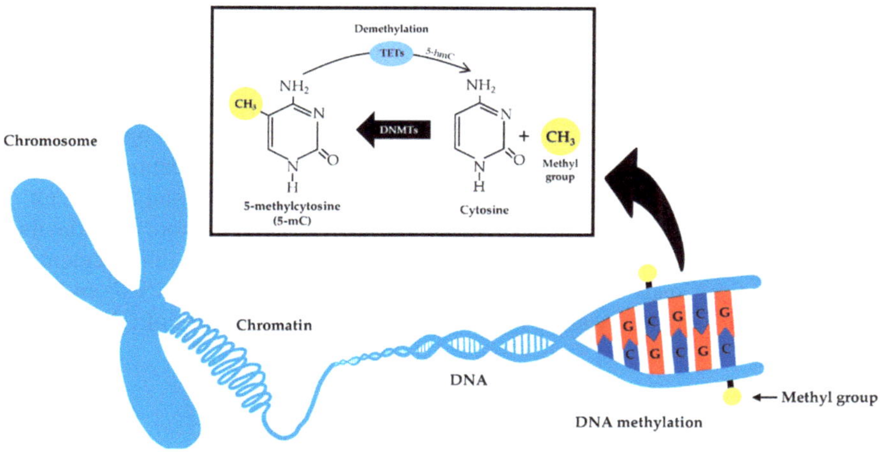

Fig. 4 Schematic representation of DNA methylation and demethylation molecular mechanisms. This file is licensed under the Creative Commons Attribution 4.0 International license

standards reveal significantly stronger bands at 1335 cm^{-1}, 1379 cm^{-1}, and 1579 cm^{-1} in the methylated DNA. The 1379 cm^{-1} enhancement is tied to a CH_3 bending mode and ring vibration (often adenine is affected by methylation status indirectly), while the 1579 cm^{-1} increase is associated with a ring vibration (C=N stretching in cytosine/guanine) that for some reason gains intensity with methylation. Essentially, methylation tends to intensify Raman lines at the edges of the fingerprint region (1200–1700 cm^{-1}), likely because the presence of 5mC slightly redistributes how bases stack and vibrate collectively.

Resonance Raman at specific UV excitations can further amplify base-specific signals. Using a 228 nm laser (which resonates with cytosine electronic transitions), researchers found that cytosine and 5mC signals are selectively enhanced. In UVRR spectra at 228 nm, hypermethylated DNA from a cancer cell line showed distinct differences from hypomethylated DNA of another source. One study noted that *global* hypermethylation led to higher overall intensities in peaks corresponding to cytosine vibrations, consistent with more 5mC content [41, 42].

7.6 SERS Detection of Methylation (Guanine Example)

Surface-Enhanced Raman Scattering (SERS) has been applied to detect methylation at specific guanine positions. Methylation of guanine can occur at the O_6 or N_7 positions (e.g., due to certain carcinogens or chemotherapy drugs). SERS, with the help of metal nanoparticle substrates, can greatly enhance Raman signals of DNA bases and even distinguish isomeric modifications. A striking finding is that a Raman band at 1323 cm^{-1} appears only when guanine is methylated at the O6 position. In contrast, the same band is absent for N_7-methylguanine samples. This provides a clear spectral handle to differentiate N_7 vs O_6 methylation in guanine, a difference of great toxicological importance, since O6-methylguanine is highly mutagenic. Additionally, SERS of methylated guanines identified bands at 1554, 647, and 700 cm^{-1} associated with specific adducts (N_7-methyl, N_7-hydroxyethyl, etc., each with slight variations). Through such band assignments and by using computational calculations, one can in principle identify not just that guanine is methylated, but also at which atom and with what group. This level of detail is difficult to obtain by other label-free means. It's worth noting SERS's limitation: reproducibility. SERS intensities can vary by ±15% due to variable hot-spot enhancement, so careful statistical analysis (PCA, etc.) is used. Nevertheless, SERS's sensitivity allowed detection of methylation in femtomole to picomole ranges of DNA.

7.7 Raman Micro-Spectroscopy in Living Cells

A major breakthrough has been applying Raman microspectroscopy to monitor global DNA methylation in live cells without labels. In a recent study, researchers compared cells with high vs low DNA methylation (e.g., cancer cells vs DNMT1 knockout cells, or stem cells in different media) and found reproducible spectral differences. Using multivariate analysis, they identified a set of Raman bands whose intensities differ systematically with methylation level. Key bands that increased with higher methylation include 786 cm^{-1}, 1257 cm^{-1}, 1317–1330 cm^{-1}, 1342 cm^{-1}, 1579 cm^{-1}, 1605–1610 cm^{-1}, and 1662 cm^{-1}. For instance, wild-type colon carcinoma cells (with normal methylation) showed stronger peaks at 786, 1257, 1330, and 1579 cm^{-1} compared to DNMT1-deficient (hypomethylated) cells. Conversely, a band in the 815–830 cm^{-1} range (related to the DNA backbone and possibly the O–P–O stretch or adenine) was relatively lower in intensity in high-methylation cells. This suggests that methylation (which often correlates with more condensed, heterochromatic DNA) might reduce exposure or freedom of certain backbone modes. Indeed, Raman imaging at high resolution even revealed that highly methylated DNA tends to concentrate in heterochromatic foci within nuclei, which correlate with intense Raman signals; essentially, one can "see" DNA methylation in a cell's Raman image as bright spots corresponding to densely packed, methyl-rich DNA (confirmed by comparison to immunostaining of 5mC). These demonstrations show that Raman microspectroscopy can be developed into a nondestructive assay for epigenetic state, potentially useful for assessing tumor cell populations or stem cell differentiation status in real time [9].

7.8 Advantages of Vibrational Approaches for Methylation

Traditional DNA methylation analysis (bisulfite sequencing, ELISA, etc.) is time-consuming and destructive and requires DNA extraction or chemical treatment. Vibrational spectroscopy offers a label-free alternative. It probes the intrinsic vibrations of the methyl groups and the perturbation they induce in the DNA bases. Methods like SERS can detect extremely low concentrations of methylated bases due to signal enhancement. UV resonance Raman can selectively enhance one base at a time by tuning the wavelength (e.g., 272 nm for thymine, 250 nm for guanine); thus, one could separately examine methylation on different bases by appropriate laser choice. FTIR, while less specific to individual bases, is rapid and can be used to monitor global methylation through markers like the 1429 cm^{-1} vs 1473 cm^{-1} intensity ratio or the appearance of new bands due to methyl C–H bending. In one study, an IR band at 1265 cm^{-1} appeared upon Z-DNA formation, which could be related to 5mC, and an increase in a band at 1124 cm^{-1} was seen (possibly due to methylated base vibrations). More directly, an IR study on blood cells noted that the relative intensity of the 1713 cm^{-1} band (base-paired DNA) can reflect global DNA content, which might indirectly indicate global hypomethylation in diseased states (since heavily diseased cells often have less DNA or more single-stranded regions due to nicks).

In summary, vibrational spectroscopy encompassing FTIR, Raman, UVRR, and SERS has emerged as a versatile set of tools for epigenetic fingerprinting. They combine simplicity (no elaborate sample prep), non-destructiveness, and the ability to work even in live cells or native tissues. With continued refinement, these techniques hold promise for clinical diagnostics (e.g., identifying cancers by their methylation patterns spectroscopically) and for basic research (e.g., observing dynamic methylation changes in cells in real time).

8 Limitations and Challenges of FTIR/Raman for DNA Analysis

While FTIR and Raman spectroscopy provide rich information on DNA structure and interactions, there are several limitations and practical challenges to consider:

8.1 Spectral Congestion and Band Assignments

DNA's vibrational spectrum, especially in the fingerprint region (600–1800 cm^{-1}), is highly congested. The overlapping of many normal modes means individual bands often represent contributions from multiple molecular groups. This makes *unambiguous assignment* difficult. For example, an IR band around 1660 cm^{-1} could involve contributions from thymine C=O stretch, cytosine C=N stretch, and even amino group vibrations. In our discussions we assign "major" contributors, but in truth most bands are composite. Theoretical calculations of DNA normal modes are complex and not fully resolved, so much assignment work relies on empirical comparisons to model compounds (mononucleotides, nucleosides, etc.). This semi-empirical approach can sometimes mislead; hence, some DNA band assignments remain tentative or conventional. A particular example is the 1710 cm^{-1} band: it's clearly associated with base pairing, but it actually arises from coupled C=O and N–H vibrations of multiple bases. Without isotopic labeling or advanced simulation, one cannot cleanly separate their contributions. Another complication is isotope exchange: if one records spectra in D_2O (often done to avoid H_2O IR absorption), many bands shift (e.g., those involving N–H or O–H vibrations). In deuterated DNA, most base and sugar bands move 10–40 cm^{-1} lower, further crowding the spectrum and complicating assignments. Thus, while we can qualitatively monitor changes (shifts, intensity variations), attributing each to a precise molecular motion can be challenging. This complexity sometimes limits the depth of conclusions. For instance, we might detect that an unknown drug causes a band at 1490 cm^{-1} to increase, but knowing whether that's due to an adenine ring mode or a cytosine amino bend might require additional evidence beyond the IR data alone. Researchers mitigate this by combining IR/Raman with other methods (NMR, X-ray, or mutational analysis) to solidify interpretations [43].

8.2 Sample State and Preparation Influences

DNA spectra can look different depending on whether the sample is a dry film, hydrated fiber, solution, or crystalline form. Water, in particular, is a double-edged sword: in solution, water causes heavy IR background and can mask some DNA bands, so often DNA is measured as a film or pellet to minimize water. But drying DNA can induce structural changes (like transition to A-form or even partial denaturation in high salt). For example, if one precipitates DNA with excess salt and dries it for IR, the sample may no longer be in the native B-form; the IR spectrum will show broadened lines and shifts indicative of altered hydrogen bonding. Similarly, too much exposure to UV laser in Raman can heat the sample, potentially causing local denaturation or damage (though UVRR at low power mitigates this). Thus, sample handling must ensure that what's observed is intrinsic and not an artifact of the preparation. The physical state also matters: oriented films give information on dichroism (useful for structure), whereas powder or solution spectra are averaged over all orientations. Spectra of DNA in solution tend to have broader peaks than those of well-ordered fibers, due to conformational heterogeneity. Achieving sufficiently high DNA concentration for a good SNR in IR can be tricky (since DNA has limited solubility at very high concentrations and water absorbs strongly). ATR-FTIR can be used to look at aqueous DNA with water subtraction, but even slight baseline variations can affect analysis of weak bands. Additionally, many biological DNA samples (e.g., genomic DNA from cells) come with contaminants (proteins, buffer salts, etc., which have their own spectral signatures). Protein amide bands (around 1650 and 1550 cm^{-1}) can overlap DNA bands. In UVRR spectra of extracted DNA, residual histones or phenol (from purification) can cause extraneous bands or fluorescence background. Careful sample purification or using difference spectroscopy (subtracting the spectrum of the contaminant) is often needed. Still, complete removal of tightly bound nuclear proteins is hard, and their amide signals might ride along with the DNA spectrum. This can complicate interpretation, although sometimes it provides complementary info (like detecting histone still bound to DNA via its amide I band) [23].

8.3 *Quantification and Reproducibility*

While vibrational spectroscopy can be quantitative (Beer's law for IR absorbance, etc.), there are challenges. For one, molar absorptivities and Raman cross-sections vary by functional group. In a mixture like DNA (where bases, sugars, phosphates each contribute), the composite band intensity isn't directly proportional to one component's concentration unless that component dominates the band. Thus, quantifying an "amount of damage" or "percent methylation" from raw IR/Raman intensities requires constructing calibration curves or empirical ratios correlated to reference methods. In our earlier example, the linear correlation of Raman 1050/1095 with DFI was established by calibrating against flow cytometry. Without such calibration, the absolute intensities could be misleading (because Raman signals also depend on focus, local concentration, etc.). Reproducibility is particularly an issue for SERS: two SERS spectra of the same DNA sample might differ if the distribution of nanoparticles (and hence hot spots) differs, even if ensemble-averaged trends are consistent. Efforts like averaging many spectra or using reproducible nanostructured substrates are employed to overcome this. In Raman microspectroscopy of cells, one also faces cell-to-cell variability—dozens of cells must be measured to get statistically robust differences for methylation states, for example. Another quantitative challenge is normalization: with whole-cell spectra, one often normalizes to a specific peak (like phenylalanine at 1004 cm^{-1} or the PO_4 symmetric stretch) to account for variations in cell size or DNA amount. If the normalization peak is itself affected by the condition (say, DNA content changes), interpretation needs caution [44].

8.4 *Instrumental and Practical Limits*

FTIR has a spatial resolution limited by IR wavelengths (>5 μm), so it's not ideal for imaging small cellular structures (though IR microspectroscopy exists, its resolution is lower than optical microscopy). Raman can achieve sub-micron resolution (even 250 nm with confocal), which is great for cell imaging, but spontaneous Raman signals are weak, requiring longer acquisition times or higher laser powers. Higher laser power can risk sample damage (especially UV lasers causing photochemistry). There's also the matter of spectral interference: water in IR, as mentioned, or fluorescence in Raman (which can swamp the Raman scattering if the sample or impurities fluoresce). Fluorescence is a frequent issue for visible excitation Raman on biological samples, though it can be reduced by using UV (where many molecules don't fluoresce strongly or the emission is outside the detection range). Using UV, however, demands more expensive lasers and careful handling to avoid UV damage.

8.5 Data Analysis Complexity

Extracting meaningful info from spectra often requires advanced analysis, from simple baseline corrections to multivariate analysis (PCA, LDA, machine learning). For example, distinguishing the subtle spectral differences between methylated and unmethylated DNA in cells needed PCA to identify which combination of band changes was significant. This means expertise and computing are as necessary as the spectral acquisition. Moreover, large spectral datasets (like Raman maps of nuclei with thousands of spectra) require processing that is time-consuming and not yet fully automated in a routine sense.

In conclusion, while FTIR and Raman spectroscopy offer rich, multiplexed data about DNA, one must be mindful of their limitations. Overlapping bands can obscure detailed assignments; sample prep and conditions can inadvertently alter DNA; quantification often needs calibration; and techniques like SERS, though sensitive, may suffer from reproducibility issues if not carefully controlled. Researchers address these by combining vibrational spectroscopy with orthogonal methods (for validation) and by improving techniques (such as developing more uniform SERS substrates or using stable isotope labeling to separate overlapping bands). The continued advancement in instrumentation (like FTIR imaging arrays, better lasers, and detectors) and analysis methods (like machine learning on spectra) is helping to mitigate these challenges, making vibrational spectroscopy increasingly robust for DNA characterization.

9 Conclusion

Despite the challenges, Fourier Transform Infrared (FTIR) and Raman spectroscopy stand out as powerful tools for DNA analysis. They are non-destructive and label-free, requiring minimal sample preparation; one can examine DNA in a native-like state (from solutions to intact cells) without adding dyes or probes. Both techniques yield a unique "molecular fingerprint" of DNA by measuring its intrinsic vibrational modes. This is immensely valuable because any change in DNA (be it a conformational transition, damage, or binding event) leaves a signature in this vibrational fingerprint. Over the years, FTIR and Raman have provided key insights: for example, identifying the B, A, and Z forms of DNA via characteristic bands, detecting the melting of DNA by the shift of the 1705 to 1660 cm^{-1} band, or revealing interactions like drug intercalation by marker band perturbations. These techniques effectively complement more structure-specific methods like X-ray crystallography or NMR by being applicable in broader conditions (non-crystalline samples, mixtures, even live cells). They are also rapid; an IR spectrum of a sample can be collected in seconds to minutes, enabling high-throughput screening in some cases.

FTIR and Raman are often complementary because they are sensitive to different molecular vibrations and selection rules. FTIR relies on a change in dipole moment; thus vibrations that involve polar functional groups (like the phosphate stretches, carbonyl stretches of bases, and –OH and –NH bends) are usually very strong in IR. Indeed, FTIR spectra excel at highlighting DNA's phosphate backbone (the $\nu asPO_2^-$ and νsPO_2^- are among the most intense bands) and the base carbonyl and ring stretching vibrations. As such, FTIR is superb for assessing backbone conformation and DNA's interactions with charged ligands. For instance, it clearly shows the difference between B-form vs. A-form in the phosphate and sugar region, and it can monitor complexation by shifts in those backbone bands. Moreover, IR is very sensitive to the presence of double-stranded vs. single-stranded DNA (through the base-pairing band near 1700 cm^{-1}). FTIR has been successfully used to identify and quantify DNA damage in situ (like distinguishing healthy vs. cancerous tissue based on DNA and protein spectral differences).

Raman spectroscopy, on the other hand, relies on a change in polarizability; thus, it excels with vibrations of non-polar bonds or symmetric vibrations. Raman spectra of DNA showcase strong signals from ring stretching modes of bases (C=C, C=N in purine and pyrimidine rings) and C–H bending, which are sometimes less prominent in IR. Raman lines like 1485 cm^{-1} or 1575 cm^{-1} (ring modes) are often sharper and easier to resolve than in IR. Also, Raman is sensitive to backbone conformations in a different way: some backbone modes (like O–P–O bending or the deoxyribose vibrations) give distinct polarized Raman bands that correlate with sugar pucker and glycosidic bond angle. Because water has weak Raman scattering, Raman spectroscopy can be done in aqueous solutions without the massive background that plagues IR, enabling *in vitro* and even *in vivo* measurements. Another big advantage of Raman is the ability to use resonance and surface enhancement to boost specific signals, e.g., UV resonance Raman to target particular bases or SERS to detect trace modifications. These enhancements allow Raman to detect things like low-abundance damage products or subtle epigenetic marks that would be invisible in IR. For example, SERS could discern N_7 vs O_6 methylguanine by a unique 1323 cm^{-1} band, and UVRR could selectively amplify the vibration of a methylated cytosine ring. FTIR cannot offer that kind of selectivity; it gives a holistic picture, which is a strength for global analysis but a weakness for pinpointing a minor component.

In many cases, information from both FTIR and Raman is combined to get a fuller understanding. A classic example is in the study of Z-DNA: IR identified certain marker bands (like the 1220 to 1215 cm^{-1} shift and the new 930 cm^{-1} band), while Raman identified others (like the appearance of 745 cm^{-1} and the shift of 781 to 772 cm^{-1}), together confirming the Z-form and offering insight into changes in both backbone and base geometry.

In conclusion, FTIR and Raman spectroscopy greatly enrich the toolkit for DNA analysis. They offer a unique combination of speed, non-destructiveness, and comprehensive sensing of molecular bonds. Rather than replacing methods like sequencing or crystallography, they add a layer of functional insight—revealing how DNA's bonds are vibrating can tell us if it's paired or not, supercoiled or relaxed, methylated or unmethylated, free or protein-bound. This vibrational "music" of DNA is now being deciphered with increasing clarity, enabling scientists to detect and understand subtle changes in DNA that underlie biological function and disease. As technology advances, we anticipate these spectroscopic methods will play an even larger role, possibly moving into clinical diagnostics (as compact Raman or IR devices) and continuing to uncover new aspects of DNA chemistry and biology in tandem with other emerging techniques.

References

1. Han Y, Han L, Yao Y, Li Y, Liu X (2018) Key factors in FTIR spectroscopic analysis of DNA: the sampling technique, pretreatment temperature and sample concentration. Anal Methods 10:2436–2443. https://doi.org/10.1039/c8ay00386f
2. Zhizhina GP, Oieinik EF (1972) Infrared Spectroscopy of Nucleic Acids. Russian Chem Rev 41:258
3. Szyc Ł, Yang M, Nibbering ETJ, Elsaesser T (2010) Ultrafast vibrational dynamics and local interactions of hvdrated DNA. Angew Chem Int Ed 49:3598–3610
4. Almassalha LM, Carignano M, Liwag EP, Li WS, Gong R, Acosta N, Dunton CL, Gonzalez PC, Carter LM, Kakkaramadam R, Kröger M, MacQuarrie KL, Frederick J, Ye IC, Su P, Kuo T, Medina KI, Pritchard JA, Skol A, Nap R, Kanemaki M, Dravid V, Szleifer I, Backman V (2025) Chromatin conformation, gene transcription, and nucleosome remodeling as an emergent system. Sci Adv 11. https://doi.org/10.1126/sciadv.adq6652
5. Kelly JG, Najand GM, Martin FL (2011) Characterisation of DNA methylation status using spectroscopy (mid-IR versus Raman) with multivariate analysis. J Biophotonics 4:345–354. https://doi.org/10.1002/jbio.201000085
6. Liu KZ, Xu M, Scott DA (2007) Biomolecular characterisation of leucocytes by infrared spectroscopy. Br J Haematol 136:713–722
7. Chen J, Lu J, Liu J, Fang J, Zhong X, Song J (2023) DNA conformational dynamics in the context-dependent non-CG CHH methylation by plant methyltransferase DRM2. J Biol Chem 299:105433. https://doi.org/10.1016/j.jbc.2023.105433
8. D'Amico F, Zucchiatti P, Latella K, Pachetti M, Gessini A, Masciovecchio C, Vaccari L, Pascolo L (2020) Investigation of genomic DNA methylation by ultraviolet resonant Raman spectroscopy. J Biophotonics 13. https://doi.org/10.1002/jbio.202000150
9. Abid Hasan SM, He Y, Chang TW, Wang J, Gartia MR (2019) Detecting DNA methylation using surface-enhanced Raman spectroscopy. J Phys Chem C 123:698–709. https://doi.org/10.1021/acs.jpcc.8b10178
10. Sudhakaran G, Kesavan D, Kandaswamy K, Guru A, Arockiaraj J (2024) Unravelling the epigenetic impact: oxidative stress and its role in male infertility-associated sperm dysfunction. Reprod Toxicol 124:108531. https://doi.org/10.1016/j.reprotox.2023.108531
11. Safar W, Azziz A, Edely M, Lamy de la Chapelle M (2023) Conventional Raman, SERS and TERS studies of DNA compounds. Chem 11:399. https://doi.org/10.3390/chemosensors11070399

12. Duan M, Li Y, Zhang F, Huang Q (2023) Assessing B-Z DNA transitions in solutions via infrared spectroscopy. Biomolecules 13:964. https://doi.org/10.3390/biom13060964
13. Szabó A, Shi B, Lee SA, Rupprecht A (1996) A Raman study of the time variability for the A-to-B transition in wet-spun films of calf-thymus DNA. J Biomol Struct Dyn 13:1029–1033. https://doi.org/10.1080/07391102.1996.10508917
14. D'Amico F, Graziano R, D'Aria F, Russomanno P, Di Fonzo S, Amato J, Pagano B (2023) Cytosine epigenetic modifications and conformational changes in G-quadruplex DNA: an ultraviolet resonance Raman spectroscopy study. Spectrochim Acta A Mol Biomol Spectrosc 300:122901. https://doi.org/10.1016/j.saa.2023.122901
15. Marty R, N'soukpoe-Kossi CN, Charbonneau D, Weinert CM, Kreplak L, Tajmir-Riahi H-A (2009) Structural analysis of DNA complexation with cationic lipids. Nucleic Acids Res 37:849–857. https://doi.org/10.1093/nar/gkn1003
16. Guchhait B, Liu Y, Siebert T, Elsaesser T (2016) Ultrafast vibrational dynamics of the DNA backbone at different hydration levels mapped by two-dimensional infrared spectroscopy. Struct Dyn 3. https://doi.org/10.1063/1.4936567
17. Fick RJ, Liu AY, Nussbaumer F, Kreutz C, Rangadurai A, Xu Y, Sommer RD, Shi H, Scheiner S, Stelling AL (2021) Probing the hydrogen-bonding environment of individual bases in DNA duplexes with isotope-edited infrared spectroscopy. J Phys Chem B 125:7613–7627. https://doi.org/10.1021/acs.jpcb.1c01351
18. Banyay M, Sarkar M, Gräslund A (2003) A library of IR bands of nucleic acids in solution. Biophys Chem 104:477–488. https://doi.org/10.1016/S0301-4622(03)00035-8
19. Mello MLS, Vidal BC (2012) Changes in the infrared microspectroscopic characteristics of DNA caused by cationic elements, Different Base richness and single-stranded form. PLoS One 7:e43169. https://doi.org/10.1371/journal.pone.0043169
20. Kuhar N, Sil S, Verma T, Umapathy S (2018) Challenges in application of Raman spectroscopy to biology and materials. RSC Adv 8:25888–25908. https://doi.org/10.1039/C8RA04491K
21. Benevides JM, Wang AHJ, Van der Marel GA, Van Boom JH, Thomas GJ (1988) Crystal and solution structures of the B-DNA dodecamer d(CGCAAATTTGCG) probed by Raman spectroscopy: heterogeneity in the crystal structure does not persist in the solution structure. Biochemistry 27:931–938. https://doi.org/10.1021/bi00403a014
22. Ghosh D, Dey SK, Saha C (2014) Mutation induced conformational changes in genomic DNA from cancerous K562 cells influence drug-DNA binding modes. PLoS One 9:e84880. https://doi.org/10.1371/journal.pone.0084880
23. Serec K, Babić SD, Podgornik R, Tomić S (2016) Effect of magnesium ions on the structure of DNA thin films: an infrared spectroscopy study. Nucleic Acids Res 44:8456–8464. https://doi.org/10.1093/nar/gkw696
24. Ruiz-Chica AJ (2004) On the interpretation of Raman spectra of 1-aminooxy-spermine/DNA complexes. Nucleic Acids Res 32:579–589. https://doi.org/10.1093/nar/gkh232
25. Whelan DR, Bambery KR, Heraud P, Tobin MJ, Diem M, McNaughton D, Wood BR (2011) Monitoring the reversible B to A-like transition of DNA in eukaryotic cells using Fourier transform infrared spectroscopy. Nucleic Acids Res 39:5439–5448. https://doi.org/10.1093/nar/gkr175
26. Zhang F, Huang Q (2023) Characterization of Z-DNA by infrared spectroscopy. Methods Mol Biol 2651:53–58
27. BBC-16-06
28. Taboury JA, Liquier J, Taillandier E (1985) Characterization of DNA structures by infrared spectroscopy: double helical forms of poly(dG-dC) poly(dG-dC), poly(dD8G-dC) poly(dD8G-dC), and poly(dG-dm5C) poly(dG-dm5C). Canadian J Chem. https://doi.org/10.1139/v85-315
29. Jain S, Thomas A, Zhuo G-Y, Mazumder N (2025) Fourier transform infrared (FTIR) spectroscopy of biomolecules. In: Recent advances in infrared spectroscopy and its applications in biotechnology. IntechOpen, London
30. Duguid JG, Bloomfield VA, Benevides JM, Thomas GJ (1996) DNA melting investigated by differential scanning calorimetry and Raman spectroscopy. Biophys J 71:3350–3360. https://doi.org/10.1016/S0006-3495(96)79528-0

31. Bonnet J, Colotte M, Coudy D, Couallier V, Portier J, Morin B, Tuffet S (2010) Chain and conformation stability of solid-state DNA: implications for room temperature storage. Nucleic Acids Res 38:1531–1546. https://doi.org/10.1093/nar/gkp1060
32. Ashwood B, Jones MS, Radakovic A, Khanna S, Lee Y, Sachleben JR, Szostak JW, Ferguson AL, Tokmakoff A (2023) Direct monitoring of the thermodynamics and kinetics of DNA and RNA dinucleotide dehybridization from gaps and overhangs. bioRxiv. https://doi.org/10.1101/2023.04.10.536266
33. Neault JF, Tajmir-Riahi HA (1999) Structural analysis of DNA-chlorophyll complexes by Fourier transform infrared difference spectroscopy. Biophys J 76:2177–2182. https://doi.org/10.1016/S0006-3495(99)77372-8
34. Hithell G, Shaw DJ, Donaldson PM, Greetham GM, Towrie M, Burley GA, Parker AW, Hunt NT (2016) Long-range vibrational dynamics are directed by Watson-crick base pairing in duplex DNA. J Phys Chem B 120:4009–4018. https://doi.org/10.1021/acs.jpcb.6b02112
35. Neault JF, Naoui M, Manfait M, Tajmir-Riahi HA (1996) Aspirin-DNA interaction studied by FTIR and laser Raman difference spectroscopy. FEBS Lett 382:26–30. https://doi.org/10.1016/0014-5793(96)00093-2
36. Zhuang Z, Cheng J, Wang X, Zhao B, Han X, Luo Y (2007) Surface-enhanced Raman spectroscopy and density functional theory study on 4,4′-bipyridine molecule. Spectrochim Acta A Mol Biomol Spectrosc 67:509–516. https://doi.org/10.1016/j.saa.2006.08.008
37. Webb SJ, Bhorjee JS (1968) Infrared studies of DNA, water, and inositol associations. Can J Biochem 46:691–695. https://doi.org/10.1139/o68-106
38. Sánchez V, Redmann K, Wistuba J, Wübbeling F, Burger M, Oldenhof H, Wolkers WF, Kliesch S, Schlatt S, Mallidis C (2012) Oxidative DNA damage in human sperm can be detected by Raman microspectroscopy. Fertil Steril 98:1124–1129.e3. https://doi.org/10.1016/j.fertnstert.2012.07.1059
39. Zupanič Pajnič I, Leskovar T, Jerman I (2021) Bone fragment or bone powder? ATR-FTIR spectroscopy–based comparison of chemical composition and DNA preservation of bones after 10 years in a freezer. Int J Legal Med 135:1695–1707. https://doi.org/10.1007/s00414-021-02620-0
40. Lu Y, Dong Y, Qin J (2016) Porous pcu-type Zn(II) framework material with high adsorption selectivity for CO 2 over N 2. J Mol Struct 1107:66–69. https://doi.org/10.1016/j.molstruc.2015.11.045
41. Benguerba Y, Alnashef IM, Ertc A, Balsamo M, Ernst B (2019) A quantitative prediction of the viscosity of amine based DESs using Sσ-profile molecular descriptors. J Mol Struct 1184:357–363. https://doi.org/10.1016/j.molstruc.2019.02.052
42. Wu Y, Wu J, Jiao L, Xu W, Wang H, Wei X, Gu W, Ren G, Zhang N, Zhang Q, Huang L, Gu L, Zhu C (2020) Cascade reaction system integrating single-atom Nanozymes with abundant cu sites for enhanced biosensing. Anal Chem 92:3373–3379. https://doi.org/10.1021/acs.analchem.9b05437
43. Naim N, Ennahli N, Hanine H, Lahlali R, Tahiri A, Fauconnier M-L, Madani I, Ennahli S (2022) ATR-FTIR spectroscopy combined with DNA barcoding and GC-MS to assess the quality and purity of saffron (Crocus sativus L.). Vib Spectrosc 123:103446. https://doi.org/10.1016/j.vibspec.2022.103446
44. Alom KM, Tukova A, Lyu N, Rodger A, Wang Y (2025) Label-free surface-enhanced Raman scattering for genomic DNA cytosine methylation Reading. Molecules 30:403. https://doi.org/10.3390/molecules30020403

Raman Imaging for Biopolymers Analysis

Jiro Karlo and Surya Pratap Singh

Abstract Biopolymers form the structural and functional backbone of life, ranging from nucleic acids and proteins to cellulose, lignin, chitin, and others. Their complexity and diversity make them central not only to biological systems but also to agriculture, food technology, medicine, and materials science. Studying them, however, has always posed a challenge. Many traditional analytical techniques require destructive processing or chemical labeling, which can obscure the very properties researchers aim to measure. Raman imaging provides a powerful alternative. Capturing the vibrational fingerprints of molecular bonds, it enables non-invasive, label-free, and chemically specific visualization of biopolymers in their native environments. This chapter explores how Raman imaging has been applied across different classes of biopolymers. The technique's unique ability to generate spatially resolved chemical maps makes it particularly valuable for understanding distribution, localization, and heterogeneity across the area of interest. We have also discussed the challenges, current limitations and future perspective of Raman imaging in biopolymer analysis. Weak scattering signals, interference from autofluorescence, and overlapping spectra can limit sensitivity and clarity, while high instrumentation costs have slowed wider adoption. Advances in enhanced Raman methods, machine learning, multimodal imaging, and portable instrumentation are beginning to address these obstacles. Taken together, Raman imaging is emerging as a versatile and transformative approach to biopolymer analysis. Its strength lies in connecting molecular-level insights providing knowledge that can advance both fundamental research and practical applications in sustainable materials, biotechnology, and health.

Keywords Raman imaging · Biopolymer · Spectroscopy

J. Karlo · S. P. Singh (✉)
Department of Biosciences and Bioengineering, Indian Institute of Technology Dharwad, Dharwad, Karnataka, India
e-mail: ssingh@iitdh.ac.in

N. Mazumder et al. (eds.), *Vibrational Spectroscopy for Biopolymer Characterization*, Biological and Medical Physics, Biomedical Engineering,
https://doi.org/10.1007/978-3-032-15306-7_5

1 Introduction

Nature has been producing polymers long before humans learned to shape synthetic polymers such as plastics. Biopolymers are the polymers that are naturally synthesized or derived from organisms. It includes nucleic acids such as DNA and RNA, proteins, inulin, casein, amylopectin, chitosan, gelatin, pectin, lignin, chitin, cellulose, hemicellulose and others [1–8]. In recent decades, the need for sustainability and finding biodegradable alternatives has drawn interest in biopolymers. Although it is synthesized in nature, it can be obtained efficiently and on a large scale from sources such as plants, animals, microorganisms, waste (agro-waste, paper-waste, green waste, wood waste), directly or indirectly [9]. Nucleic acids and proteins are the primary biopolymers synthesized by all living organisms, and they are entirely carbon-neutral [9]. Gelatin, a biopolymer obtained from the denaturation of collagen, is used in coating drugs in capsules, casing of sausages, and other [10, 11]. Other well-known protein-based biopolymers are legumin and albumin. Microbial polyester is being used for biomedical applications and is a potential alternative for synthetic polymers/plastics [11, 12]. Further, many other naturally occurring biomolecules can be converted into biodegradable polymers, such as lactic acid which can be polymerised to poly(lactic acid) and triglycerides can also be polymerised for bio-based plastics [11].

Overall, the study of biopolymers sits at the intersection of sustainability, biology, and advanced materials science. Their origin, structural diversity, and functional adaptability make them indispensable for addressing modern challenges such as plastic pollution, biomedical innovation, and green energy solutions. From naturally occurring nucleic acids and proteins to engineered derivatives like poly(lactic acid), biopolymers provide opportunities not only to replace conventional synthetic materials but also to develop novel systems with superior biocompatibility, biodegradability, and environmental safety.

2 Conventional Imaging Method for Polymer Analysis

The characterization of biopolymers ranging from nucleic acids and proteins to polysaccharides and other natural polymers relies heavily on imaging techniques that reveal both structure and composition. Traditionally, methods such as optical microscopy, fluorescence microscopy, electron microscopy (EM), and chemical staining have been widely used. Optical microscopy is commonly employed to observe the morphology of polymers [13–16]. However, the contrast of most biopolymers is low, necessitating stains or labels for clear visualization. Fluorescence microscopy improves contrast by tagging molecules with fluorophores, enabling the visualization of biopolymers [17, 18]. Despite its widespread use, fluorescence microscopy requires tagging or even suffers from photobleaching. Transmission and scanning electron microscopy (TEM and SEM) provide ultra-high spatial resolution with high quality images down to the nanometer scale, allowing detailed structural analysis of fibers, fibrils, or aggregates. Electron microscopy is especially valuable in biopolymer characterization applications. For surface as well as fracture analysis, SEM is used which gives information about structure and surface of biopolymers, whereas TEM is mostly used for microstructural analysis of biopolymers. However, this lacks direct chemical specificity; differentiating biopolymers in a complex matrix often necessitates additional labelling, sample preparation or complementary techniques. Biopolymers are inherently non-conductive, so preparing them for SEM analysis typically requires coating with a thin metallic layer. The metals most commonly used for this purpose include gold, gold–palladium alloys, platinum, and chromium [19–28]. Second harmonic generation (SHG) imaging is a nonlinear optical method increasingly used to investigate biopolymers, especially fibrillar structures such as collagen in tissues or cellulose in plants [29, 30]. Atomic force microscopy (AFM) is a high-resolution technique that maps sample surface topography using a nanoscale cantilever. It provides detailed morphological information at the nanometer scale. AFM has been widely used to assess surface roughness and structural features of biopolymer films [31–33].

Raman spectroscopy and imaging a powerful vibrational spectroscopy technique that allows non-destructive analysis with minimal sample preparation. While the instrumentation is relatively affordable, its application in the study of membranes and biopolymer films remains limited. Nevertheless, several notable studies have demonstrated its potential as an insightful tool for understanding material structure and system morphology [20, 34–37]. For example, Garms et al. highlighted how Raman spectroscopy was crucial in elucidating the interactions drug molecules with natural rubber latex, while another study revealed that marine shell fibres are enveloped by an organic membrane sheath [38, 39]. Beyond conventional spectroscopy, Raman can be leveraged to generate detailed chemical images, which is a central focus of this chapter. In one study, composites made from thermoplastic biopolymers reinforced with purified cellulose fibers were examined, and Raman imaging revealed uniform dispersion of fibers within the polymer matrices for all three biopolymers tested [20]. Overall, Raman imaging plays a crucial role in biopolymer research by providing label-free, non-destructive, chemically specific, and multiplexed imaging capabilities. This allows researchers to explore the molecular organization, dynamics, and interactions of various biopolymer systems, revealing insights that are often hard to obtain using conventional imaging techniques. This chapter explores the principles of Raman imaging combined with multivariate analysis and for biopolymer analysis in detail. Raman imaging offers a bridge between fundamental understanding and practical applications, positioning biopolymers as key players in shaping a sustainable and technologically advanced future.

3 Raman Spectroscopy and Imaging Analysis

Characterizing and visualizing the distribution of biopolymers is vital yet remains analytically challenging due to the complexity and heterogeneity of samples in their native environment. Raman imaging, a hybrid technique that combines Raman spectroscopy with optical microscopy and digital imaging, has emerged as a powerful and label-free tool for visualising and characterising such biopolymers [20, 34, 40]. Raman spectroscopy is built on the Raman effect, an inelastic light-scattering process that probes molecular vibrations of chemical bonds. The Raman effect was first discovered by Sir CV Raman [41]. Through inherently weak (~1 in every million photons is inelastically scattered), the Raman spectrum carries rich qualitative and quasi-quantitative information. Importantly, water produces only a minor Raman signal, making the method particularly attractive for analysing biological specimens. Overall, Raman spectroscopy is celebrated as a stain-free, minimally destructive, non-invasive technique that requires little sample preparation. The spontaneous Raman effect is a two-photon event where an incident photon induces deformation of the electron cloud of a molecule (polarizability) with respect to its vibrational motion. This distortion leads to the formation of a transient dipole in a so-called virtual energy state. To return to its original ground state, the molecule instantly releases a photon, giving rise to the Raman scattered photon. If the Raman scattered photon is low in energy, it is known as Stokes Raman scattering, and if it is high in energy, it is known as anti-Stokes Raman scattering [34, 40].

Raman imaging extends the capability and applicability of Raman spectroscopy by generating spatial information of the sample. A typical Raman image is a pseudo--coloured image made of pixels, with each pixel a complete Raman spectrum. Therefore, each pixel is in the form of a hyperspectral data cube with *x* and *y* spatial dimensions and a wavenumber (Raman shift) dimension. Further targeting a specific or a range of wavenumber, corresponding to a molecular bond, it provides spatially resolved biochemical information by assigning colour to all pixels based on their spectral intensity values, based on pixel location. To obtain a Raman image, a computer-controlled Raman microscope with motorised *x*-, *y*-, and/or *z*-high precision sample stages is used to control the lateral and axial coordinates. The region of interest in the sample is first defined, then scanned point-by-point (most common in the context of biopolymers) while optimising multiple parameters such as objective lens, focus, step size, acquisition time and laser power to ensure high-quality chemical images [34, 42]. For the label-free Raman imaging, intrinsic Raman contrast from molecular bonds such as C–H, C=O, N–H, O–H, C–C and others is the basis of the Raman imaging. As spontaneous Raman scattering is a weak phenomenon, it requires longer exposure time per pixel or higher laser power. However, the increase in laser power and/or long exposure time can overwhelm Raman signals by the autofluorescence of the sample or increase the chance of local photo-induced damage [42]. Recent advancements in spectrometers, chemometric analysis tools, microscope automation, and laser systems have revolutionised Raman imaging, enabling one-visualisation of chemical composition, gradients, homogeneity, purity, and other properties.

Alongside the spontaneous Raman scattering-based imaging, different classes of nonlinear Raman imaging techniques are also there, most notably coherent anti--Stokes Raman scattering (CARS) and stimulated Raman scattering (SRS). These approaches provide a significant boost in signal strength compared to spontaneous Raman scattering by exploiting the nonlinear optical response of a sample excited with two laser frequencies. When the frequency difference between the excitation beams matches a vibrational mode of the molecules, strong resonance enhancement occurs. One of the major advantages of nonlinear Raman imaging is its ability to deliver rapid image acquisition. However, this speed comes with a trade-off that is, at any given moment, only a single vibrational wavenumber can be monitored, meaning that constructing a full spectroscopic dataset requires tuning the excitation frequency step by step across the desired spectral window. The resulting images can then be assembled to produce a hyperspectral dataset comparable to those obtained through spontaneous Raman microscopy. Therefore, careful selection of the spectral range to be scanned is critical, as it balances the amount of information gained against the total acquisition time. In practice, many nonlinear Raman studies, particularly in biology, have relied on univariate imaging of just a handful of vibrational bands. This can be used for biopolymer analysis in the future. Recent advancement of CARS and SRS have demonstrated feasibility for hyperspectral imaging coupled with advanced multivariate analysis. Looking ahead, such strategies hold the promise to extract composition in best detail in rapid manner, which can be used for biopolymer imaging in future.

4 Multivariate Analysis for Raman Imaging

The simplest known approach to generate Raman images from a sample is either selecting a specific vibrational peak or integrating in a range that corresponds to a compound of interest, which in our case is biopolymers, and mapping its signal intensity across the sample. This method is fast and computationally light, making it attractive for quick visualization. Multivariate analysis is another approach that have been increasingly applied for analysing complex Raman imaging data sets. In this approach, rather than focusing on a single peak, multivariate methods consider the entire spectral profile across all wavelengths simultaneously. Approaches such as Principal Component Analysis (PCA), K-means clustering, Multivariate Curve Resolution–Alternating Least Squares (MCR-ALS), and Vertex Component Analysis (VCA) decompose the dataset into underlying components, enabling separation of overlapping signals and identification of subtle spectral variations [43–51]. This makes multivariate methods especially powerful in heterogeneous environments where many biomolecules coexist and interact. As a result, researchers can obtain highly detailed chemical information with fine spatial resolution. These methods extract hidden patterns, separate overlapping signals, and reveal the underlying chemical distribution more reliably. Today, several toolboxes provide access to a variety of such multivariate analysis techniques, making advanced data interpretation increasingly accessible to researchers [52–54].

In this chapter, we have discussed two multivariate imaging techniques which has been used in previous studies discussed in application section for biopolymer imaging analysis—(1) multivariate curve resolution alternating least squares and (2) k-means clustering.

4.1 Multivariate Curve Resolution: Alternating Least Squares

MCR-ALS has been used for the biopolymer analysis combined with Raman spectroscopy which follows the nonnegative constraint approach designed to decompose hyperspectral Raman imaging datasets into chemically meaningful components. In this framework, the original Raman imaging data matrix R with dimensions: spatial pixels XY × spectra are decomposed into two smaller matrices: the abundance matrix A ($XY \times P$) and the spectral matrix S^T, where P represents the number of pure components. The relationship can be written as: $R = A\ S^T + E$; Here, matrix E denotes the residual error, having the same dimensions as R, and represents the part of the signal that the model does not capture. The optimization in MCR happens in an iterative manner using the alternating least squares (ALS) algorithm, where A and S are refined step by step to minimize the Frobenius norm $\|R—AS\|^2$. Importantly, nonnegativity constraints ($A \geq 0$, $S \geq 0$) are imposed, ensuring that both abundance maps and spectral intensities remain physically interpretable, as negative concentrations or intensities would have no chemical meaning [44].

4.2 k-means Clustering

K-means clustering is another powerful approach for uncovering hidden patterns in Raman hyperspectral data which has been previously used for biopolymer analysis. Although it does not yield chemically pure spectra like MCR-ALS, it remains a valuable approach for rapid classification and visualization of hyperspectral Raman imaging data sets as a partitioning method for unsupervised data analysis. In essence, clustering aims to divide a dataset into a predefined number of groups (clusters) in such a way that the spectra within a cluster are as similar as possible, while differences between clusters are maximized. This separation is achieved by using a distance measure, with the Euclidean distance being particularly effective for Raman spectroscopy data. The procedure begins by randomly selecting K spectra to serve as the initial cluster centroids. A centroid represents the central or average position of a cluster within a multidimensional space. The first step involves calculating the distance between each spectrum and every centroid, after which each spectrum is assigned to the cluster whose centroid is closest. Once all spectra have been tentatively grouped, the centroids are updated by computing the mean position of the spectra in each cluster. This iterative procedure of assigning spectra to the nearest centroid and then recalculating the centroid positions is repeated continuously. The process continues until convergence is achieved, meaning that either the cluster assignments remain unchanged between iterations or further adjustments no longer significantly improve the clustering results. Through this repetition, the clusters progressively stabilize, and the centroids accurately reflect the central tendencies of the grouped spectra. In practice, a maximum number of iterations may also be defined as a stopping criterion. For visualization, the clustering results are commonly represented as false-color maps. In these images, each cluster is assigned a distinct color, providing an intuitive picture of the spatial distribution of chemically similar regions in the sample. To further improve robustness, strategies such as multiple seeding (running the algorithm with several random starting points and selecting the solution with the lowest variance) or informed initialization (using cluster seeds derived from another method like hierarchical cluster analysis, HCA) can be applied [43].

5 Applications of Raman Imaging for Biopolymer Analysis

Raman imaging is a versatile bio-analytical tool for biopolymer characterization, yet its application has been largely confined to nucleic acids and proteins, with comparatively limited attention given to other biopolymers such as lignin and cellulose. In this section, we have listed out some studies showing the potential applications of Raman Imaging for biopolymer analysis. Direct imaging of nucleic acid biopolymers, such as nucleic acids (RNA and DNA) at the single-cell level in a label-free manner remains a challenge.

Samuel et al. have spatially mapped the RNA and DNA in the cell and liquid–liquid phase separated membrane less organelles using Raman imaging. They successfully resolved two spectral components specific to RNA and DNA from the complex biological signal matrix, Raman imaging data of the cell. This was done using a powerful unmixing algorithm—multivariate curve resolution analysis. RNA shows characteristic dual peaks at ~783 cm^{-1} and ~813 cm^{-1}, broader bands at 1245 cm^{-1} and 1318 cm^{-1}, and a PO_2^- stretching band at 1093–1101 cm^{-1}. In contrast, DNA displays a single peak near 785 cm^{-1}, sharper peaks at 1340 cm^{-1} and 1375 cm^{-1}, a weaker 1256 cm^{-1} band, and a PO_2^- signal at 1093 cm^{-1}. Additional bands at 1481–1487 cm^{-1} and 1571–1576 cm^{-1} correspond to purine base vibrations in both nucleic acids. Notably, RNA-rich domains were spatially aligned with protein-dense regions, suggesting the presence of liquid–liquid phase-separated organelles; disappearance of these domains during mitosis supported both the spectral assignments and methodological reliability. DNA mapping revealed characteristically condensed chromosomes in mitotic cells without major alterations in helix structure. This Raman imaging strategy, coupled with multivariate analysis offers a promising tool to for studying naturally occurring nucleic acid biopolymer (Fig. 1

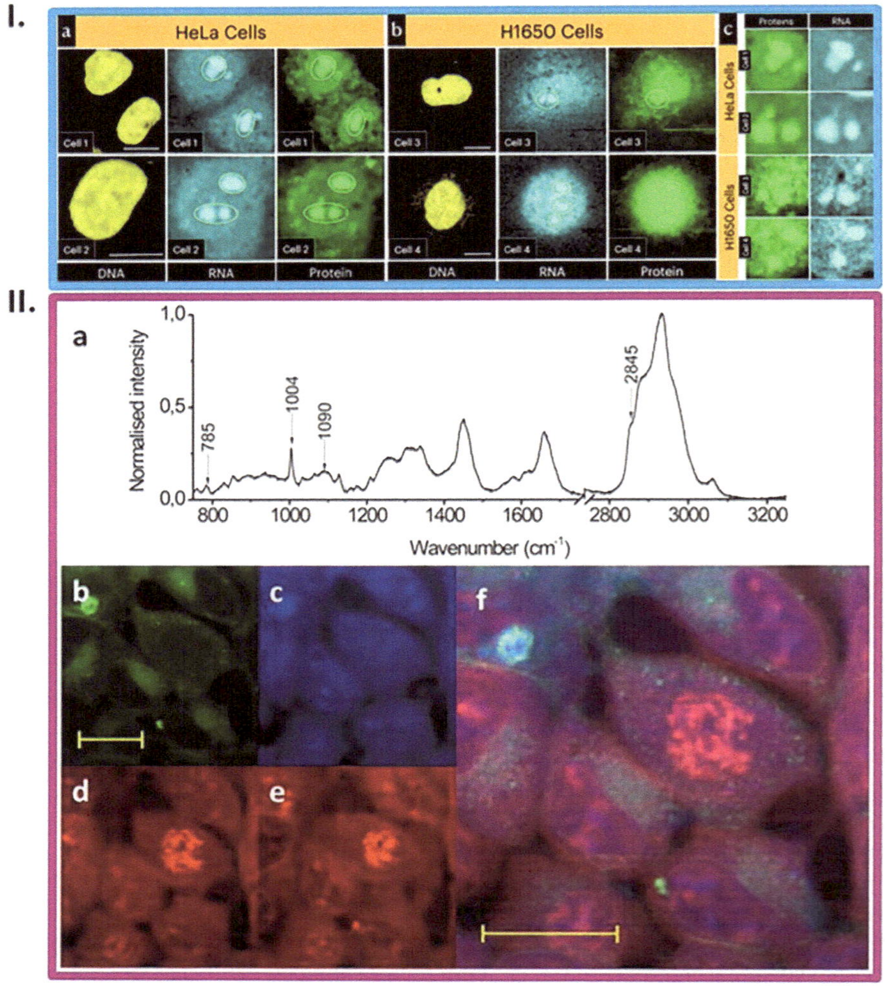

Fig. 1 (I) (a) HeLa cells, I (b) H1650 cells. Nucleoli are circled yellow and Cajal bodies are circled red. I (c) Zoomed images showing RNA and proteins aggregates observed inside nuclear regions. Identical shapes confirm their assignments to nuclear LLPS domains. In cell 3, however, the LLPS image correlation is comparatively poor. Scale bars 10 μm. (II) (a) Raman spectrum of a HEK-293 cell pellet. II(b)-II (f) SRS images of HEK-293 cells at II(b) 2845 cm^{-1}, primarily lipid II (c) 1004 cm^{-1}, phenylalanine II (d) 785 cm^{-1}, nucleic acid II (e) 1090 cm^{-1}, primarily nucleic acid. II (f) multicolor overlay of II(b), (c), (d). Scale bar is 20 μm [55, 56]

(I)) [55]. Zhang et al. employed a highly sensitive imaging approach based on stimulated Raman scattering (SRS) to visualize nucleic acid biopolymers in live cells without the need for labels. Using this technique, they were able to image individual polytene chromosomes within the salivary gland cells of *Drosophila melanogaster*. The imaging specifically targeted the vibrational modes at 785 cm^{-1} and 1090 cm^{-1}

to capture nucleic acid signals within the nucleus. Their observations revealed that variability in nucleic acid intensity between cells corresponds to differences in cell-cycle phases. During interphase, nucleoli exhibited stronger nucleic acid signals, reflecting elevated ribosomal RNA production. As cells progressed into mitosis, the Raman profiles changed in line with chromatin condensation: for example, cells in prophase displayed enhanced nuclear contrast as DNA began to condense, whereas metaphase cells showed distinct, band-like nucleic acid signals corresponding to the alignment of condensed chromosomes along the metaphase plate. These characteristic spectral signatures allow active, dividing cells to be distinguished from those not undergoing division (Fig. 1(II)) [56].

Sanaeifar et al. have shown that the applicability of confocal Raman microspectroscopy-based chemical imaging as a powerful analytical tool for monitoring biopolymer dynamics in plants during pathogen invasion. They reported chemical imaging by integrating spectral intensity of wavenumber specific to lignin at range 1589 to 1607 cm^{-1}, cellulose (1087–1100 cm^{-1}) and C–H for all polymer cellulose, hemicellulose, pectin, and lignin (2980–2995 cm^{-1}). They presented Raman chemical images of tea leaf cells affected by blight, examining four sections that represented distinct stages of infection. At the earliest stage, lignin was primarily confined to the vascular bundles and the corners of bundle sheath cell walls, while cellulose was distributed throughout the entire bundle sheath. Additionally, the overall biopolymer signal appeared more widespread, reflecting contributions from hemicellulose and pectin. With increasing infection, lignin and cellulose distributions became more widespread and irregular, extending across whole cell walls and into intracellular regions, alongside evident vascular bundle damage. In advanced infection stages, intracellular accumulation of lignin/phenolics and cellulose was observed, and total biopolymer content markedly increased, especially within vascular bundles, indicative of tylose formation to prevent the further development of pathogens (Fig. 2) [57]. Gierlinger et al. demonstrated that Raman imaging can serve as a non-destructive technique for directly visualizing variations in lignin content without requiring any staining or chemical treatment of plant cell walls. The method also enables detailed characterization of cellulose orientation, allowing even the S1 layer, which is smaller than 0.5 μm, to be resolved. Overall, this study offers valuable insights into the spatial distribution and alignment of essential cell wall components such as lignin and cellulose, achieving high-resolution mapping that would be difficult with conventional methods [59].

Inulin (Fig. 3I) is a versatile biopolymer that is used as a bifidogenic prebiotic. This compound is widely recognized for its beneficial effects on gut health, primarily because it is not broken down or absorbed in the upper regions of the digestive tract. Due to these properties, it functions effectively as a dietary fiber, and in June 2018, the Food and Drug Administration officially approved its use as a fiber ingredient in food products. Expanding on its functional applications, Deylaghian et al. conducted detailed studies to explore how inulin, a naturally occurring biopolymer, interacts with soil components, focusing on the formation of bonds that influence its behavior and stability. They reported that the Raman spectra from the soil sample show a peak from quartz and kaolinite minerals. They observed the appearance of

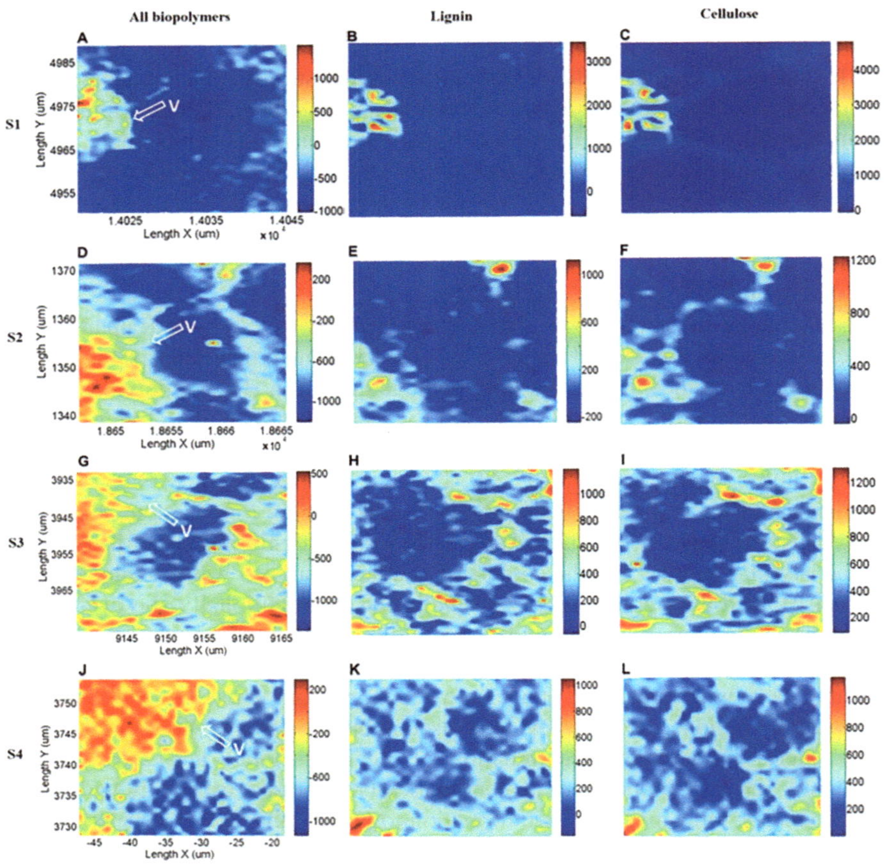

Fig. 2 Raman chemical images of four sections: S1 (**A–C**), S2 (**D–F**), S3 (**G–I**), and S4 (**J–L**). (**A**, **D**, **G**, **J**): all polymers band intensity, (**B**, **E**, **H**, **K**): lignin band intensity, and (**C**, **F**, **I**, **L**): cellulose band intensity [57]

new Raman peaks upon the addition of inulin to the soil. Peaks appearing between 3400 and 3700 cm^{-1} were attributed to direct hydrogen bonding between the biopolymer and soil components, while signals around 3800 cm^{-1} suggested the formation of ionic bonds with silica oxide. The Raman spectra also highlighted characteristic carbohydrate vibrations of the biopolymer, particularly in the regions of 900–1100 cm^{-1} and 2800–3000 cm^{-1}. In addition, Raman mapping was employed to visualize how the biopolymer was distributed across the soil surface [60]. Letocha et al. focused on developing high-quality microsphere carriers using tapioca flour, which were further coated with biopolymers such as hyaluronic acid, chitosan, or gelatin to encapsulate *Lactobacillus casei* probiotic bacteria. Their study demonstrated that Raman imaging could effectively verify the presence of these biopolymer coatings without causing visible damage to the microspheres. For the hyaluronic

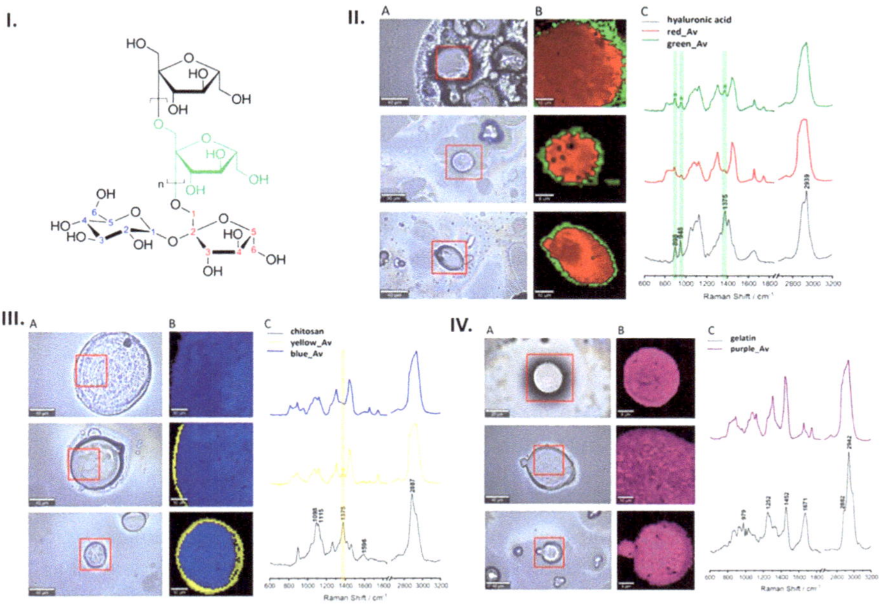

Fig. 3 (I) Structure of Inulin (II) Raman mapping results for selected microspheres coated with hyaluronic acid: images with measurement areas marked II (A); with K-means cluster analysis II (B); and average spectra for classes (spectral colors correspond to class colors in KMC images) compared with the standard spectrum II (C). III. Raman mapping results for selected chitosan-coated microspheres: images with measurement areas marked III (A); with K-means cluster analysis III (B); and average spectra for the classes (the colors of the spectra correspond to the colors of the classes in the K-means cluster images) compared with the standard spectrum III (C). IV. Raman mapping results for selected gelatin-coated microspheres: images with measurement areas marked IV (A); along with KMC analysis IV (B); and the average spectrum of the microsphere compared to the gelatin standard IV (C) [58]

acid-coated microspheres, k-means cluster analysis was applied, revealing two distinct clusters: one at the central region and another along the boundary edges. The edge cluster was identified as hyaluronic acid, as the component spectra exhibited strong peaks at 890, 950, and 1375 cm^{-1}, which correspond to characteristic vibrations of hyaluronic acid. Similarly, chitosan-coated microspheres showed a comparable pattern, with the 1375 cm^{-1} peak clearly visible at the edge, confirming the presence of chitosan. In contrast, gelatin-coated microspheres produced only a single cluster in the k-means analysis, and no distinct spectral bands corresponding to the standard gelatin spectrum were observed in the Raman spectral profiles (Fig. 3II) [58]. Kalisz et al. carried out a comprehensive comparison of several fixation and embedding techniques optimization for zebrafish larvae, including paraformaldehyde, glutaraldehyde, low-temperature fixation, as well as embedding in gelatin and agarose using advanced vibrational imaging methods. Their Raman analyses revealed a noticeably higher protein-associated signal intensity in gelatin compared to those agarose, likely due to inherent biochemical differences between the two supporting substrates. Although both gelatin and agarose showed relatively lower intensities in the phosphate stretching region, these bands were still consistently detected and well distributed throughout the tissue, enabling reliable visualization of phosphate-containing molecular components [61].

Casein and pectin are important biopolymers commonly employed in food products and drug delivery systems due to their functional properties [62, 63]. In earlier studies, casein was utilized to encapsulate hydrophobic bioactive substances, such as vitamin D, for improved stability and delivery [63, 64]. Zhuang et al. utilized confocal Raman imaging to examine the structural organization of individual casein microparticles. They generated spatial distribution maps of the Amide I band by integrating the spectral range between 1646 and 1686 cm^{-1}. Their results showed no noticeable differences in the amide region when comparing the Raman spectra of casein microparticles with those of native casein micelles. Additionally, the Raman imaging confirmed that casein molecules are distributed uniformly throughout the microparticles, indicating an even internal composition [65].

Hossain et al. demonstrated the direct measurement of amylose and amylopectin within individual starch granules. By combining Raman imaging with multivariate curve resolution, they found that amylopectin was predominantly concentrated in the central region of the granules, while amylose was more abundant near the periphery [66]. Tian et al. explored how cell wall biopolymers change in kiwifruit stems following infection by *Pseudomonas syringae* pv. *actinidiae*, the bacterium that causes bacterial canker. They employed confocal Raman imaging together with multivariate curve resolution–alternating least squares (MCR-ALS) to analyze hyperspectral data and generate detailed molecular maps of major cell wall components, including high- and low-methylated pectin, cellulose, hemicellulose, and lignin. Their results showed that, 3 days post-infection, the levels of cellulose and high-methylated pectin increased, while cellulose, hemicellulose, and lignin largely remained unchanged. By day five, however, all examined biopolymers—high- and low-methylated pectin, cellulose, hemicellulose, and lignin—had declined markedly, resulting in significant cell wall degradation (Fig. 4) [67]. Agarwal et al. demonstrated the use of Raman imaging to map the spatial distribution of cell wall components, specifically cellulose and lignin, in cross-sections of black spruce wood. The spectral region from 1519 to 1712 cm^{-1} was used to assess lignin distribution, while the range from 978 to 1178 cm^{-1} was targeted for cellulose. Their results showed that lignin concentration was highest at the cell corners, with levels in the compound middle lamella comparable to those in the secondary wall. In contrast, cellulose was found at lower levels in the cell corners and compound middle lamella, but its abundance was greater in the secondary wall [68].

Li et al. examined both the economically significant crop and model plant species *Nicotiana tabacum*, aiming to understand the spatiotemporal distribution of leaf cell wall components. They utilized confocal Raman imaging in combination with multivariate curve resolution analysis to generate detailed chemical maps. Their results indicated that cellulose and pectin were enriched in the cell walls of the midrib in mature leaves, with similar accumulation patterns observed in the lateral vein cell walls. Furthermore, they observed that during the flue-curing

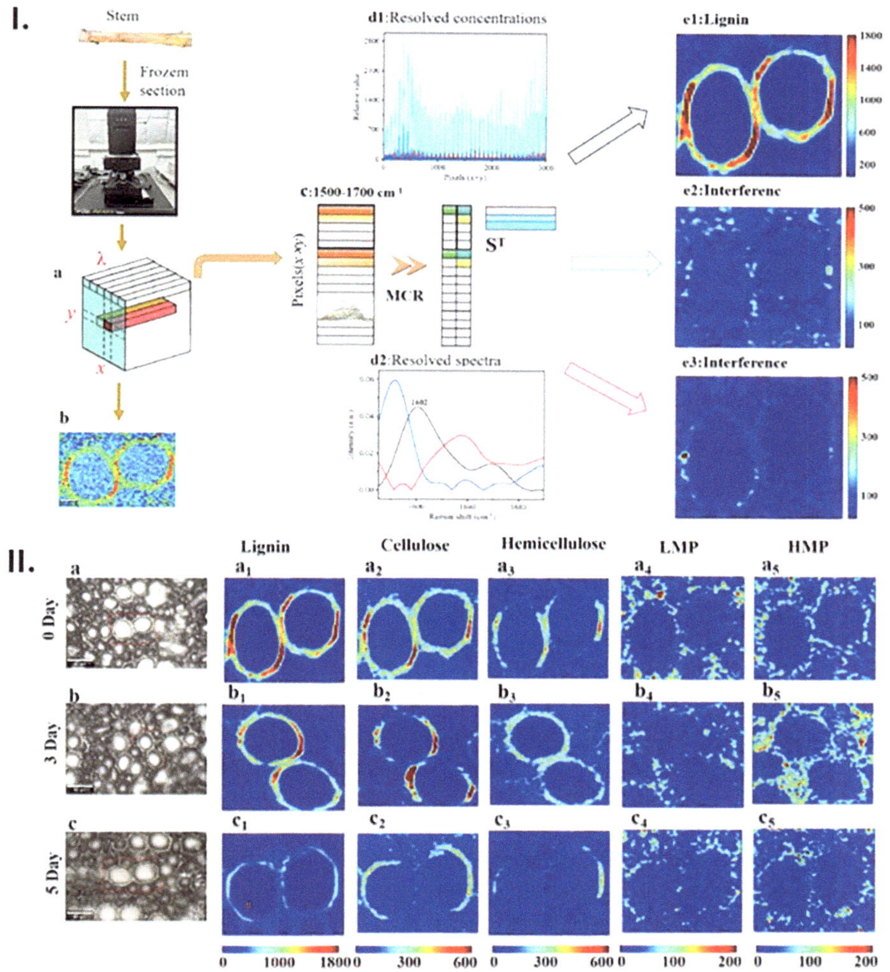

Fig. 4 (I) MCR-ALS resolution schematic and results for the lignin fingerprint region in healthy kiwifruit stems. (a) The original imaging data array $T_{X \times Y \times Z3}$ obtained by Raman confocal imaging of the surface of the kiwifruit stem section; (b) Cellulose imaging map obtained by traditional peak fitting analysis of the region within the imaging data (note that this image contains interference signals); (c) The preprocessed $T_{X \times Y \times Z3}$ data matrix is expanded to form an augmented matrix, which is then decomposed using the MCR-ALS ($N = 3$) algorithm into the expanded concentration matrix (d1) and spectral matrix (d2); (e1–e3) represent the processing of the three obtained components, with the molecular imaging distribution maps of the three pure substances being plotted separately. (II) MCR-ALS resolved comparative analysis of molecular imaging maps of five types of analytes in the cell wall of different infected kiwifruit stems. Images a, b and c are Raman microscopic images at 50× magnification; (a_1, b_2, c_3) represent lignin; (a_2, b_2, c_2) represent cellulose; (a_3, b_3, c_3) represent hemicellulose; (a_5, b_5, c_5) represent HMP; and (a_4, b_4, c_4) represent LMP. In the Raman images, pixels with redder colors correspond to stronger spectral signals and higher content, while pixels with bluer colors indicate weaker spectral signals and lower content [67]

process, both cellulose and highly methylated pectin experienced noticeable degradation [69].

In summary, Raman imaging has demonstrated remarkable versatility in probing a wide spectrum of biopolymers from nucleic acids and proteins to plant cell wall polymers, polysaccharides, and other biopolymers. Yet, despite these advances, the application of Raman imaging in biopolymer research remains unexplored, with nucleic acids and proteins far more extensively investigated than other biopolymers such lignin, cellulose, starch, or inulin. Future work will likely focus on expanding Raman imaging-based methodologies, integrating them with advanced chemometric and multivariate tools, and applying them to more complex systems to achieve a deeper understanding of biopolymer organization, interactions, and dynamics.

6 Limitations and Future Perspectives

Across diverse fields, from food and plant-derived materials to microbial and cellular matrices, Raman spectroscopy and imaging have proven to be powerful techniques for polymer analysis. However, still application of Raman imaging for biopolymers is in its nascent phase. Looking forward, its future perspectives are interesting yet grounded by technical and other limitations that should be carefully acknowledged. One of the most promising directions to overcome the limitations in biopolymer research is by integrating Raman imaging with advanced computational approaches such as artificial intelligence and machine learning. These AI/ML algorithms can be used to process hyperspectral Raman imaging datasets obtained from biopolymers for classification, identification, unmixing, localization, and quantification. In agricultural, plant, and food sciences, machine learning-driven Raman imaging could possibly be used for real-time monitoring of complex biopolymers synthesis and degradation. Similarly, in tissue or cellular level, AI-based approaches may help in unmixing overlapping Raman peaks of nucleic acids from proteins or lipids from complex biological matrix, providing insights into cellular physiology, stress responses, and disease states. An exciting direction for the future is creating new SERS (surface enhanced Raman scattering) substrates specifically designed for targeting different biopolymers, which could enhance the Raman signal from biopolymers and make Raman imaging faster and more efficient for biopolymer analysis. By engineering nanostructured surfaces, researchers can amplify weak Raman signals of low-abundance biopolymers. Coupled with microfluidic platforms, portable SERS-based Raman sensors could be fabricated for on-site detection of biopolymer contamination or bioprocess monitoring, bridging laboratory research with field applications. Moreover, label-free Raman imaging of biopolymers is expected to benefit from continuous improvements in instrumentation. Therefore, advancements in technologies of laser sources, detectors, and optical filters will gradually increase acquisition speed, spatial resolution, and spectral quality while minimizing photodamage. Faster 2D as well as volumetric imaging will open new doors to studying dynamic processes such as microbial production of biopolymers or the biopolymer distribution across diverse samples in near real-time. Combining Raman imaging of biopolymers with complementary techniques such as CARS and SRS could enable multimodal systems that provide more sensitive and context-rich visualization of biochemical distributions. This can enhance sensitivity, the rapid imaging capabilities with more reliable quantification with less background. Ultimately, these approaches enhance both spatial and chemical resolution, providing deeper insight into biopolymer organization and function.

Despite these exciting future perspectives, Raman imaging still faces several limitations that must be addressed. First and foremost is its inherently weak scattering cross-section, making it challenging to study lowly abundant or weakly scattering biopolymers without long acquisition times or specialized enhancement techniques. This limitation is particularly evident in bulky food matrices and turbid biological tissues, where signal attenuation and scattering prevent deep imaging beyond superficial layers. Spectral complexity and peak overlap remain additional problems. Many biopolymers unless pure, share many Raman peaks in the Raman spectrum at the same or nearly adjacent wavenumbers, leading to complex multicomponent spectra. Although advanced chemometric-based multivariate statistical methods help in extracting the individual components corresponding to a biopolymer, but sill full deconvolution of Raman spectra in heterogeneous complex matrices is rarely straightforward. Another practical limitation is photodamage and fluorescence background. Many natural biopolymers (especially plant-derived ones) may contain chromophoric impurities or phenolic compounds that give rise to strong autofluorescence under laser excitation. This broad fluorescence background can mask the Raman signals, particularly under visible excitation wavelengths. Though near-infrared (785 or 830 nm) wavelength lasers reduce autofluorescence, they also lead to weaker Raman scattering, creating a sensitivity trade-off. Prolonged exposure to focused laser beams can also induce photochemical degradation, denaturation, or thermal damage in sensitive biopolymers like proteins and nucleic acids. Cost and accessibility also represent non-negligible constraints. High-end Raman microscopes for imaging equipped with confocal optics, multiple laser lines, and sensitive detectors are expensive and require skilled operators, and come with a high cost. As a result, their routine uses in industrial or field settings remains limited. Although portable or handheld Raman devices are becoming more popular, their sensitivity and resolution are insufficient for detailed biopolymer imaging and are typically only suitable for bulk composition analysis.

Ultimately, the future of Raman imaging in biopolymer analysis rests on making the technology more accessible, pushing the boundaries of sensitivity, rapid imaging and unlocking deeper layers of meaning from the data it generates. For Raman imaging to become a mainstream tool for wider adoption in biopolymer analysis, the path forward will depend on creating standardized protocols, user-friendly open-source software, openly available Raman databases for biopolymers, and cost--effective instruments designed with the needs of diverse communities in mind, from food technologists and plant biologists to microbiologists and materials scientists. With ongoing interdisciplinary collaborations and rapid technological progress, Raman imaging is set to establish itself as a vital tool for non-invasive, non-destructive, and label-free characterization of biopolymers in their native environments, opening new frontiers in our understanding of both natural systems and engineered biomaterials.

References

1. Liu L, Liang W, Zhang Y, Fu Q (2023) Nanoencapsulation in polymeric materials: weaving magical coats for microorganisms. Nano Today 52:101973
2. Hassabo A, Zayed M, Bakr M, Othman H (2022) The utilisation of Gelatin biopolymer in textile wet processing. J Textiles Color Polym Sci. https://doi.org/10.21608/jtcps.2022.128913.1113
3. Picot-Allain MCN, Neergheen VS (2023) Pectin a multifaceted biopolymer in the management of cancer: a review. Heliyon 9:e22236
4. Kurita K (2006) Chitin and chitosan: functional biopolymers from marine crustaceans. Mar Biotechnol 8:203–226
5. Gupta P, Nayak KK (2015) Characteristics of protein-based biopolymer and its application. Polym Eng Sci 55:485–498
6. Vásquez-Garay F, Carrillo-Varela I, Vidal C, Reyes-Contreras P, Faccini M, Teixeira Mendonça R (2021) A review on the lignin biopolymer and its integration in the elaboration of sustainable materials. Sustainability 13:2697
7. Sharma S, Kumar A (2019) Keratin as a Protein Biopolymer. Springer, Cham. https://doi.org/10.1007/978-3-030-02901-2
8. Liu Y, Ahmed S, Sameen DE, Wang Y, Lu R, Dai J, Li S, Qin W (2021) A review of cellulose and its derivatives in biopolymer-based for food packaging application. Trends Food Sci Technol 112:532–546
9. Baranwal J, Barse B, Fais A, Delogu GL, Kumar A (2022) Biopolymer: a sustainable material for food and medical applications. Polymers 14:983
10. Ahmad MI, Li Y, Pan J, Liu F, Dai H, Fu Y, Huang T, Farooq S, Zhang H (2024) Collagen and gelatin: structure, properties, and applications in food industry. Int J Biol Macromol 254:128037
11. what_are_biopolymers. https://www.sda-uk.org/materials/popups/plastics/what_are_biopolymers.html
12. Cammas-Marion S, Martínez-Barbosa ME (2021) Microbial polyesters: synthesis and applications. In: Fundamentals of natural fibres and textiles. Elsevier, Amsterdam, pp 515–555
13. Liakos I, Rizzello L, Scurr DJ, Pompa PP, Bayer IS, Athanassiou A (2014) All-natural composite wound dressing films of essential oils encapsulated in sodium alginate with antimicrobial properties. Int J Pharm 463:137–145
14. Martins GC, Penido ES, Alvarenga IFS, Teodoro JC, Bianchi ML, Guilherme LRG (2018) Amending potential of organic and industrial by-products applied to heavy metal-rich mining soils. Ecotoxicol Environ Saf 162:581–590
15. Ebnesajjad S, Ebnesajjad CF (2014) Preface. In: Surface treatment of materials for adhesive bonding. Elsevier, Amsterdam, pp xiii–xiv
16. Vimala K, Mohan YM, Sivudu KS, Varaprasad K, Ravindra S, Reddy NN, Padma Y, Sreedhar B, MohanaRaju K (2010) Fabrication of porous chitosan films impregnated with silver nanoparticles: a facile approach for superior antibacterial application. Colloids Surf B Biointerfaces 76:248–258
17. Valdman D, Lopez BJ, Valentine MT, Atzberger PJ (2013) Force spectroscopy of complex biopolymers with heterogeneous elasticity. Soft Matter 9:772–778
18. Wang X, Li M, Hou Y et al (2020) Tumor-microenvironment-activated in situ self-assembly of sequentially responsive biopolymer for targeted photodynamic therapy. Adv Funct Mater. https://doi.org/10.1002/adfm.202000229

19. Molinari G, Parlanti P, Aliotta L, Lazzeri A, Gemmi M (2024) TEM morphological analysis of biopolymers: the case of poly (lactic acid) (PLA). Mater Today Commun 38:107868
20. Tomoda BT, Yassue-Cordeiro PH, Ernesto JV, Lopes PS, Péres LO, da Silva CF, de Moraes MA (2020) Characterization of biopolymer membranes and films: physicochemical, mechanical, barrier, and biological properties. In: Biopolymer membranes and films. Elsevier, Amsterdam, pp 67–95
21. Barbosa GP, Debone HS, Severino P, Souto EB, da Silva CF (2016) Design and characterization of chitosan/zeolite composite films — effect of zeolite type and zeolite dose on the film properties. Mater Sci Eng C 60:246–254
22. Das MM, Ramasamy P (2016) Sample preparations for scanning electron microscopy—life sciences. In: Modern electron microscopy in physical and life sciences. IntechOpen, London. https://doi.org/10.5772/61720
23. Pereira R, Tojeira A, Vaz DC, Mendes A, Bártolo P (2011) Preparation and characterization of films based on alginate and aloe Vera. Int J Polym Anal Charact 16:449–464
24. Venkatesan J, Lee J-Y, Kang DS, Anil S, Kim S-K, Shim MS, Kim DG (2017) Antimicrobial and anticancer activities of porous chitosan-alginate biosynthesized silver nanoparticles. Int J Biol Macromol 98:515–525
25. Dong Z, Wang Q, Du Y (2006) Alginate/gelatin blend films and their properties for drug controlled release. J Memb Sci 280:37–44
26. Mandal BB, Mann JK, Kundu SC (2009) Silk fibroin/gelatin multilayered films as a model system for controlled drug release. Eur J Pharm Sci 37:160–171
27. Denavi G, Tapia-Blácido DR, Añón MC, Sobral PJA, Mauri AN, Menegalli FC (2009) Effects of drying conditions on some physical properties of soy protein films. J Food Eng 90:341–349
28. Noishiki Y, Nishiyama Y, Wada M, Kuga S, Magoshi J (2002) Mechanical properties of silk fibroin–microcrystalline cellulose composite films. J Appl Polym Sci 86:3425–3429
29. Williams RM, Zipfel WR, Webb WW (2005) Interpreting second-harmonic generation images of collagen I fibrils. Biophys J 88:1377–1386
30. Cisek R, Tokarz D (2024) Second harmonic generation microscopy of polysaccharide macrostructures. In: Advanced biophysical techniques for polysaccharides characterization. Elsevier, Amsterdam, pp 81–120
31. Lawrence BD, Omenetto F, Chui K, Kaplan DL (2008) Processing methods to control silk fibroin film biomaterial features. J Mater Sci 43:6967–6985
32. Wiley (2008) Characterization analysis of polymers. Wiley, Hoboken
33. Silva SS, Goodfellow BJ, Benesch J, Rocha J, Mano JF, Reis RL (2007) Morphology and miscibility of chitosan/soy protein blended membranes. Carbohydr Polym 70:25–31
34. Karlo J, Dhillon AK, Razi SS, Siddhanta S, Singh SP (2024) Imaging based Raman spectroscopy. In: Raman spectroscopy. Springer, Cham, pp 349–375
35. Schrader B (1995) Tools for infrared and Raman spectroscopy. In: Infrared and Raman Spectroscopy. Wiley, Hoboken, pp 63–188
36. Bista RK, Bruch RF, Covington AM (2010) Vibrational spectroscopic studies of newly developed synthetic biopolymers. Biopolymers 93:403–417
37. Bista RK, Bruch RF, Covington AM (2011) Infrared spectroscopic study of thermotropic phase behavior of newly developed synthetic biopolymers. Spectrochim Acta A Mol Biomol Spectrosc 81:583–589
38. Schmahl WW, Griesshaber E, Kelm K, Goetz A, Jordan G, Ball A, Xu D, Merkel C, Brand U (2012) Hierarchical structure of marine shell biomaterials: biomechanical functionalization of calcite by brachiopods. Z Kristallogr Cryst Mater 227:793–804
39. Garms B, Borges F, Santos R et al (2017) Characterization and microbiological application of ciprofloxacin loaded in natural rubber latex membranes. Br J Pharm Res 15:1–10
40. Stewart S, Priore RJ, Nelson MP, Treado PJ (2012) Raman Imaging. Annu Rev Anal Chem 5:337–360
41. Raman CV, Krishnan KS (1928) A new type of secondary radiation. Nature 121:501–502

42. Shen Y, Hu F, Min W (2019) Raman imaging of small biomolecules. Annu Rev Biophys 48:347–369
43. Hedegaard M, Matthäus C, Hassing S, Krafft C, Diem M, Popp J (2011) Spectral unmixing and clustering algorithms for assessment of single cells by Raman microscopic imaging. Theor Chem Accounts 130:1249–1260
44. Chitra Ragupathy I, Schweikhard V, Zumbusch A (2021) Multivariate analysis of hyperspectral stimulated Raman scattering microscopy images. J Raman Spectrosc 52:1630–1642
45. Nascimento JMP, Dias JMB (2005) Vertex component analysis: a fast algorithm to unmix hyperspectral data. IEEE Trans Geosci Remote Sens 43:898–910
46. Karlo J, Gupta A, Singh SP (2024) In situ monitoring of the shikimate pathway: a combinatorial approach of Raman reverse stable isotope probing and hyperspectral imaging. Analyst. https://doi.org/10.1039/D4AN00203B
47. de Juan A, Jaumot J, Tauler R (2014) Multivariate curve resolution (MCR). Solving the mixture analysis problem. Anal Methods 6:4964–4976
48. Karlo J, Dewangan NK, Phatak A, Singh SP (2026) Sensing and hyperspectral imaging of the newly synthesized carotenoids in single-cell micro-factories. Sensors and Actuators B: Chemical. https://doi.org/10.1016/j.snb.2025.139367
49. Noothalapati H, Iwasaki K, Yamamoto T (2017) Biological and medical applications of multivariate curve resolution assisted Raman spectroscopy. Anal Sci 33:15–22
50. Kallepitis C, Bergholt MS, Mazo MM, Leonardo V, Skaalure SC, Maynard SA, Stevens MM (2017) Quantitative volumetric Raman imaging of three dimensional cell cultures. Nat Commun 8:14843
51. Karlo J, Carrasco-Navarro V, Koistinen A, Singh SP (2024) Tracking trash to treasure: *in situ* monitoring of single microbial cell oil biosynthesis from waste cooking oil using Raman spectroscopy and imaging. RSC Adv 14:33323–33331
52. Georgiev D, Pedersen SV, Xie R, Fernández-Galiana Á, Stevens MM, Barahona M (2024) RamanSPy: an open-source python package for integrative Raman spectroscopy data analysis. Anal Chem 96:8492–8500
53. Amigo JM, Babamoradi H, Elcoroaristizabal S (2015) Hyperspectral image analysis. A tutorial. Anal Chim Acta 896:34–51
54. Mobaraki N, Amigo JM (2018) HYPER-tools. A graphical user-friendly interface for hyperspectral image analysis. Chemom Intell Lab Syst 172:174–187
55. Samuel AZ, Sugiyama K, Ando M, Takeyama H (2022) Direct imaging of intracellular RNA, DNA, and liquid–liquid phase separated membraneless organelles with Raman microspectroscopy. Commun Biol 5:1383
56. Zhang X, Roeffaers MBJ, Basu S, Daniele JR, Fu D, Freudiger CW, Holtom GR, Xie XS (2012) Label-free live-cell imaging of nucleic acids using stimulated Raman scattering microscopy. ChemPhysChem 13:1054–1059
57. Sanaeifar A, Ye D, Li X, Luo L, Tang Y, He Y (2022) A spatial-temporal analysis of cellular biopolymers on leaf blight-infected tea plants using confocal Raman microspectroscopy. Front Plant Sci. https://doi.org/10.3389/fpls.2022.846484
58. Łętocha A, Michalczyk A, Miastkowska M, Sikora E (2024) Effect of encapsulation of *Lactobacillus casei* in alginate–tapioca flour microspheres coated with different biopolymers on the viability of probiotic bacteria. ACS Appl Mater Interfaces 16:52878–52893
59. Gierlinger N, Schwanninger M (2006) Chemical imaging of poplar wood cell walls by confocal Raman microscopy. Plant Physiol 140:1246–1254
60. Deylaghian S, Nikooee E, Habibagahi G, Nagel T (2024) Inulin biopolymer as a novel material for sustainable soil stabilization. Sci Rep 14:31078
61. Kalisz G, Budzynska B, Sroka-Bartnicka A (2025) The optimization of sample preparation on zebrafish larvae in vibrational spectroscopy imaging. Spectrochim Acta A Mol Biomol Spectrosc 326:125288
62. Freitas CMP, Coimbra JSR, Souza VGL, Sousa RCS (2021) Structure and applications of pectin in food, biomedical, and pharmaceutical industry: a review. Coatings 11:922

63. Wusigale LL, Luo Y (2020) Casein and pectin: structures, interactions, and applications. Trends Food Sci Technol 97:391–403
64. Poltronieri P, Čerekovič N (2018) Nanoencapsulation of vitamin D in caseins and other plant proteins as a means to stabilize and increase its bioavailability in fortified foods. In: Handbook of nanomaterials for industrial applications. Elsevier, Amsterdam, pp 235–242
65. Zhuang Y, Sterr J, Schulte A, Kulozik U, Gebhardt R (2016) Casein microparticles from blend films forming casein/α-tocopherol emulsion droplets in solution. Food Biophys 11:332–338
66. Hossain IM, Pooja N, Kondeti SSC, Yamamoto T, Mazumder N, Noothalapati H (2025) Direct estimation of amylose and amylopectin in single starch granules by machine learning assisted Raman spectroscopy. Carbohydr Polym 366:123929
67. Tian W, Luo L, Wang B, Zhao Z, Kang C, Chen D, Yan X, Long Y (2025) Raman imaging--chemometrics analysis of cell wall biopolymer dynamics in Pseudomonas syringae pv. actinidiae-infeicted kiwifruit stems. Chinese J Anal Chem 53:100556
68. Agarwal UP (2006) Raman imaging to investigate ultrastructure and composition of plant cell walls: distribution of lignin and cellulose in black spruce wood (Picea mariana). Planta 224:1141–1153
69. Li M, Wei K-S, Xue Y, Wu S-J, Liu Y-J, Chen D-M, Yan X-F, Kang C (2024) Microscopic spatiotemporal changes in cell wall cellulose and pectin during Nicotiana tabacum L. leaf growth and senescence based on label-free Raman microspectroscopic imaging combined with multivariate curve resolution. Ind Crops Pro 222:119865

Raman Spectroscopy for Bioplastic Research

Sagnika Dutta, Vedant Vikas Sarode, Budheswar Dehury, and Ajeetkumar Patil

Abstract Bioplastics derived from renewable biomass are emerging alternatives to conventional petroleum-based plastics, but their viability as replacements requires thorough molecular analysis. Raman spectroscopy (RS) has emerged as a technique that is optimal for this purpose because of its advantageous properties, such as being rapid, nondestructive, and label-free. It facilitates basic material identification for advanced analyses of the structure, namely, the crystallinity, polymorphism, molecular orientation, and influence of fillers, as well as processing and degradation from the spectral shifts and peak intensity changes that occur. During the synthesis and processing of bioplastics, RS tracks monomer-to-polymer conversion and additive dispersion with quantitative, real-time feedback and a minimal lag based on the ester and carboxyl bands. The application of the technique extends to guiding formulation strategies by quantifying plasticizers and fillers. Utilization of extensive spectral libraries, routine calibration, and integration of machine learning (ML) or deep learning (DL) algorithms to automate feature extraction as well as denoising enables univariate calibrations and multivariate chemometrics such as principal component analysis, partial least squares, linear discriminant analysis, and multivariate curve resolution. Ongoing advances, such as surface-enhanced Raman scattering, Fourier transform-Raman, and Raman tomography, pave the way for overcoming several challenges, such as low-scattering cross-sections, fluorescence interference, and overlapping bands, and, therefore, more promising, quantitative bioplastic research.

Keywords Spectrum analysis, Raman · Biopolymer · Poly(lactide) · Chemometrics · Crystallization · Polyhydroxyalkanoates

S. Dutta · V. V. Sarode · B. Dehury
Department of Bioinformatics, Manipal School of Life Sciences, Manipal Academy of Higher Education, Manipal, India

A. Patil (✉)
Manipal Institute of Applied Physics, Manipal Academy of Higher Education, Manipal, India
e-mail: ajeetkumar.p@manipal.edu

N. Mazumder et al. (eds.), *Vibrational Spectroscopy for Biopolymer Characterization*, Biological and Medical Physics, Biomedical Engineering,
https://doi.org/10.1007/978-3-032-15306-7_6

1 Introduction

The ubiquity of plastics is owed to its versatility, stability, low cost, and ease. Therefore, it is not surprising that its use has grown much faster than traditional materials such as glass, metals and wood and is replacing them. By 2050, the production of plastic is expected to be responsible for 50% of the world's oil demand. However, this indispensable material also leads to more than 30 million metric tons of it being disposed of in the environment every year, where it lies without degradation for years and causes significant damage, which necessitates the identification of alternatives to petroleum-derived polymers. Bioplastics are polymers that are made from natural raw materials and have similar physical and chemical properties to plastics derived from petroleum. The viability of these materials as replacements for fossil fuel-based plastics depends on their performance during use, which requires detailed analysis and decreases their cost of production while monitoring their degradation rate to mitigate harm to the environment [1, 2].

In this chapter, we review RS as a versatile tool for analyzing bioplastics. We outline the major types of bioplastics and their applications and introduce the basic principles of RS, highlighting its advantages. The subsequent sections discuss how Raman spectra identify specific biopolymers and reveal key material properties, such as crystallinity and molecular orientation, while also monitoring bioplastic synthesis and degradation processes. We close by examining Raman imaging and chemometric methods for blends and composites and summarizing the current limitations of this technique and how to overcome them in advancing sustainable bioplastic development. The International Union of Pure and Applied Chemistry (IUPAC) describes bioplastics as follows: "a bioplastic is derived from biomass or monomers derived from biomass and which, at some stage in its processing into finished products, can be shaped by flow" [3]. The important advantages of bioplastics include reduced pressure on nonrenewable resources, increased waste disposal options, and reduced greenhouse gas emissions [1]. They can be divided into biodegradable and nonbiodegradable, where the former can be easily decomposed and broken down in the environment, whereas the latter can be recycled only in industrial settings [4]. Bioplastics such as polyhydroxyalkanoates (PHA), polylactic acid (PLA), polybutylene succinate (PBS), and cellulose-, lignin- and starch-based polymers are biodegradable, whereas biobased-polyethylene, biobased polyamide, and biobased-polyethylene terephthalate are not biodegradable despite being produced from bioresources [5]. A few of the numerous applications that bioplastics have as replacements for petroleum-based plastics are listed in Table 1.

Table 1 Applications of common bioplastics in various industries [5, 6]

Bioplastic	Industry	Applications
PLA	Healthcare	Gene delivery, tissue engineering, drug delivery
	Packaging	Food packaging
	Electrical and electronic	Three-dimensional printing filament
PHA	Healthcare	Implants, biomedical devices, drug carrier
	Packaging	Food packaging
	Agriculture	Crop protection film, carrier for insecticides
PHB	Healthcare	Cardiac valve manufacturing, medicine encapsulation
	Agriculture	Encapsulation of fertilizers
	Consumer goods	Disposable bottles, diaper
	Food	Encapsulation of food supplements, food packaging
Polyamide 11	Automobile	Car door handles and interior parts
	Electrical and electronic	Electrical cables, wires
	Healthcare	3D printing of prosthetic teeth
Cellulose-based	Electrical and electronic	Transparent dialers, screen shields, headphones
	Packaging	Food packaging
	Healthcare	Coating tablets

2 Raman Spectroscopy: A Tool for Bioplastic Analysis

RS has emerged as a powerful tool for analysis because it is nondestructive, thus allowing repetitive analysis of the same sample if needed, assisting in fast, label-free feedback, and due to its ability to extract the chemical, morphological, and crystal properties of polymers. It shows characteristic vibrations such as bending and stretching of bonds, which act as a fingerprint to identify molecules. A standard Raman spectrometer has three core components: a monochromatic laser source, a grating filter designed to remove unscattered Rayleigh radiation and a sensitive charge-coupled device (CCD) detector [7]. It performs better than FTIR because of its compatibility with aqueous environments, and is not affected by water or moisture present in the sample, resulting in better spatial resolution and wider spectral coverage. RS helps to understand bioplastic properties such as crystallinity and degradation. Identification of bonds and chemical composition is carried out by analyzing the frequency change occurring after irradiation of the sample using a laser. The frequency change is caused by the movement of the molecule in virtual state and inelastic scattering of light [8–10]. Beyond RS, several established analytical techniques have been used to evaluate bioplastics, albeit with their own constraints.

Fourier transform infrared (FTIR) spectroscopy, nuclear magnetic resonance (NMR) spectroscopy, X-ray diffraction (XRD), thermogravimetric analysis (TGA), differential scanning calorimetry (DSC), and pyrolysis–gas chromatography–mass spectrometry (Pyr-GC/MS) are other techniques that help characterize the physico-chemical properties of bioplastics [11–13]. To analyze samples of size 10 μm and above, μFTIR is used. It is a nondestructive method and gathers information on functional groups based on the vibration of molecules. RS extends the capabilities of FTIR since it is suitable for smaller sample sizes and is not affected by water. NMR, a phenomenon that occurs as a result of exposure to electromagnetic radiation on specific nuclei placed in a magnetic field, has been used to determine the average molecular weight and structural properties of PBSA bioplastics. This technique is valued for its precision and rapid data acquisition. However, they require large sample sizes and need to be dissolved in a solvent, causing important characteristics such as size and color to be lost, making their use with polymers having high crystallinity such as PBS and PHB difficult because of their low solubility [11, 14]. Thermal methods can extract meaningful structure–property information from small sample sizes, and these techniques, such as Py-GC/MS and TGA, have other advantages, such as requiring little to no preparation of the sample. They measured the change in the properties with the change in temperature via DSC, which is a technique for determining information such as the thermal degradation temperature, heat capacity, and melting point of polymers and has been used on PBSA films to determine their crystallinity [11]. Pyr-GC/MS works by degrading the sample thermally without air and then using mass spectrometry for analysis of the gaseous reaction products, and it has been used successfully on both pure bioplastic samples and those with additives. However, these thermal methods do have the disadvantage of destroying the sample during analysis [12, 15, 16].

3 Raman Spectroscopy for Polymer Characterization

The peaks that are observed in the Raman spectra are associated with specific functional groups present in the material, which aid in its identification. The molecular structures of the four bioplastics covered here are shown in Fig. 1. PLA is a polymer that is obtained by extracting sugar from starch feedstocks and then converting it to lactic acid; then, the dimer lactide is subsequently polymerized to polylactic acid [20]. For PLA, the Raman spectra show bands at 2951 cm^{-1} for the stretching of CH, 1768 cm^{-1} for C=O carbonyl stretching, 1450 cm^{-1} for asymmetric C-H bending, and 1382 cm^{-1} and 1130 cm^{-1} for the symmetric and asymmetric bending vibrations of CH3, 872 cm^{-1} for C–COO stretching and 735 cm^{-1} for C=O stretching, respectively [21].

PHA granules are produced by bacteria from biomass by biosynthesis in the presence of excess carbon. This carbon is generally supplied from ecological wastes such as fermented molasses and from paper and palm oil mills. Furthermore, various shapes can be formed with it on application of heat for use in different

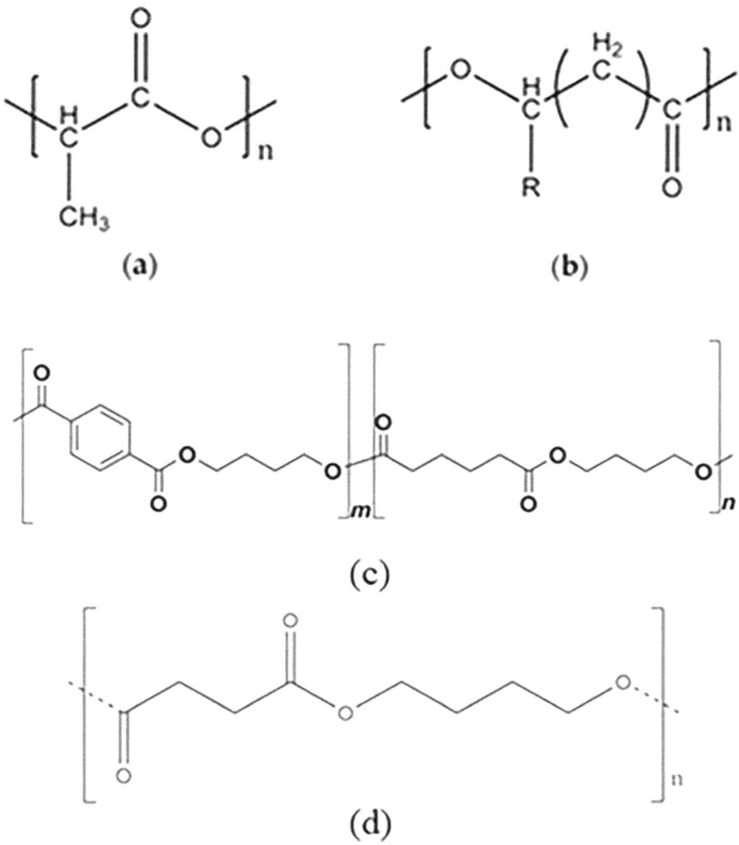

Fig. 1 Chemical structures of (**a**) PLA (**b**) PHA (reused with permission from [17]), (**c**) PBAT (reused with permission from [18]), and (**d**) PBS (reused with permission from [19])

industries, for example medicine, agriculture, and packaging [2]. For PHA, the Raman spectrum shows bands at 2998 cm^{-1} for the asymmetric stretching of CH3; at 2993 cm^{-1} for the symmetric stretching of CH2; at 2877 cm^{-1} for CH stretching; at 1730 cm^{-1} for carboxylic group stretching; at 1455 cm^{-1} for the asymmetric deformation of CH_3; at 1361 cm^{-1} for the symmetric deformation of CH_2; at 1100 cm^{-1} for the symmetric stretching of C-O-C and in-plane bending of CH_3; at 1055 cm^{-1} for the C–CH_3 bond symmetric stretching; at 835 cm^{-1} for the ester bond stretching; at 610 cm^{-1} for the deformation of C–CO and C–CH_3 bonds; and at 433 cm^{-1} for the skeletal deformation of the C–C bond [17].

PBS is produced from the esterification of succinic acid, obtained from the fermentation of renewable resources, with 1,4-butanediol, obtained through biosynthesis [22]. For PBS, the peaks at 2928 cm^{-1} and 1303 cm^{-1} were associated with the asymmetric and symmetric deformational vibrations of CH2 groups in the main chain; at 1723 cm^{-1}, the peak was due to ester carbonyl C=O stretching; at

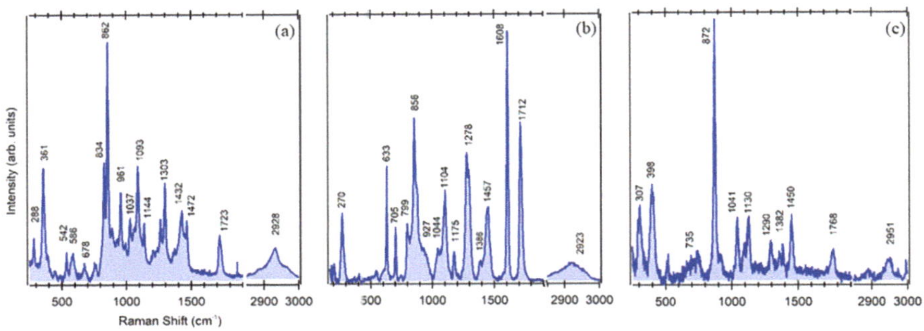

Fig. 2 Raman spectra of (**a**) PBS, (**b**) PBAT, (**c**) PLA [reused with permission from 21]

1144–1260 cm^{-1}, the peak was due to C–O–C group stretching in the ester linkages; at 1037 cm^{-1}, for O-C-C stretching vibrations; at 961 cm^{-1}, for C–OH bending in the carboxylic acid groups; and at 834 and 862 cm^{-1}, for C=O stretching; and at 288 cm^{-1}, for bending vibrations along the carbon, carbon, and oxygen backbone vibrations [21]. The characteristic spectra of PBS, PBAT, and PLA are shown in Fig. 2.

PBAT is a polymer that combines aliphatic and aromatic polyesters and is obtained by esterification of adipic acid and terephthalic acid with 1,4-butanediol via a condensation reaction followed by melt polymerization to remove impurities [23]. For PBAT, the peak at 2923 cm^{-1} is due to methyl vibration, the peak at 1712 cm^{-1} corresponds to carbonyl stretching, the peak at 1608 cm^{-1} corresponds to C–C stretching in the benzene ring, and the peak at 1457 cm^{-1} corresponds to in-plane bending of CH_2. The peaks at 1278 cm^{-1} and 1104 cm^{-1} correspond to C–O stretching in the benzene ring, helping in PBAT identification. The peak at 856 cm^{-1} was associated with the out-of-plane bending mode of =C–H in the benzene ring [21].

3.1 *Structural and Morphological Analysis*

The crystallinity of a material determines its mechanical and structural properties and this ultimately determines the long-term viability of the material. In a study on polyethylene plastic using RS, differentiation of crystalline, interfacial and amorphous states was done [24]. Raman spectra capture the changes in bonds as crystallization occurs. This involves understanding the sequence of changes with the help of 2D-COS, as demonstrated in studies on the bioplastic PHBHx (a polymer that belongs to the PHA family of bioplastics) which compared Raman spectra in the amorphous state and the crystalline state. Increased interaction of CH with carbonyl groups was shown by bands at 3000 cm^{-1} and 3010 cm^{-1}. The downshift of the carbonyl band from 1737 cm^{-1} to 1726 cm^{-1} as well as the appearance of a phonon band at 80 cm^{-1} indicated a crystal structure formation [25]. Addition of a filler like

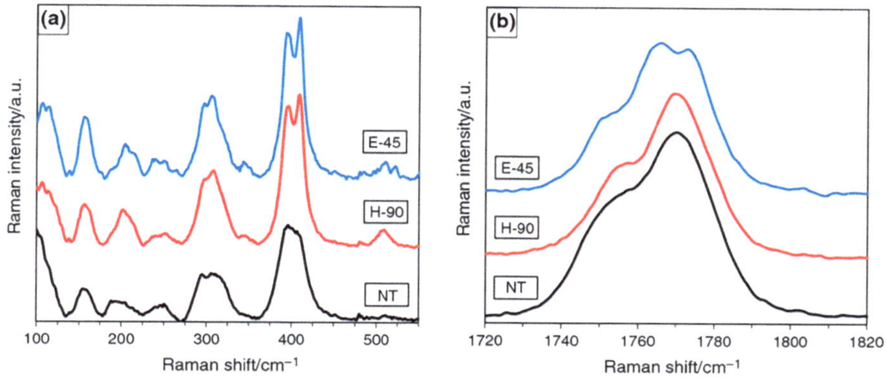

Fig. 3 Differentiation between α (E-45) and α′ (H-90) crystalline structures via Raman spectroscopy: (**a**) 100–550 cm^{-1} and (**b**) 1720–1820 cm^{-1} ranges (reused with permission from [27])

cellulose microfibers can increase the crystallinity of PHAs, a property that can be controlled by heat and quenching [26]. RS can also be used to distinguish between polymorphs (different crystalline forms of the same material) of bioplastic, such as the α and α' crystalline forms of PLA. In a study analysing PLA fibers, it was confirmed that two different postcrystallization steps resulted in different crystalline structures. In the ethanol-treated fiber, a triplet in the carbonyl stretching region indicated an abundance of α crystals, whereas in the thermally annealed fiber, the peak normally observed at 200 cm^{-1} decreased, indicating a more disordered α' structure, as shown in Fig. 3. The findings obtained via RS were consistent with the results obtained via DSC and XRD measurements [27]. RS was also used to observe the conformational changes that occurred upon deformation in the amorphous state of PLA by identifying bands linked to different isomeric states and shifts in the skeletal deformation bands (peaks due to bending of backbone atoms) to understand the conformation distribution [28].

Polarized RS demonstrates its usefulness in the study of molecular orientation of polymer chains such as in PLLA. Crystalline and amorphous regions are distinguished based on the analysis of changes in morphology-dependent peaks caused by heat. By determining the Raman tensor direction for each peak, measuring the tilt angle relative to the polymer chain and deriving the orientation parameters, the structural reorganization of polymer chains during deformation can be quantitatively understood [29, 30].

3.2 *Monitoring Bioplastic Degradation Mechanisms*

RS is a versatile tool for studying degradation pathways in bioplastics. It does so by tracking bond modifications which inform assessments of long-term stability and environmental fate [21]. By providing a detailed "molecular fingerprint" of

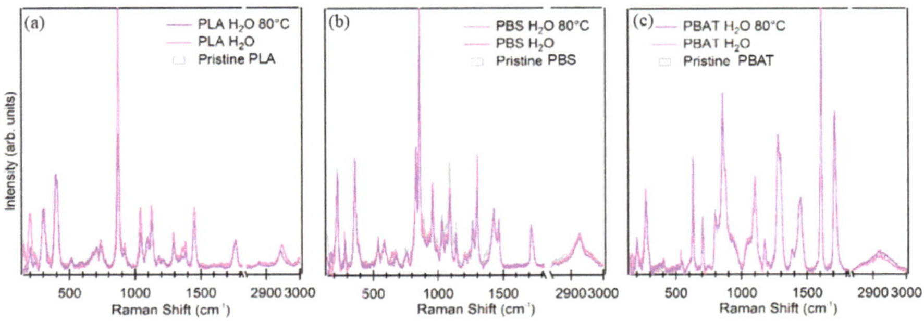

Fig. 4 Raman spectra of the single-layer bioplastics after water treatment at RT for 12 h (pink line) and 80 °C for 3 h (violet line) compared with those of the pristine material (gray spectra): (**a**) PLA, (**b**) PBS, and (**c**) PBAT [reused with permission from 21]

functional groups and bonds, RS offers a nondestructive way to monitor chemical changes during bioplastic degradation. Because of the insensitivity of RS to water, it works well for studying degradation under aqueous conditions, which infrared methods often struggle with [31–33]. Figure 4 shows how water treatment at room temperature (12 h) and 80 °C (3 h) altered the Raman spectra of PLA, PBS, and PBAT. As a sample ages, its Raman spectrum shifts, and peaks change in intensity, indicating chemical alterations. For example, when polyethylene terephthalate (PET) is irradiated, the C=C stretch shifts from 1608.5 cm^{-1} to 1610 cm^{-1}, and the peak broadens, indicating radiation damage; chain breaks in a similar way appear as a reduction in the intensity of diagnostic bands [33]. In polyanhydride networks, the carbonyl peak at 1815 cm^{-1} gradually disappears during hydrolysis, revealing how the anhydride groups erode from the surface inward [33]. Water–polymer interactions and polymer concentrations in different regions of hydrogel systems are indicated on the basis of the stretching of the CH_x band near 2893 cm^{-1} [32]. Unique Raman signatures are observed for different bioplastics, such as PLA, PBAT, PHBV and PBS, which makes it straightforward to differentiate between them [32, 34]. The presence of a particular bond and its quantification on the basis of the peak areas, as well as the mapping of those peaks in 2D or 3D, to understand degradation regions is made possible by modern Raman microscopy. For example, when fluorinated polyurethane is aged, a clear decrease in the C–H and C–F signals of the outer layers is observed, confirming greater degradation of the surface than of the interior [33].

The raman chemical hyperspectral imaging (RHI) method shows the in situ distribution of molecular species across a surface, which is made possible by the high temporal (<1 s) and spatial (~1 μm^2) resolutions of RS. This allows for quick two-dimensional chemical mapping [32, 33, 35]. Without impact detail, automated mapping further speeds up the collection of data over large areas [31]. To create three-dimensional spectral images for depth profiling, confocal Raman microscopy acquires serial optical sections at various focal planes. These images can show how aging patterns, such as "surface eroding" effects, progress beneath the surface [32, 33]. When probing down to several millimeters, spatially offset RS (SORS) shifts the collection point away from the laser spot to capture subsurface signals, enabling detection to a depth of 5 mm by collecting spectral contributions from different sample depths. For building volumetric models of internal features, Raman tomography stacks 2D cross-sectional images (or slices) obtained through methods such as optical sectioning with confocal Raman microscopes or physical sectioning [32].

For example, 3D Raman maps of aged fluorinated polyurethane show the greatest loss of C–H (2500–3400 cm^{-1}) and C–F (1280–1400 cm^{-1}) signals at the outermost layers after 1800 h of weathering, confirming a surface-eroding pattern [33]. Polyanhydride networks show a depth-dependent decline in the 1815 cm^{-1} carbonyl band, tracing hydrolysis from the surface inward [33]. Confocal Raman imaging was used to track intravenous graphene in mice over a period of 8–90 days in a biological setting. Changes in D (~1360 cm^{-1}) and G (~1585 cm^{-1}) and the emergence of D′ (~1620 cm^{-1}) bands indicate structural disorder driven by macrophages [32]. Similarly, in isotactic polypropylene composites with multiwalled carbon nanotubes, accelerated aging was mapped by 2D Raman images of the 1716 cm^{-1} carbonyl peak [33]. These two- and three-dimensional Raman techniques are crucial for determining the time and location of molecular breakdown in bioplastics, where controlled degradation often serves as a design goal in both laboratory and real-world settings [33].

3.3 *Monitoring of Bioplastic Synthesis and Processing*

RS, which is integrated with chemometric tools, has emerged as an important analytical method for real-time monitoring of bioplastic synthesis and processing. During the polymerization steps, monomer-to-polymer conversion can be understood by looking for changes in the characteristic ester and carboxyl bands, thus determining esterification percentages in polyesters [36]. These molecular-level observations support the optimization of the reaction conditions and offer insights into kinetic mechanisms, such as those observed in cyanoacrylate-based adhesives, where the rate of polymer chain propagation is closely influenced by the monomer concentration [37].

In continuous bioprocessing, RS is widely employed as a process analytical technology (PAT) owing to its ability to perform noninvasive, inline measurements even in aqueous and optically complex media. Through the application of near-infrared excitation sources (typically at 785 nm or 830 nm) and multivariate chemometric models such as principal component analysis (PCA) and partial least squares (PLS), it becomes possible to minimize fluorescence interference and derive quantitative data on key parameters, including chemical composition and biomass density, with significantly reduced analytical lag times [38–40].

In regard to film formation and melt processing, Raman spectra can be used to identify various bioplastic types, such as PLA, PBAT, PBS, and PHBV, through their unique vibrational fingerprints [32, 34], and Raman imaging can be used to understand the spatial distribution of molecular constituents within composites, molecular mixing and crystallinity variations present in interfacial boundaries, as shown in PLA/PHA laminates [40, 41]. With time, aging and degradation results in subtle functional group changes and this can be monitored [31, 33, 40, 42, 43]. Combining RS with density functional theory (DFT) or Hartree–Fock (HF) models allows analysis of additives like plasticizer content in polymers. This is extremely important to us, as it allows us to understand the quality of materials by checking the concentration of additives, which is important for regulatory purposes [44]. These applications demonstrate the role of RS in maintaining bioplastic quality and reproducibility across the production lifecycle [32, 33].

3.4 *Characterization of the Bioplastic Blends and Composites*

Bioplastics are currently not as efficient as petroleum-derived plastics when cost performance is considered; therefore, the addition of fillers or other polymers is essential for strengthening toughness, improving gas and moisture barriers, and increasing softening points [33]. Here, RS is used as a remarkably sensitive probe where each additive reveals its vibrational fingerprint, making it possible to spot and measure components directly in the matrix [32, 45]. Plasticizers are small, polar molecules that slip between polymer chains, weaken hydrogen bonds, increase spacing, and thus decrease the glass transition temperature, which improves film flexibility and stretching [33]. Fillers do the opposite by filling voids, resulting in higher tensile strength and Young's modulus, and sometimes, they even gain conductivity or antimicrobial traits [33]. Phthalic acid esters show Raman peaks at approximately 400, 650, 1040, 1120, 1160, 1284, 1450, 1580, 1600 and 1726 cm^{-1}, with dimethyl phthalate marked by extra bands at 818 and 964 cm^{-1} [44]. Graphene oxide is flagged by its D band (~1345–1360 cm^{-1}) and G band (~1570–1585 cm^{-1}), where shifts in the I_D/I_G ratio hint at defect levels [32, 46]. Starch and cellulose nanocrystals can transform mechanical and barrier properties, but signals tend to hide behind the polymer bands [33, 42]. Even so, PBAT stands out in complex blends because of its unique benzene-ring peaks, which do not overlap with those of PLA, PBS or PHBV [32, 34, 42]. Ultimately, whether standard Raman or advanced O-PTIR is used, these techniques are invaluable for choosing and mapping additives—and they remind us how much more we need in terms of lignocellulosic spectral libraries [41, 42].

3.5 *Quantitative Analysis and Chemometrics in Bioplastic Raman Research*

Because the intensity of a particular Raman band is directly proportional to the number of corresponding molecular vibrations present in the sample, analysis via RS correlates peak intensities with concentrations [38, 47]. Single-band features, such as area, intensity, or band center, are focused on when monitoring polymer properties or degradation products via univariate approaches, and properties such as band ratios indicate changes in crystallinity and/or mechanical characteristics. An example here is the comparison of the symmetric and asymmetric CH_2 stretching modes, which is demonstrated for polyethylene [38, 39]. Efficient quantification by surface-enhanced Raman methods using isotopolog tracers, where internal standardization is generally applied to address heterogeneity in samples and fluctuations in instruments, has been applied in biomedical assays [48].

Along with good instrumentation, quantitative Raman analysis also requires extensive, well-curated spectral libraries and calibration routines that are carefully designed so that compound identities are understood with promising confidence when the measured spectra are compared against databases such as Wiley's KnowItAll Raman Library [8, 31, 39, 43]. The reduction of false positives while dealing with complex samples is achieved by including spectra from withered materials and biological sources in repositories that are open-access, and the results are comparable when different time points or instruments are used for routine calibration of both wavenumber and intensity [43, 47].

Several studies illustrate practical applications of Raman and allied imaging techniques. Submicron maps of molecular distributions at the PHA–PLA laminate interface were generated by combining optical photothermal infrared (O-PTIR) with Raman and analyzing the results through two-dimensional correlation spectroscopy [41]. A study used electrospun PEO–PCL–hyaluronic acid nanofibers, applying singular value decomposition to assess polymer homogeneity, and reported that shifts in HA content seemed to influence mixing efficiency in a predictable way [32]. A study dealt with elaborate double-walled microspheres made of PLGA, PLLA, and PCL. They employed the band target entropy minimization (BTEM) algorithm to extract pure component spectra, offering a clear window into the contribution of each polymer [32]. Raman imaging of compostable bioplastic bags under biodegradation captures the evolving spatial patterns of new surface functional groups. This approach provides a foundation for the direct quantification of fragmentation products [42].

Chemometric approaches plays the central role in turning complex Raman spectra into clear quantitative insights [40, 47, 49]. Principal component analysis (PCA) extracts variants that identify different sample types in addition to clarity, whereas partial least squares (PLS) regression links spectral peaks with analyte concentrations, improving quantitative accuracy [32, 35, 38, 40, 47].

For classification tasks, linear discriminant analysis (LDA) offers separation of categories, and multivariate curve resolution (MCR) helps to unmix overlapping bands so individual component contributions can be estimated [32, 40]. ML and DL algorithms are incorporated into workflows to automate feature extraction, filter out noise, and reduce interpretive mistakes [8, 38, 45, 49].

When coupled with Raman hyperspectral imaging, these multivariate strategies form a robust workflow for detecting and quantifying microplastics and nanoplastics within complex environmental matrices. This can solve fluorescence interference and band overlap, and pollutants can be detected at submicron scales [35].

4 Challenges and Future Directions

Despite its molecular specificity, RS encounters several obstacles in bioplastic characterization. Its scattering cross-section is inherently weak, six to ten orders of magnitude smaller than that of fluorescence, resulting in low signal-to-noise ratios for components such as starch, whose signals may be masked by stronger polymer bands. Therefore, the acquisition time is increased, and the sensitivity is limited to early-stage aging [32, 33, 35, 42]. Bond identification becomes complex, and accuracy is also reduced, as fluorescence from environmental contaminants, pigments, and/or biofilms tends to mask original Raman spectra [31–33, 45]. Spatial resolution is also affected by diffraction to the micrometer scale, and penetration depths typically reach only a few hundred micrometers, thus limiting analyses of features of smaller or thicker samples [32, 42]. Moreover, the use of serial point measurements for mapping large areas remains slow and expensive because of expensive instrumentation and the lack of automated, high-throughput platforms [31, 42, 45]. The risk of identifying a polymer incorrectly is increased because complex bioplastic matrices have overlapping peaks due to additives and the presence of chemically similar polymers when combined with incomplete reference libraries and limited peak assignments [31, 43, 44, 50].

As a method to reduce background interference and enhance sensitivity, the development of surface-enhanced Raman spectroscopy (SERS) when used on plasmonic substrates can amplify signals by up to 10^8 times, whereas Fourier-transformed Raman (FT-Raman) spectroscopy using near-infrared excitation can reduce fluorescence [8, 32, 35]. Time-domain and spectral-domain fluorescence suppression techniques, such as shifted-excitation Raman difference spectroscopy and chemo-photobleaching, are methods offering signal clarity [32]. Advancements in imaging, such as confocal Raman imaging (CRI), spatially offset Raman spectroscopy (SORS), and emerging Raman tomography, can now enable depth profiling up to millimeters and three-dimensional chemical reconstructions [32, 35]. The integration of optical photothermal infrared (O-PTIR) for simultaneous IR and Raman acquisition, the use of portable Raman systems for field use, and the development of automated particle selection workflows can improve the outcome [8, 31, 45]. On the computational side, DL is being used to denoise spectra, and chemometric approaches such as PCA and MCR-ALS simplify band assignment in complex spectra while supporting improved spatial resolution [32, 35]. Including Raman spectra of plasticizers in open-access reference libraries, supported by DFT-based peak assignments, helps reduce false positives and improves confidence in identification [43, 44]. Adjusting laser power carefully prevents sample damage during measurement [42]. Taken together, these measures work toward a more accurate and sensitive approach for addressing current challenges in bioplastic research.

5 Conclusion

Polymers such as bioplastics, which are derived from biomass, have properties similar to those of conventional plastics and can be biodegradable or nonbiodegradable. The versatile properties of these materials make them useful in various industries, highlighting the need for detailed characterization. RS provides rapid, label-free analysis of polymers, and its laser-based vibrational spectra act as unique "fingerprints" of the chemical bonds while tolerating water, unlike infrared methods, making it useful in aqueous bioplastic samples. Different bioplastics have different Raman spectra (e.g., PLA shows a strong carbonyl stretch at approximately 1768 cm^{-1}, whereas PBS shows a carbonyl band near 1723 cm^{-1}), which enables the unambiguous identification of polymers and helps in monitoring chemical modifications, and structural properties such as crystallinity, polymorphic forms, and different crystal phases can be identified on the basis of shifts in key bands and monitoring changes in molecular orientation. Polymer chain alignment and conformational distributions can be revealed with polarized Raman measurements. Additionally, the degradation of bioplastics can be tracked by specific vibrational bands during aging (e.g., ester linkage hydrolysis by disappearance of the carbonyl band), and advanced techniques such as confocal Raman mapping and hyperspectral imaging help spatially resolve degradation, i.e., deterioration progress through the material. It enables real-time reaction monitoring, can follow monomer-to-polymer conversion, and can detect evolving functional groups, which results in the optimization of reaction conditions and quality control in polymer manufacturing. The identification of plasticizers, fillers and other additives on the basis of their unique vibrational signatures via Raman spectroscopy (e.g., the D and G bands of graphene-based fillers) helps in understanding the additive distribution, composition of composites, and interactions between the matrix and fillers. In addition, the quantification of polymer fractions or microplastic content can be derived by correlating band intensities with concentrations. The major advantages of RS for bioplastic research include its noninvasive nature, minimal sample preparation, and compatibility with aqueous environments, whereas low signal levels, longer acquisition times and fluorescence from contaminants or additives obscuring Raman peaks are challenges. Some of these challenges can be addressed via the use of emerging techniques, such as surface-enhanced Raman spectroscopy (SERS), near-infrared excitation, and confocal, spatially offset, and tomographic RS, to increase sensitivity, reduce fluorescence, improve depth profiling, and provide 3D mapping capabilities. Band assignments and particle selection can be made easier with the integration of DL and the expansion of spectral databases.

References

1. Abina A, Korošec T, Puc U, Zidanšek A (2023) Review of bioplastics characterisation by terahertz techniques in the view of ensuring a circular economy. Photonics 10(8):883. https://doi.org/10.3390/photonics10080883
2. Acharjee SA, Bharali P, Gogoi B, Sorhie V, Walling B, Alemtoshi. (2023) PHA-based bioplastic: a potential alternative to address microplastic pollution. Water Air Soil Pollut 234(1). https://doi.org/10.1007/s11270-022-06029-2
3. Saharan R, Kharb J (2022) Exploration of bioplastics: a review. Orient J Chem 38(4):840–854. https://doi.org/10.13005/ojc/380403
4. Getino L, Martín JL, Chamizo-Ampudia A (2024) A review of Polyhydroxyalkanoates: characterization, production, and application from waste. Microorganisms 12(10):2028. https://doi.org/10.3390/microorganisms12102028
5. Nanda S, Patra BR, Patel R, Bakos J, Dalai AK (2022) Innovations in applications and prospects of bioplastics and biopolymers: a review. Environ Chem Lett 20(1):379–395. https://doi.org/10.1007/s10311-021-01334-4
6. Kong U, Mohammad Rawi NF, Tay GS (2023) The potential applications of reinforced bioplastics in various industries: a review. Polymers 15(10):2399. https://doi.org/10.3390/polym15102399
7. Asamoah BO, Uurasjärvi E, Räty J, Koistinen A, Roussey M, Peiponen K-E (2021) Towards the development of portable and in situ optical devices for detection of micro-and Nanoplastics in water: a review on the current status. Polymers 13(5):730. https://doi.org/10.3390/polym13050730
8. Chakraborty I, Banik S, Biswas R, Yamamoto T, Noothalapati H, Mazumder N (2023) Raman spectroscopy for microplastic detection in water sources: a systematic review. Int J Environ Sci Technol 20(9):10435–10448. https://doi.org/10.1007/s13762-022-04505-0
9. Gieroba B, Kalisz G, Krysa M, Khalavka M, Przekora A (2023) Application of vibrational spectroscopic techniques in the study of the natural polysaccharides and their cross-linking process. Int J Mol Sci 24(3):2630. https://doi.org/10.3390/ijms24032630
10. He W, Liu R, Fei F, Xi S, Du Z, Luan Z, Sun C, Zhang X (2024) In situ real-time pathway to study the polyethylene long-term degradation process by a marine fungus through confocal Raman quantitative imaging. Sci Total Environ 939:173582. https://doi.org/10.1016/j.scitotenv.2024.173582
11. Baidurah S (2022) Methods of analyses for biodegradable polymers: a review. Polymers 14(22):4928. https://doi.org/10.3390/polym14224928
12. Censi V, Saiano F, Bongiorno D, Indelicato S, Napoli A, Piazzese D (2022) Bioplastics: a new analytical challenge. Front Chem 10. https://doi.org/10.3389/fchem.2022.971792
13. Zhou Z, Duan X, Yu Y, Ma L, Moreno A, Xia Y, Chen L, Ye S, Cong R (2025) Recent advances and applications of NMR techniques in plastic characterizations. Anal Chem 97(11):5847–5865. https://doi.org/10.1021/acs.analchem.4c05230
14. Schmidt J, Haave M, Underhaug J, Wang W (2024) Unlocking the potential of NMR spectroscopy for precise and efficient quantification of microplastics. Microplast Nanoplast 4(1). https://doi.org/10.1186/s43591-024-00095-5
15. Fojt J, Románeková I, Procházková P, David J, Brtnický M, Kučerík J (2022) A simple method for quantification of Polyhydroxybutyrate and Polylactic acid micro-bioplastics in soils by evolved gas analysis. Molecules 27(6):1898. https://doi.org/10.3390/molecules27061898
16. Okoffo ED, Chan CM, Rauert C, Kaserzon S, Thomas KV (2022) Identification and quantification of micro-bioplastics in environmental samples by pyrolysis–gas chromatography–mass spectrometry. Environ Sci Technol 56(19):13774–13785. https://doi.org/10.1021/acs.est.2c04091

17. Mondragón-Herrera LI, Vargas-Coronado RF, Carrillo-Escalante H, Cauich-Rodríguez JV, Hernández-Sánchez F, Velasco-Santos C, Avilés F (2024) Mechanical, thermal, and physicochemical properties of filaments of poly (lactic acid), Polyhydroxyalkanoates and their blend for additive manufacturing. Polymers 16(8):1062. https://doi.org/10.3390/polym16081062
18. Šerá J, Kadlečková M, Fayyazbakhsh A, Kučabová V, Koutný M (2020) Occurrence and analysis of thermophilic poly(butylene adipate-co-terephthalate)-degrading microorganisms in temperate zone soils. Int J Mol Sci 21(21):7857. https://doi.org/10.3390/ijms21217857
19. Aliotta L, Seggiani M, Lazzeri A, Gigante V, Cinelli P (2022) A brief review of poly (butylene succinate) (PBS) and its Main copolymers: synthesis, blends, composites, biodegradability, and applications. Polymers 14(4):844. https://doi.org/10.3390/polym14040844
20. Jiménez L, Mena M, Prendiz J, Vega-Baudrit J (2019) Polylactic acid (PLA) as a bioplastic and its possible applications in the food industry. Food Sci Nutr 5(2):1–6. https://doi.org/10.24966/fsn-1076/100048
21. Ambrosio G, Faglia G, Tagliabue S, Baratto C (2021) Study of the degradation of biobased plastic after stress tests in water. Coatings 11(11):1330. https://doi.org/10.3390/coatings11111330
22. Barrino F, De La Rosa-Ramírez H, Schiraldi C, López-Martínez J, Samper MD (2023) Preparation and characterization of new bioplastics based on polybutylene succinate (PBS). Polymers 15(5):1212. https://doi.org/10.3390/polym15051212
23. Roy S, Ghosh T, Zhang W, Rhim J-W (2024) Recent progress in PBAT-based films and food packaging applications: a mini-review. Food Chem 437:137822. https://doi.org/10.1016/j.foodchem.2023.137822
24. Lin W, Cossar M, Dang V, Teh J (2007) The application of Raman spectroscopy to three-phase characterization of polyethylene crystallinity. Polym Test 26(6):814–821. https://doi.org/10.1016/j.polymertesting.2007.05.004
25. Adar F, Street R, Noda I (2023) Isothermal crystallization of polyhydroxyalkanoate (PHA) utilizing Raman spectroscopy to follow chain packing as well as molecular motion. Spectrochim Acta A Mol Biomol Spectrosc 285:121861. https://doi.org/10.1016/j.saa.2022.121861
26. Mármol G, Gauss C, Fangueiro R (2020) Potential of cellulose microfibers for PHA and PLA biopolymers reinforcement. Molecules 25(20):4653. https://doi.org/10.3390/molecules25204653
27. Vadas D, Nagy ZK, Csontos I, Marosi G, Bocz K (2020) Effects of thermal annealing and solvent-induced crystallization on the structure and properties of poly(lactic acid) microfibres produced by high-speed electrospinning. J Therm Anal Calorim 142(2):581–594. https://doi.org/10.1007/s10973-019-09191-8
28. Yang X, Kang S, Yang Y, Aou K, Hsu SL (2004) Raman spectroscopic study of conformational changes in the amorphous phase of poly(lactic acid) during deformation. Polymer 45(12):4241–4248. https://doi.org/10.1016/j.polymer.2004.03.107
29. Park MS, Wong YS, Park JO, Venkatraman SS, Srinivasarao M (2011) A simple method for obtaining the information of orientation distribution using polarized Raman spectroscopy: orientation study of structural units in poly(lactic acid). Macromolecules 44(7):2120–2131. https://doi.org/10.1021/ma101553v
30. Tanaka M, Young RJ (2006) Molecular orientation distributions in Uniaxially oriented poly(L-lactic acid) films determined by polarized Raman spectroscopy. Macromolecules 39(9):3312–3321. https://doi.org/10.1021/ma0526286
31. Birch QT, Potter PM, Pinto PX, Dionysiou DD, Al-Abed SR (2021) Isotope ratio mass spectrometry and spectroscopic techniques for microplastics characterization. Talanta 224:121743. https://doi.org/10.1016/j.talanta.2020.121743
32. Fernández-Galiana Á, Bibikova O, Vilms Pedersen S, Stevens MM (2024) Fundamentals and applications of Raman-based techniques for the design and development of active biomedical materials. Adv Mater 36(43). https://doi.org/10.1002/adma.202210807

33. Tian R, Li K, Lin Y, Lu C, Duan X (2023) Characterization techniques of polymer aging: from beginning to end. Chem Rev 123(6):3007–3088. https://doi.org/10.1021/acs.chemrev.2c00750
34. Cai Y, Lv J, Feng J (2013) Spectral characterization of four kinds of biodegradable plastics: poly (lactic acid), poly (Butylenes Adipate-co-terephthalate), poly (Hydroxybutyrate-co-Hydroxyvalerate) and poly (Butylenes succinate) with FTIR and Raman spectroscopy. J Polym Environ 21(1):108–114. https://doi.org/10.1007/s10924-012-0534-2
35. Ly NH, Kim M-K, Lee H, Lee C, Son SJ, Zoh K-D, Vasseghian Y, Joo S-W (2022) Advanced microplastic monitoring using Raman spectroscopy with a combination of nanostructure-based substrates. J Nanostructure Chem 12(5):865–888. https://doi.org/10.1007/s40097-022-00506-0
36. Chang C, Feng L-F, Gu X-P, Zhang C-L, Dai L-K, Chen X, Hu G-H (2022) In situ Raman spectroscopy real-time monitoring of a polyester polymerization process for subsequent process optimization and control. Ind Eng Chem Res 61(49):17993–18003. https://doi.org/10.1021/acs.iecr.2c02933
37. Li Y, Xiao L, Wang Z, Chen K, Xue C, Yu M, Wang Y, Kong F, Liu K, Qin K (2021) Raman spectroscopic characterization of polymerization kinetics of cyanoacrylate embolic glues for vascular embolization. Polymers 13(19):3362. https://doi.org/10.3390/polym13193362
38. Esmonde-White KA, Cuellar M, Uerpmann C, Lenain B, Lewis IR (2017) Raman spectroscopy as a process analytical technology for pharmaceutical manufacturing and bioprocessing. Anal Bioanal Chem 409(3):637–649. https://doi.org/10.1007/s00216-016-9824-1
39. Rytelewska S, Dąbrowska A (2022) The Raman spectroscopy approach to different freshwater microplastics and quantitative characterization of polyethylene aged in the environment. Microplastics 1(2):263–281. https://doi.org/10.3390/microplastics1020019
40. Yogurtcu B, Cebi N, Koçer AT, Erarslan A (2024) A review of non-destructive Raman spectroscopy and Chemometric techniques in the analysis of cultural heritage. Molecules 29(22):5324. https://doi.org/10.3390/molecules29225324
41. Marcott C, Kansiz M, Dillon E, Cook D, Mang MN, Noda I (2020) Two-dimensional correlation analysis of highly spatially resolved simultaneous IR and Raman spectral imaging of bioplastics composite using optical photothermal infrared and Raman spectroscopy. J Mol Struct 1210:128045. https://doi.org/10.1016/j.molstruc.2020.128045
42. Fang C, Zhang X, Zhang Z, Naidu R (2024) Characterising fragmentation of compostable bioplastic: releasing microplastics or small bioplastic debris. Environ Sci Europe 36(1). https://doi.org/10.1186/s12302-024-00946-1
43. Miller EA, Yamahara KM, French C, Spingarn N, Birch JM, Van Houtan KS (2022) A Raman spectral reference library of potential anthropogenic and biological ocean polymers. Sci Data 9(1). https://doi.org/10.1038/s41597-022-01883-5
44. Sun T, Wang Y, Li M, Hu D (2023) Raman spectroscopic study of five typical plasticizers based on DFT and HF theoretical calculation. Foods 12(15):2888. https://doi.org/10.3390/foods12152888
45. Böke JS, Popp J, Krafft C (2022) Optical photothermal infrared spectroscopy with simultaneously acquired Raman spectroscopy for two-dimensional microplastic identification. Sci Rep 12(1). https://doi.org/10.1038/s41598-022-23318-2
46. Bokobza L, Bruneel J-L, Couzi M (2013) Raman spectroscopic investigation of carbon-based materials and their composites. Comparison between carbon nanotubes and carbon black. Chem Phys Lett 590:153–159. https://doi.org/10.1016/j.cplett.2013.10.071
47. Guo S, Popp J, Bocklitz T (2021) Chemometric analysis in Raman spectroscopy from experimental design to machine learning–based modeling. Nat Protoc 16(12):5426–5459. https://doi.org/10.1038/s41596-021-00620-3

48. Langer J, Jimenez De Aberasturi D, Aizpurua J, Alvarez-Puebla RA, Auguié B, Baumberg JJ, Bazan GC, Bell SEJ, Boisen A, Brolo AG, Choo J, Cialla-May D, Deckert V, Fabris L, Faulds K, García De Abajo FJ, Goodacre R, Graham D, Haes AJ, Liz-Marzán LM (2020) Present and future of surface-enhanced Raman scattering. ACS Nano 14(1):28–117. https://doi.org/10.1021/acsnano.9b04224
49. Chandra A, Kumar V, Garnaik UC, Dada R, Qamar I, Goel VK, Agarwal S (2024) Unveiling the molecular secrets: a comprehensive review of Raman spectroscopy in biological research. ACS Omega 9(51):50049–50063. https://doi.org/10.1021/acsomega.4c00591
50. Wang JL, Schubert SH, Dragone NB, Ciechanowicz SM, Hynek BM (2024) Applications of microbial bioplastic polyhydroxyalkanoates as biosignatures for astrobiological detection. Int J Astrobiol 23. https://doi.org/10.1017/s1473550424000193

Fourier Transform Infrared (FTIR) Spectroscopy Analysis of Starch-Based Bioplastics for Cosmetic Facial Sheet Mask Applications

Siya Jain and Nirmal Mazumder

Abstract The escalating demand for sustainable and safe cosmetic products has catalyzed the adoption of bioplastic substrates, particularly those based on starch, in the development of facial sheet masks. This chapter provides a comprehensive review of the use of Fourier Transform Infrared (FTIR) spectroscopy for the structural and compositional characterization of starch-based bioplastics within cosmetic applications. FTIR is highlighted as a key analytical method, capable of monitoring functional group changes, molecular interactions, and the impact of additives such as plasticizers and fillers on film performance. Multiple case studies demonstrate how FTIR distinguishes between native and modified starches, elucidates effects on moisture resistance, flexibility, and bioactive incorporation. These insights reinforce FTIR as an indispensable quality control and material development tool for next-generation cosmetics, ultimately supporting the industry's environmental and safety objectives.

Keywords Bioplastic · Cosmetic · FTIR spectroscopy · Biodegradable · Sustainable · Biopolymer

Abbreviations

FTIR	Fourier transform infrared
GRAS	Generally regarded as safe
IR	Infrared
PHA	Polyhydroxyalkanoates
PLA	Polylactic acid
TPS	Thermoplastic starch

S. Jain · N. Mazumder (✉)
Department of Biophysics, Manipal School of Life Sciences, Manipal Academy of Higher Education, Manipal, India
e-mail: nirmal.mazumder@manipal.edu

N. Mazumder et al. (eds.), *Vibrational Spectroscopy for Biopolymer Characterization*, Biological and Medical Physics, Biomedical Engineering,
https://doi.org/10.1007/978-3-032-15306-7_7

1 Introduction

The cosmetics and personal care industry is undergoing a significant transformation toward the use of sustainable and biodegradable materials, responding to a growing demand for environmentally responsible and functional products. Facial sheet masks, now a prominent feature in skincare, are prized for their targeted delivery of active ingredients through an easy-to-use, occlusive format that enhances skin hydration, detoxification, and anti-aging effects [1]. Despite their widespread adoption, most commercial sheet masks are made from synthetic, non-biodegradable fibers such as rayon, polypropylene, and other petrochemical-derived polymers. These materials present major environmental concerns due to their persistence in landfills and contribution to plastic waste accumulation. In addition, prolonged use of synthetic substrates has raised dermatological issues, including skin irritation and potential disruption of the skin barrier [2].

Amid these issues, research attention has turned to the development of eco--friendly, biocompatible cosmetic materials. Natural biopolymers, particularly starch, are emerging as strong candidates owing to their cost-effectiveness, renewability, and excellent film-forming capabilities. Starch, comprised of amylose and amylopectin, can be derived from sources such as corn, rice, potato, and cassava. Rice starch, in particular, is valued for its small granule size and high amylopectin content, attributes that yield desirable film properties and a pleasant skin feel [3, 4]. Generally regarded as safe (GRAS), starch-based films have proven to be effective in delivering both lipophilic and hydrophilic cosmetic active ingredients. However, inherent drawbacks such as brittleness and low moisture resistance in native starch films typically necessitate modification through plasticizers, cross-linkers, or bioactive additive incorporation to achieve the required functional properties for cosmetic applications.

Fourier Transform Infrared (FTIR) spectroscopy is a vital analytical technique used to identify and characterize the chemical structures of materials, including polymers and bioplastics. FTIR works by measuring the absorption of infrared radiation by functional groups within a sample, producing a molecular fingerprint that reveals compositional and structural information. In starch-based bioplastics, FTIR enables researchers to monitor key functional groups, study molecular interactions, and evaluate the impact of plasticizers, cross-linkers, or bioactive additives on the polymer matrix [5, 6]. Its rapid and non-destructive nature makes it particularly valuable for quality control and formulation optimization in cosmetic applications, ensuring that bioplastic substrates possess the required properties for safe and effective facial mask development [3, 4]. Multiple studies have demonstrated FTIR's ability to differentiate between native and modified starches, detect impurities, and elucidate material degradation or enhancement mechanisms.

2 Bioplastics in Cosmetics

Starch is hydrophilic and has thermoplastic characteristics that are comparable to those of synthetic polymers. Typically, native starch is plasticized at a high temperature under high shear to yield thermoplastic starch (TPS) [7]. Granular starch can be converted to thermoplastic starch (TPS) by kneading, extruding, injection moulding, compression moulding, heating, and casting in an excessive water solution with the help of plasticizers including water, glycerol, or urea, and additives including lecithin, citric acid, titanium dioxide, silicon dioxide, and monoglycerides. Surface tension, retrogradation of starch, the behaviour of moisture sorption, and primarily thermal, mechanical, and barrier qualities make TPS an appealing choice. The starch sources, additives processing variables, and chemical treatment of starch all influence the behaviour of TPS. The glass-transition temperature (Tg), additives, crystallinity, and the amylose-to-amylopectin ratio all affect the tensile strength of starch [8]. Plasticizers are the most prevalent and important additives for TPS structure and performance, and have drawn the most attention because they may alter the tensile capabilities of TPS.

Rice starch is a particularly promising material for cosmetic masks. Its fine granule size and favorable amylose content produce films that feel smooth on the skin and have good drying kinetics. Rice starch is also naturally hypoallergenic (low protein content) and forms whitish films, desirable for facial products [9]. In sum, starch films in cosmetics serve as sustainable carriers and functional mask substrates. Their moisturizing and adhesive properties, combined with the ability to load actives (peptides, vitamins, acids), make them attractive for modern "clean" skincare. The synergy of starch with other cosmetic ingredients provides hydration, improved ingredient contact, and a gentle modality of delivery.

Cosmetic packaging including cases for powders, creams, and lipsticks, is now commonly produced from cellulose, starch, and polylactic acid-based bioplastics, offering both excellent mechanical properties and reduced environmental impact [10]. Additionally, bioplastics have been incorporated in cosmetic formulations as carriers or encapsulation materials for active compounds leveraging their biocompatibility and ability to form efficient delivery systems for skin applications [11]. For example, polylactic acid (PLA) and polyhydroxyalkanoates (PHA) feature prominently in biodegradable containers, microbeads in exfoliating scrubs, and facial sheet masks designed for single-use but rapid decomposition [10, 12]. Innovations also include the use of protein-based and lipid-based biopolymers, which not only enhance the hygienic properties of cosmetic carriers but also provide antioxidant, antibacterial, and moisture-retaining benefits, further aligning with consumer expectations for safe, eco-friendly beauty products [12]. Figure 1 shows the lifecycle of a bioplastics as biodegradable polymer with varied amounts of application in industries such as packaging, and cosmetic.

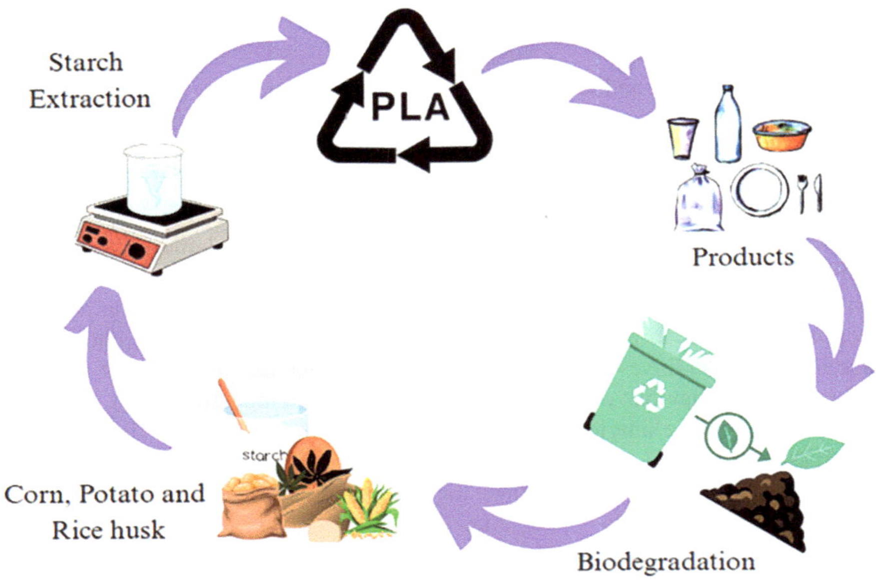

Fig. 1 The life cycle of bioplastics [13]

3 FTIR Analysis of Cosmetic-Based Bioplastics

FTIR is widely acknowledged as an efficient tool for elucidating the chemical structure and modifications in starch-based bioplastics, including those designed for cosmetic facial masks. Gurunathan et al. [13] conducted a comprehensive study on corn and potato starch blends for bioplastic materials using FTIR to monitor the effect of plasticizers and fillers. As shown in Fig. 2, characteristic peaks at approximately 2920–3010 cm^{-1} (C–H stretch) were indicative of the starch backbone, while plasticized samples exhibited broader O–H absorption bands at 3290–3316 cm^{-1} due to glycerol and sorbitol incorporation. The addition of fillers like calcium carbonate impacted peak intensity without generating new bands, illustrating improved mechanical properties and changed water solubility attributes essential for cosmetic mask performance [13, 14].

Recent work by Coltelli et al. [14] explored skin-compatible beauty masks using polyhydroxyalkanoate (PHA)/starch and biopolyester blends. FTIR confirmed effective plasticization and interaction between starch and polyesters, with observable shifts in the O–H and C–O stretch regions as seen in Fig. 3. Process conditions such as extrusion temperature were optimized by correlating FTIR data with morphological and mechanical properties, demonstrating how fine spectral shifts can guide formulation and scalability in cosmetic mask manufacturing [14].

Abotbina et al. [15] extended the FTIR analysis to cornstarch-based films containing various plasticizers. All samples, including control and plasticized films,

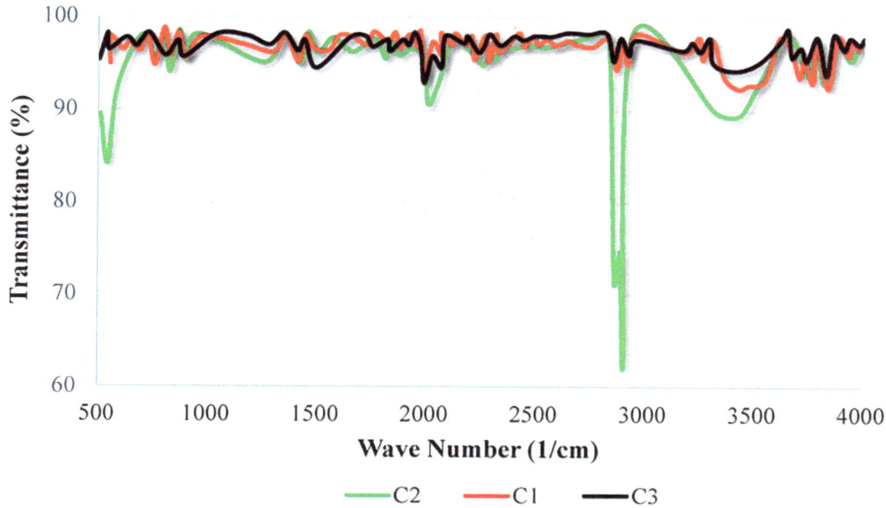

Fig. 2 The FTIR results of bioplastic film of compositions 1, 2, and 3 [13]

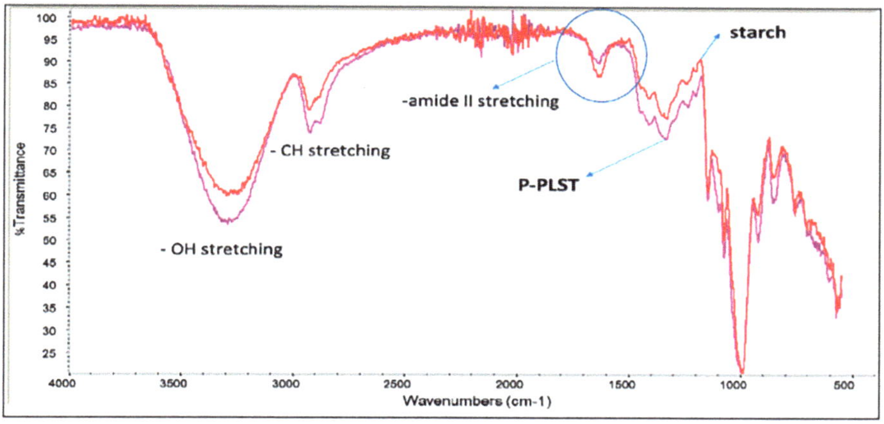

Fig. 3 FTIR spectra of starch (as red) and plasticized starch (as purple) [14]

displayed similar key FTIR peaks, like O–H stretching around 3358 cm^{-1} and C–H stretching at 2942 cm^{-1}, representing polysaccharide structure (shown in Fig. 4). The uniformity in spectra was attributed to the polyol nature of both starch and plasticizers, suggesting that major intermolecular rearrangements do not always yield new spectral features but rather modify existing intensities and positions [15].

Pooja et al. [7] did a FTIR analysis for native starch and its plasticized biopolymer films. The native starch from both potato and corn sources were evaluated. They reported both native starch and starch biopolymer film's FTIR spectra (Fig. 5)

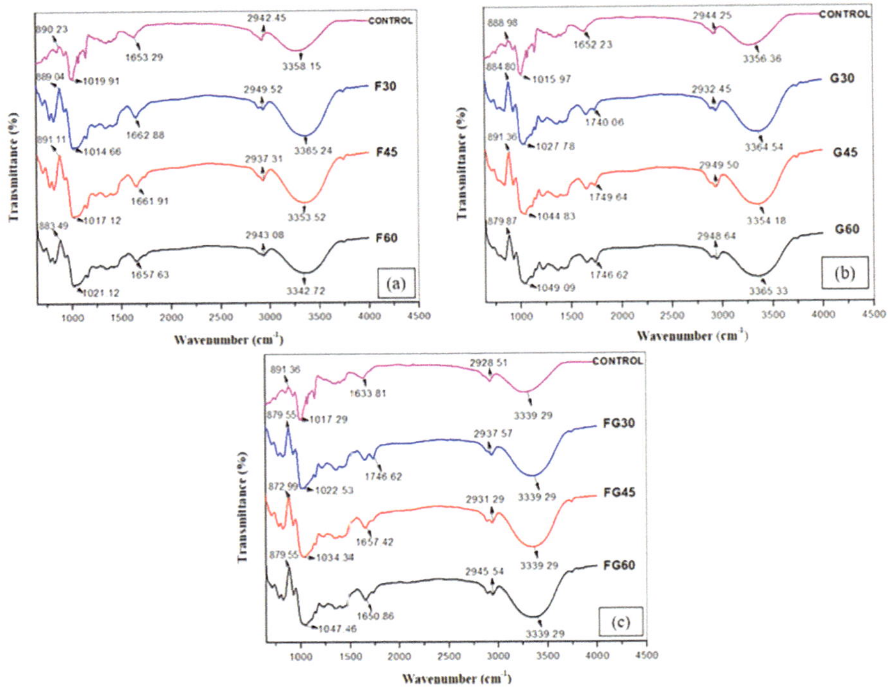

Fig. 4 FTIR spectra of starch plasticized films with plasticizer (**a**) fructose, (**b**) glycerol, (**c**) fructose/glycerol [15]

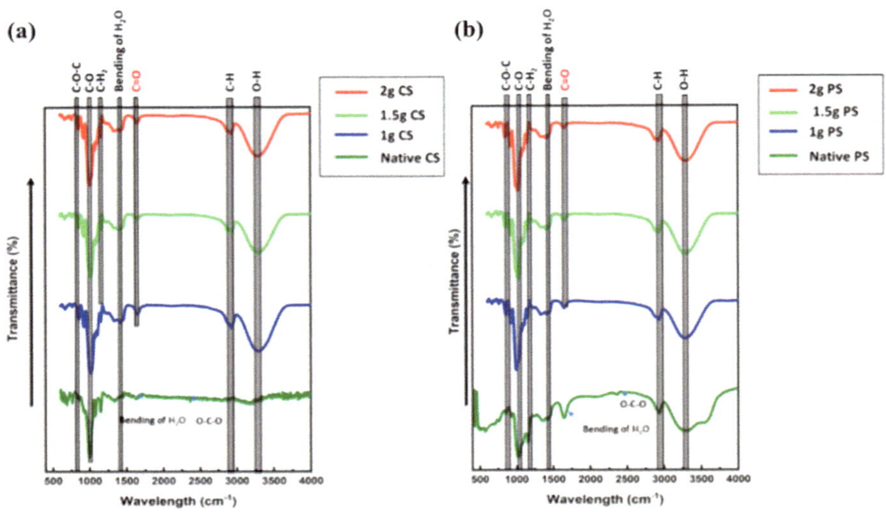

Fig. 5 FTIR spectra of (**a**) corn starch and its biopolymer films (**b**) potato starch and its biopolymer films [7]

revealed four principal vibrational peaks associated with water bending, pyranose, C–H bending, and O–H stretching ring. It was discovered that the pyranose ring peak (C–O–C) between 900 and 920 cm^{-1} had low or negligible strength in the spectra of natural starch. But because glycerol was added to the biopolymer films, larger peaks were found in the same area. At 1600 cm^{-1}, the C=O stretching band is more intense found in the films of starch polymers, in contrast to native starch powder, most likely as a result of structural alterations brought about by the film-forming process.

Collectively, these studies tell us about FTIR's versatility for cosmetic bioplastics from screening for chemical purity and additive compatibility to connecting spectral data with real-world performance attributes like moisture retention, flexibility, and skin safety.

4 Challenges and Future Prospects

The application of FTIR characterization in starch-based bioplastics for cosmetic masks faces several key challenges. One major issue is the inherent hydrophilicity and low mechanical strength of native starch, which can result in film instability and poor barrier properties when exposed to moisture [13] critical for cosmetic mask performance. The integration of plasticizers, fillers, or bioactive additives, while aiming to enhance mechanical and functional properties, often complicates the FTIR spectral analysis due to overlapping peaks and subtle structural changes. Additionally, sample heterogeneity arising from composite blends (eg, starch with PHA or chitosan) can reduce reproducibility, and environmental factors such as humidity and UV exposure can further influence FTIR outcomes, making it challenging to standardize protocols across studies [14]. Another technical challenge involves distinguishing between signals from the bioplastic matrix and those from incorporated cosmetic actives or antioxidants, which may have similar IR absorptions and can mask or shift key regions of interest [14, 15].

Despite these challenges, the future prospects for FTIR in cosmetic bioplastic research are highly promising. Advances in FTIR imaging and real-time monitoring are enabling more precise mapping of molecular interactions, including the dynamic release of bioactives and changes in functional groups during use [14]. Emerging composite materials, such as thermoplastic starch reinforced with microcrystalline cellulose or chitosan, show improved mechanical and barrier properties, as revealed by FTIR studies that precisely track interaction and dispersion. The development of fully biobased and smart beauty masks capable of responding to environmental cues will also benefit from sophisticated FTIR-based protocols that link spectral signatures directly to performance functionality [13]. As industry standards evolve and sustainability becomes the new trend, FTIR will remain essential in verifying the safety, authenticity, and degradation profiles of next-generation cosmetic materials.

5 Conclusion

FTIR spectroscopy has emerged as an indispensable characterization technique for starch-based bioplastics in cosmetic applications, allowing for molecular-level insights into the structure, compatibility, and performance of innovative sustainable materials. Addressing current analytical and formulation challenges through methodological advancements, FTIR will continue to enable rigorous evaluation and optimization, supporting the cosmetics industry's shift toward safer, eco-friendly, and high-performance products.

References

1. Nilforoushzadeh MA, Amirkhani MA, Zarrintaj P, Salehi Moghaddam A, Mehrabi T, Alavi S, Mollapour Sisakht M (2018) Skin care and rejuvenation by cosmeceutical facial mask. J Cosmet Dermatol 17(5):693–702
2. Latif S, Saha M, Morshed MM, Rahman MA, Akter M (2022) Toxicological and ecological risks associated with personal care products: an overview. Environ Res 204(Part B):1
3. Jayalath TK, Samaraweera H, Samarasinghe A (2024) Physicochemical properties and application potential of rice starch films in the cosmetic industry. Int J Biol Macromol 246:125920
4. Trevisol TC, Henriques RO, Souza AJA, Cesca K, Furigo A (2023) Starch- and carboxymethyl cellulose-based films as active beauty masks with papain incorporation. Int J Biol Macromol 231:123258
5. Coltelli M-B, Danti S, Trombi L, Morganti P, Donnarumma G, Baroni A, Fusco A, Lazzeri A (2018) Preparation of innovative skin compatible films to release polysaccharides for biobased beauty masks. Cosmetics 5(4):70
6. Yahia R, Owda ME, Abou-Zeid RE, Abdelhai F, El-Gamil HY, Abdo AM, Ali AA (2024) Biodegradable, UV absorber and thermal stable bioplastic films from waxy corn starch/polyvinyl alcohol blends. Biomass Convers Biorefinery 14(22):27989–28006
7. Pooja N, Banik S, Chakraborty I, Sudeeksha HC, Mal SS, Srisungsitthisunti P, Patil A, Mahato KK, Mazumder N (2024) Comparative analysis of biopolymer films derived from corn and potato starch with insights into morphological, structural and thermal properties. Discov Sustain 5(1):3
8. Zhang Y, Rempel C, Liu Q (2014) Thermoplastic starch processing and characteristics—a review. Crit Rev Food Sci Nutr 54(10):1353
9. Shoukat R, Cappai M, Pilia L, Pia G (2025) Rice starch chemistry, functional properties, and industrial applications: a review. Polymers 17(1):110
10. Nanda S, Patra BR, Patel R, Bakos J, Dalai AK (2021) Innovations in applications and prospects of bioplastics and biopolymers: a review. Polymers 13(22):379–395
11. Narancic T, Cerrone F, Beagan N, O'Connor KE (2020) Recent advances in bioplastics: application and biodegradation. Polymers 12(4):920
12. Abang S, Wong F, Sarbatly R, Sariau J, Baini R, Besar NA (2023) Bioplastic classifications and innovations in antibacterial, antifungal, and antioxidant applications. J Bioresour Bioprod 8(4):361–387
13. Gurunathan MK, Navasingh RJH, Selvam JDR, Čep R (2025) Development and characterization of starch bioplastics as a sustainable alternative for packaging. Sci Rep 15(1):15264

14. Coltelli M-B, Panariello L, Morganti P, Danti S, Baroni A, Lazzeri A, Fusco A, Donnarumma G (2020) Skin-compatible biobased beauty masks prepared by extrusion. J Funct Biomater 11(2):23
15. Abotbina W, Sapuan SM, Sultan MTH, Alkbir MFM, Ilyas RA (2021) Development and characterization of cornstarch-based bioplastics packaging film using a combination of different plasticizers. Polymers 13(20):3487

Raman Spectroscopy for Microplastic Research

Vedant Vikas Sarode, Sagnika Dutta, P. V. Lahari, Muralidhara Nitheesh Beliraya, and Budheswar Dehury

Abstract Raman spectroscopy (RS) is presented as a non-destructive, label-free analytical platform for the identification and characterization of microplastics of a size ranging between 1 μm and 5 mm, and nanoplastics, having a size less than 1 μm in diverse environmental and biological matrices. The chapter details the physical principles and instrumentation of different Raman methods such as micro-Raman spectroscopy, confocal Raman mapping and surface-enhanced Raman spectroscopy (SERS) and summarizes their application in water, sediment, soil, air, and tissue samples when coupled with microscopy, image analysis and chemometric workflows. The evidence from literature demonstrates polymer fingerprinting (for polyethylene, polypropylene, polystyrene, polyethylene terephthalate, polyvinyl chloride, polyamide and others), particle-level metrics like size, abundance, spatial distribution, crystallinity and routine detection in lower micrometer range. SERS and laboratory-scale tip-enhanced approaches extend sensitivity toward nanoscale but remain largely non-routine. Fluorescence and pigment interference altered spectra by weathering and biofouling, limited throughput for hyperspectral mapping, and the absence of harmonized reference libraries are the persistent barriers identified. Recent changes were surveyed, which included portable instruments, automated high-throughput mapping, and artificial intelligence and/or machine learning classifiers. Coordinated standardization, expansion of aged/biofouled spectral libraries and validated, automated AI-assisted pipelines are necessary to translate Raman methods into scalable, reproducible microplastic monitoring.

Keywords Microplastics · Spectrum analysis, Raman · Polystyrenes · Polyethylene terephthalates · Machine learning

V. V. Sarode · S. Dutta · P. V. Lahari · M. N. Beliraya · B. Dehury (✉)
Department of Bioinformatics, Manipal School of Life Sciences, Manipal Academy of Higher Education, Manipal, Karnataka, India
e-mail: budheswar.dehury@manipal.edu

N. Mazumder et al. (eds.), *Vibrational Spectroscopy for Biopolymer Characterization*, Biological and Medical Physics, Biomedical Engineering,
https://doi.org/10.1007/978-3-032-15306-7_8

1 Introduction

Plastics have become omnipresent, and their widespread adoption is attributed to their durability, versatility, inexpensiveness and ease of use. Given these advantages, they have outpaced and replaced the more traditional materials such as wood, glass and metal from various industries. Plastics consist of various polymers, out of which the most common are polyethylene (PE), polyester (PET), polystyrene (PS), polyvinyl chloride (PVC), polypropylene (PP) and polyurethane (PUR), with their rankings in terms of harmfulness labelled under the 'plastic spectrum' [1, 2]. Apart from these, various other polymers present in the environment are polycarbonate (PC), polytetrafluoroethylene (PTFE), polyamide (PA), poly(methyl methacrylate) (PMMA), and acrylonitrile butadiene styrene (ABS) (Fig. 1).

Each year, more than 30 million metric tons of this indispensable material is dumped in the environment, where it persists for years without degrading resulting in worldwide pollution damaging the planet [3, 4]. Plastic pollution comes in different sizes, such as nanoplastics (<1 μm), microplastics (1 μm to 5 mm) and mesoplastics (<25 mm). The larger sized plastics, however, can be reused and recycled with the help of recent technological advancements, so the greater damage lies because of the smaller microplastic contamination [5, 6]. Microplastics are defined as fragments of synthetic polymer materials that are of sizes 1 μm to 5 mm [7]. Depending on the source of their formation, microplastics can be subdivided into primary and secondary microplastics. Primary microplastics are the plastics that are

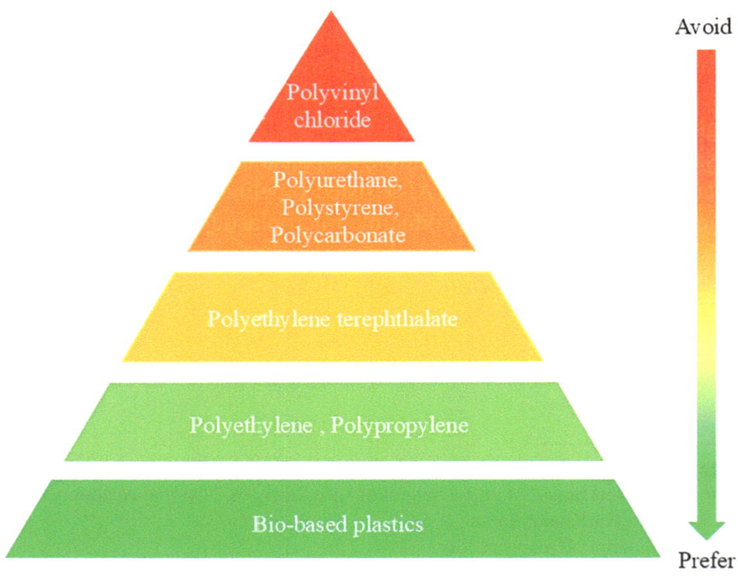

Fig. 1 The plastic spectrum pyramid ranked from the most harmful plastics to avoid to better choices for the environment. (Reproduced with permission from [1])

produced in small sizes whereas secondary microplastics are the ones that occur from chemical or biological decomposition of larger plastic fragments. Primary microplastics are generally introduced into the biosphere from discharge of households, such as PE beads present in cosmetic products like face scrubs, vehicle paints. Wearing down of tires from friction is one of the largest sources of secondary microplastics present in the ocean, apart from this, textile fibres, generally PET, PP that arise from washes are also common secondary microplastics [8, 9]. Microplastics are not only ubiquitous in both aquatic and land environments owing to their small size, they are also readily ingested by organisms without notice, affecting their survival. Ingested PE were shown to affect the spermatogenesis process in earthworms, intestinal tract of microorganisms can be blocked and presence of microplastics in soil were also shown to inactivate soil enzymes. Zebrafishes have shown differences in their behavioural patterns on exposure to PS, other microplastics have also been found in edible commercial fish and shrimp as well which ultimately are taken in my humans. It has been found that presence of microplastics of size 150 μm can cross the cell membrane in the human body and result in oxidative stress, reproductive as well as metabolic issues. Microplastic contamination has even been found in saliva, breastmilk and placenta, highlighting the seriousness of the issue [6, 8, 10, 11]. All this reinforces the need for collection of both quantitative and qualitative data about microplastics, their sources, distribution to get an idea about the correlation between the environmental damage and microplastics to understand the necessary actions to take [7, 8]. Raman spectroscopy has proven to be useful for this purpose.

In this chapter, we present the fundamentals and instrumentation of Raman spectroscopy including micro-Raman, confocal Raman mapping and surface-enhanced Raman spectroscopy, review their applications in water, sediment, soil, air and biological tissues, evaluate particle-level detection and quantitation i.e, routine detection into the lower micrometre range and the limits of nanoscale analysis, compare Raman with complementary techniques, identify practical challenges notably - fluorescence and pigment interference, effects of weathering and biofouling, limited throughput and reproducibility, and the absence of harmonised reference libraries and inter-laboratory standards, survey recent advances ranging from portable and automated high-throughput mapping, integration with spectral databases and artificial-intelligence/machine-learning classifiers, and recommend coordinated standardisation, expansion of aged/biofouled spectral libraries and validated automated, AI-assisted pipelines to enable scalable, reproducible microplastic monitoring.

2 Fundamentals of Raman Spectroscopy

Sir C.V. Raman and K.S. Krishnan recognized Raman spectroscopy as a subcategory of vibrational spectroscopy in 1928. The 1970s saw the first use of Raman spectroscopy in catalysis. The vibrational properties of molecular species could be studied due to Raman spectroscopy's broad spectrum range and high resolution

[12]. The two main categories of scattering events are elastic and inelastic. Rayleigh scattering and other forms of elastic scattering produce photons with the same wavelength as the incident light [13] whereas Raman scattering is an inelastic phenomenon in which the energy of the incident and radiated photons differs, causing the molecule to either be excited to a higher vibrational or rotational level or degraded to a lower level [14]. Raman spectroscopy's fundamental principle is the inelastic scattering of light at molecular level. When a sample is exposed to an external laser beam, such as monochromatic ultraviolet, photons will excite the molecules. These excited photons stimulate the transition of the molecule's vibrational energy levels, i.e., from the ground state to an unstable excited state, and then the excited state returns to a lower energy state through photon emission [15]. The excitation source, sample illumination/light collection system, sample holder, and detecting system make up the four elements of the Raman spectroscopy equipment. Typically, a laser beam is used to illuminate a sample. The scattered light is then collected using a lens or optical fiber and transmitted to the detector via a spectrometer or an interference filter to produce the Raman spectrum [12]. This spectrum represents scattering intensity (photons per second) on the y-axis and wavelength or Raman shift on the x-axis. Every band in a Raman spectrum represents a molecule's functional groups and/or chemical bonds [16]. The bands that make up the Raman spectrum are positioned according to the sample components' vibration frequencies. The concentration of the compound in the sample can be determined using band intensity [17].

A vast variety of polymers composed of various semisynthetic or synthetic organic components are referred to as "plastics." Each particle breaks up into ever-tinier pieces as plastic degrades, creating what are known as "microplastics" (MPs, <5 mm) [18]. Since MPs are prevalent practically everywhere and are an emerging environmental problem, there is considerable worry about the risks they bring [19]. Two types of MPs have been identified: primary and secondary. Manufactured plastics, for instance those used in cleaning and personal care products, are primary MPs. Plastic particles known as secondary MPs are the result of breakdown of macroplastics. Examples for secondary microplastics are tires, road paint, and washing synthetic clothing [20].

The primary benefit of Raman is its ability to identify particles which are smaller than 10 μm, which FTIR spectroscopy and even FTIR imaging systems often overlook [21]. One of the most sensitive instruments, Surface Enhanced Raman Spectroscopy (SERS) enables the detection of low concentration analytes. The dispersed Raman signal is amplified by the particles absorbed on the metal surface, which is the basis of SERS technology. Plasmon resonance is a phenomenon where electromagnetic waves interact with metal to produce signal amplification [17]. The benefits of SERS include its high-resolution, which facilitates simultaneous multi-component analysis; its non-invasive nature, its ease of sample handling, its speed, its capacity to identify analytes in-situ and its portability [22].

3 Application of Raman Spectroscopy in Microplastic Analysis

3.1 *Analysis of Environmental Samples Using Raman Spectroscopy*

Raman spectroscopy is applied broadly to detect microplastics in various water sources, including freshwater, groundwater, drinking water, marine, and wastewater [21]. Water does not interfere with Raman analysis, but autofluorescence from organic and inorganic impurities can superpose Raman spectra and hinder particle identification, which is generally mitigated by varying the laser excitation wavelength (e.g., from 532 nm to 785 nm) or by adding delay time to photobleach the fluorescence [23–26]. A few common case studies are as follows: first, identification of PE, PP, PVC, PA, and PS, originating from anthropogenic activities like polymer bags or fishing nets, with sizes from 1.38 to 47.8 μm in coastal surface waters of Okinawa, Japan [21]. Second, detection of the chemical composition of microplastics in the lower micrometer range, like PES and PP from plastic bottles, and PE or polyolefins from beverage cartons and glass bottles, in water samples from Germany, where recycled bottles were observed to contain more microplastics than single-use bottles [21, 25]. Third, detection of 4 types of microplastics (PE, PP, PS, PET) in groundwater and surface bodies (70 m depth borehole) of SE Attica, Greece, where PE was found mainly as dark and light blue fibers (>150 μm), PP as blue fibers (>150 μm), PET as whitish irregular particles (>20 μm), PS as highly reflective, irregular blue particles and pigments like copper phthalocyanine were also identified in blue PE and PP particles [27]. Fourth, in wastewater samples collected in Israel, Raman microspectroscopy successfully detected microplastics as small as 1 μm. These microplastics included materials such as PE, PVC, PP, PC, PTFE, PS, PUR, PA, and polyolefin elastomer found in untreated wastewater. In wastewater obtained from a car service facility, Raman spectroscopy identified polystyrene (PS) and polyacrylic particles as small as 200 nm. Similarly, PS and polymethyl methacrylate (PMMA) particles in the size range of 1.2–600 μm were detected in wastewater generated during eyeglass polishing. The technique proved effective in minimizing interference from organic matter that would otherwise obscure the spectra [21].

Sediments in marine environments are widely recognized as long-term sinks for microplastics. Raman spectroscopy is frequently employed in their characterization, with traditional preparation techniques involving acid or base digestion, oxidation, or enzymatic treatment to remove pigments and organic material that complicate spectra [24, 25]. A case study from Huiquan Bay, China, reported the detection of seven polymer types—PP, PE, PTFE, PA, PS, PET, and ABS—within the size range of 5–500 μm. The notable feature of this work was its simplified, non-digestion method, which preserved the structural features of the microplastics [25].

Soil environments, where microplastic pollution is equally persistent, can also be investigated through Raman techniques. Using flotation to isolate particles and combining Raman analysis with hyperspectral unmixing and machine learning, researchers successfully detected and visualized PE, PP, and PS, even when additives like stabilizers or pigments complicated the matrix. Another direct application of Raman, coupled with Unconstrained Least Squares (Avmax-UCLS), was done to visualize and identify PE and PS from the contaminated sand samples without complex pre-processing [28].

The ingestion and accumulation of microplastics by various organisms is a serious concern. Microplastics have been recently discovered inside the human placenta, human lung tissue, lymph nodes, and blood [21, 24, 25, 29]. Raman spectroscopy was applied to identify polypropene and PVC particles of size 1 μm in freshwater bivalves (Unio pictorum) exposed to sewage treatment plant effluents, and to observe microplastic interactions with whole jellyfish in 70 nm thick resin sections to characterize PET [21, 23].

Microplastics can also be suspended in the atmosphere and pose inhalation risk for which no standard sampling or analytical methodology is available, specifically for microplastics having a size less than 10 μm. A study involved the use of a high-volume sampler equipped with a five-stage cascade impactor to collect samples and their particle-based analysis on Teflon filters without pretreatment using Raman microspectrometry, which revealed a homogenous distribution of microplastics across different slots of the filter and a concentration of 452 ± 134 microplastics/m^3 in the first stage. Morphological classification of the detected microplastics was done into fragments and fibers with a size range of 2.8–24.8 μm and 28.6–212 μm, respectively. Chemically, the microplastics were identified as carbon black (92%) and polypropylene (8%) [29]. Raman microspectroscopy has also been used in the identification of aerosol particles collected from a laboratory room and their classification into five types, which are carbon, glassy silicates, carbonates-silicates mixed compounds, and silicates with/without fluorescence background [30].

3.2 Use of Raman for Detecting Small-Sized Particles

Food filter bags contain microplastics which can be identified using Raman imaging. The study utilized a laser confocal micro-Raman system to collect Raman spectra and images. MPs were mapped using peak area imaging based on a single characteristic peak. We can analyze the distribution of MPs within an area which is indicated by colors in a Raman spectrum. For example, in three selected regions of interest (ROIs), the numbers of PET, PP, NY6, and PE MPs identified by peak area imaging were 30, 2, 0, and 3, respectively (Fig. 2). Overall, MPs accounted for 94.4% of the total samples [31].

Raman spectroscopy was applied to quantitatively and qualitatively characterize MPs in bottled water packaged in three different materials: polyethylene terephthalate (PET), recycled PET (rPET), and glass. For the analysis, every detected particle was analyzed by Raman spectroscopy to assess its polymer identity. The highest number of particles was found for glass bottles with a minimum of 28 particles and

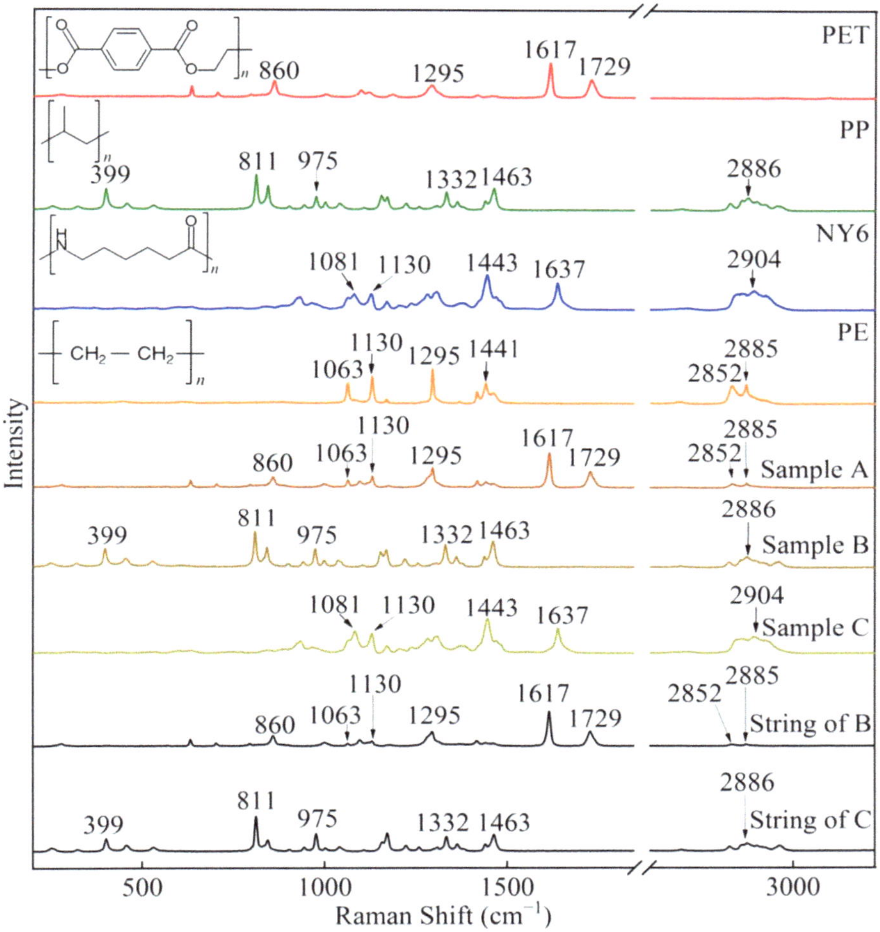

Fig. 2 Depicts the Raman spectra of the standards, filter bags, and strings. PET, PP, NY6, and PE represent standard Raman spectra in database. Sample A, sample B, sample C, string of B, and string of C represent the mean spectra. (Reused with permission from [31])

a maximum of 152 particles; for PET bottles, the range was 18–114 particles; while rPET bottles showed a minimum of 20 and a maximum of 48 particles. In terms of polymer identity, PE was the most abundant polymer overall (74.5%), comprising 26.9% of common PE and 47.6% of PE + additives, followed by PET (21.7%), PS (2.9%), and PET + PS (0.9%). Glass bottles showed the highest amount of PE particles with a total of 97.7% (38.2% of normal PE and 59.5% PE + additives), whereas very few PET (0.2%) and PET + PS (1.0%) particles were found in glass bottles. Raman spectroscopy was confirmed to be a sensible and tailored technique for MP detection [32].

Raman spectroscopy offers detailed information on a molecule's vibrational mode identifying the constituents of a compound. The sample surface is scanned using confocal Raman microscopy creating a scanning spectrum matrix by gathering the spectrum at each pixel. The spectrum intensity is mapped to produce images which identify the compound constituents. The signal-noise ratio of Raman imaging is significantly improved by averaging hundreds or thousands of Raman spectra. Principal Component Analysis (PCA) is used to study the key features of high dimensional datasets. Ideally, PCA can produce several principal components that reflect significant variances in Raman activity when used to analyze a Raman spectrum matrix. By evaluating their association to conventional plastic spectra, these primary components can be further assessed and allocated to certain polymers [33].

3.3 Qualitative and Quantitative Analysis by Raman Spectroscopy

Raman spectroscopy helps in analyzing qualitative properties such as polymer type, morphology, color, degradation, and quantitative properties such as particle size, count/abundance, concentration, and crystallinity/density.

The fingerprint of the chemical structure can be obtained by Raman spectroscopy, allowing for polymer type identification after comparing with commercial or custom-made reference libraries [34–37]. Examples of environmental samples identified are PE, PP, PS, PET, PVC, PA, PUR, PMMA, PTFE, ABS, poly(vinyl alcohol), polyester, polyacrylonitrile, and rubber [38, 39]. Non-plastic materials, such as inorganic particles (e.g., calcium carbonate, quartz), carbon, and cellulose, have also been identified [34, 36–39]. When coupled with static image analysis, the morphological characterization of microplastics based on shape descriptors such as minor and major size, particle area, diameter, perimeter, ellipse ratio, and circularity can also be done to understand the particles' fate and source. Differentiation is achieved by the random forest classifier model with a set of values derived from shape descriptors [35, 36, 39]. The color of the microplastics has also been observed, but not efficiently, since pigments and dyes can interfere with Raman analysis, leading to autofluorescence of the polymers. Some of the identified dyes include phthalocyanine blue (PB15) and titanium dioxide (TiO_2) [34, 36–39]. Insights into the degradation or ageing of microplastics can be obtained based on common indicators, which can be first, enhanced background noise and autofluorescence due to organic and inorganic matter, complex matrices, or structural disorders, second, appearance of oxygen-containing functional groups (e.g., C=O stretching at 1643 cm^{-1}) due to oxidation [34, 39].

Specialized techniques like micro-Raman detect particles as small as 1 μm, which is critical since particles keep fragmenting, resulting in an increase in particle number at smaller sizes. A study on bottled water found that 80% of these microplastics range between sizes 5–20 μm and therefore it is essential to focus on detection of these small microplastics. This detection can be enhanced by coupling with image analysis software where the minimum Feret's diameter is used for classification of particles having size as small as 1 μm [34–40]. Semi-automated and fully automated Raman micro-spectrometers can be paired with static image-analysis or particle-finder software to count microplastic particles and estimate abundance. These workflows reduce human intervention and therefore help prevent undercounting that arises from subsampling or manual sorting. For large sample sets, random window sampling (RWS) is commonly applied; RWS is tailored to smaller particles in the 1–50 μm range [34–38]. For example, ASPEx-μ-Raman analyses of North Sea surface water produced concentrations of 38–2621 particles per cubic metre for particles between 10 and 500 μm. One recurrent issue is particle clumping on filters, which can lower apparent counts; adaptive de-agglomeration methods were implemented to mitigate this bias [35, 38, 39]. Additionally, spectral indicators such as 'cryst_1' (ratio of CH_2 symmetric to asymmetric stretching) and 'cryst_2' (ratio of CH_2 bending bands) are employed to monitor density and crystallinity variations [39]. Therefore, Raman spectroscopy, when coupled with automated systems may provide a comprehensive approach to characterize microplastics by analyzing their quantitative and qualitative properties.

4 Raman Imaging and Mapping Techniques

The concepts of Raman spectroscopy and confocal microscopy give rise to a useful technique called Confocal Raman Microscopy. This technique can determine the specimen's composition as well as internal structure details. The advantage of this technique is that it lets us visualize the thin optical sections without damaging the internal structure [41]. The probe head in confocal Raman microscopy focuses the laser light on the sample via the microscope objective and the charge-coupled device's (detector) camera records the signal to produce the spectrum [17]. Micro-Raman spectroscopy is composed of an optical microscope with a Raman spectrometer. The laser beam may be concentrated on an area of about 1 μm by employing a high magnification objective [42]. Acquiring spectra at the tiny scale can improve the measurement's resolution and effectiveness, requiring fewer samples and possibly amplifying some effects over highly localized areas [43].

Raman imaging, also known as mapping, is the process of creating images with both spectral and spatial data using micro-Raman spectroscopy. It involves acquiring spectra as a spectrum matrix at a scanning location array. Raman imaging offers visual information on the size, shape, and distribution of plastic objects [44]. The laser scans the sample surface to map the image, and the Raman signal is recorded at each pixel by CCD. Different polymers emit varying intensities of Raman spectra. Using various colours, the intensities at those various peaks were mapped as images. To map the intensity signal as various image colours, multiple images were concurrently mapped from the same spectrum at various peaks. ImageJ software allows for the merging of two or more images that correspond to distinct characteristic peaks, for example, by logic-OR [45]. The size of the particle is determined by elastic scattering, whereas the material type is determined by inelastically scattered light, or Raman scattering [46]. Thus, Raman imaging enables chemical identification and spatial characterization of materials at a micro-level.

5 Other Available Techniques

Raman spectroscopy has proved to be a tool of paramount importance for analysis because it is nondestructive, thus allowing repetitive analysis of the same sample if necessary, rapidly assisting in fast, label-free feedback. Beyond this, it is also valued for its ability to extract the chemical, morphological, and crystalline properties of polymers [47].

There are various other techniques in use currently as well to analyse the physicochemical properties of microplastics acquired from the surroundings such as scanning electron microscopy (SEM), Fourier-transform infrared spectroscopy (FTIR), Quantitative Nuclear Magnetic Resonance (qNMR) Spectroscopy, Thermal Desorption Gas Chromatography-Mass Spectrometry (TDS-GC-/MS), Pyrolysis Gas Chromatography-Mass Spectroscopy (PyGC/MS), Thermogravimetric Analysis (TGA), Gel Permeation Chromatography (GPC). SEM is a method that provides the surface details of microplastics upto a size of 0.5 nm and is used for analysing weathering of microplastics in nature. However, they are limited in separation of plastics from other types of polymers and have other disadvantages such as its high cost. μFTIR is a method suited for the analysis of plastics above a size of 10 micrometres that uses infrared light to gather information about the functional groups from the vibration of molecules. Despite sharing a lot of similar features like their nondestructive nature and rapid result acquisition, Raman spectroscopy performs better than FTIR, as it is suitable for smaller sample sizes too and is compatible with aqueous environments and does not share the limitation of FTIR where sample sizes cannot be of thickness less than 150 nm [2, 48]. NMR is a phenomenon that occurs as a result of exposure to electromagnetic radiation on specific nuclei placed in a magnetic field. qNMR spectroscopy is a simple, precise and size-independent analysis technique for understanding the concentration, composition, structure of microplastic in a sample [49]. However, information pertaining to the size and shape of the sample is lost as it has to be dissolved in a solvent and pose difficulty for high-resistance polymers like PP and PE which only dissolve at high temperatures [2]. GPC can be used to track the molecular weight distribution changes with degradation in plastics [47]. Traditional spectroscopy techniques can help in identification and counting whereas thermal methods like pyrolysis can extract mass based quantitative information. These thermal techniques such as PyGC/MS measure the change in the properties with change in temperature and thus provide information on microplastic degradation in the environment [7]. PyrGC/MS works by degrading the sample thermally without air and then using mass spectrometry for analysis of the gaseous reaction products, and has been used to determine PS, PP, PET microplastics from riverbeds. They offer the advantage of no size limitations, a benefit that is countered by the disadvantage of destruction of sample [10]. The disadvantages that come with every technique may be overcome through merging them, for example, molecular spectra can identify plastics but lack provision of morphological information for the quantification of them, and in such cases combining techniques such as SEM and Raman aid in their advancements [5].

6 Challenges and Limitations in Raman Spectroscopy for Microplastic Research

6.1 Fluorescence Interference and Mitigation Strategies

Raman spectroscopy's challenge in microplastic analysis is interference by fluorescence, particularly in environmental matrices with pigments, bio-organic compounds, or dyes. Fluorescence interferes with the weak Raman signal, complicating polymer identification. Several techniques have been developed to avoid this issue, including applying longer excitation wavelengths (e.g., lasers at 785 nm or 1064 nm), photobleaching, time-gated Raman, and baseline correction algorithms. Fluorescence is still a significant limitation even after these advancements, particularly in coloured or weathered plastics [24]. Pigment or bio-organic fluorescence still blanks weak Raman signals. Even SERS and sophisticated AI models find it difficult to eliminate this interference.SpecATNet assisted in cleaning noisy spectra but was still interfered with in specific samples. Such interference constrains detection precision, particularly in complicated matrices. Fluorescence reduction is still a vital area for future innovation in Raman.

6.2 Limitations in Weathered or Bio-fouled Particle Analysis

Environmental aging mechanisms such as photodegradation, oxidation, and biofouling alter microplastics' spectral properties and surface chemistry. These mechanisms cause spectral identification to be more challenging since degraded polymers can have expanded or shifted Raman peaks or possess surface layers that mask the polymer's signal. Biofilms can introduce additional spectral noise or fluorescence, which makes plastic discrimination harder. Effective pre-cleaning protocols and deeper spectral libraries are constantly being researched to counteract this issue [36]. Primpke et al.'s platform is effective for fresh microplastics but not yet for weathered microplastics. Real-world plastics are oxidized, UV-degraded, and fouled by microorganisms. These modifications alter spectra, rendering classification less straightforward. Most automated systems have not been tested with aged environmental particles. There is a need for biofouling-inclusive spectral libraries [50].

6.3 Limitations in Measuring Nanoplastics

Raman spectroscopy has several limitations in detecting and quantifying nano plastics (<1 μm) due to the light diffraction limit and reduced scattering signal from such tiny particles. Further, the reduced signal-to-noise ratio and increased photodegradation hazard at the nanoscale size prevent practical analysis. While tip-enhanced Raman spectroscopy (TERS) and SERS have potential, both are still limited to the laboratory. It is unsuitable for routine environmental monitoring [51]. Though nano plastics can be detected, precise quantitation remains difficult. Phenomena such as particle agglomeration and heterogeneous substrate enhancement lead to signal fluctuation. Laboratory calibration curves cannot represent actual environmental situations. Therefore, field sample data are frequently semi-quantitative. A uniform protocol for nanoscale quantitation is an urgent need [52].

6.4 Sample Throughput and Analysis Time

Raman spectroscopy, in point-by-point mapping mode, is by necessity time-consuming when scanning large environmental samples. The need for accurate focus, slow scan rates, and repeated spectral acquisition limits throughput, especially when thousands of particles must be scanned and identified per sample. While automated systems offer some relief, full automation is still in the pipeline and requires considerable data post-processing [53]. Hyperspectral Raman imaging gives precise particle information but is time-consuming. It takes skilled interpretation and high computing capacity to process large datasets. AI assists, but complete automation of spectral authentication has still not been achieved. This slows down analysis and restricts the volume of samples in environmental research. Increased throughput and speedier pipelines are needed for scalability [48].

6.5 Reproducibility and Standardization Issues Across Studies

Standardized methods are a key barrier to Consistency and Standardization in Raman-based microplastic research. Sample preparation, instrumental conditions, spectral libraries, and particle classification cut-offs variations lead to study variations. Lack of quality control parameters that can be magnifies reproducibility. Initiatives like developing open-source databases and inter-laboratory validation studies are necessary towards standardization [54]. Different labs use different Raman instruments, substrates, and identification methods. There is no inter-study comparison to any standard reference materials or protocols. This renders its low reproducibility and polymer identification non-consistent. Cross-lab calibration and open-access spectral libraries are imperative. There is one way to do this: harmonization. Only then can results be globally reliable.

7 Recent Advances and Innovations in Raman Spectroscopy for Microplastic Research

7.1 Portable Raman Devices for Field Studies

The size reduction of Raman spectrometers has significantly transformed the in-situ detection of microplastics. Convenient and portable Raman instruments allow direct, in-field analysis of environmental samples, e.g., water bodies, beach sediments, and soil, with minimal sample preparation and laboratory shipping requirements. The instruments often employ laser sources between 785–1064 nm to minimize fluorescence interference, A common challenge in environmental samples. Despite spectral resolution and sensitivity limitations, continuous improvements are closing the performance gap between handheld and benchtop instruments [55]. Portable SERS devices allow in-field analysis of nano plastics with minimal preparation. Wang et al. showed the detection of polystyrene as low as 1 nm and 14 μg/mL with a handheld Raman spectrometer. Their substrate exhibited good reproducibility in intricate samples such as lake water. Such technologies reduce reliance on laboratory facilities to analyze plastics. Such technologies take Raman spectroscopy closer to field-friendly environmental analysis [52] in situ.

7.2 Machine Learning and AI for Spectral Classification

Machine learning (ML) and artificial intelligence (AI) techniques are extensively employed to improve the Raman spectrum classification, particularly in complex samples with overlapping bands. Techniques such as random forests, support vector machines, and neural networks are employed to accurately classify polymer types with efficiency and speed, even from Irregular data. These approaches undertake automation and increase the reproducibility of spectral interpretation, making this a component of more robust microplastic monitoring systems [56]. Deep learning models like SpecATNet improve the classification of superimposed microplastic spectra. They are capable of detecting up to six different types of polymers in combined environmental samples. Deep learning models like SpecATNet improve the classification of superimposed microplastic spectra. They are able to accurately identify as many as six forms of polymers in blended environmental samples. The procedure eliminates the pre-treatment step and reduces misclassification. Such AI devices undertake spectral analysis with great certainty and minimal exertion. This increases dependability and speeds up Raman-based plastic identification [57].

7.3 *Development of Automated Raman Systems*

Automation of Raman spectroscopy has made it possible to create high-throughput analysis systems that can scan entire filter membranes or microscope slides of microplastics. These systems integrate Raman mapping, image processing, and autofocus algorithms to identify and detect thousands of particles per sample. Automation minimizes la significantly minimizes labor and analysis time, making large-scale environmental monitoring possible bor and analysis timeem that can automatically screen thousands of microplastics per sample. This system reduces labor and increases the reproducibility of environmental surveys. Large-scale scanning of the filter and slide is allowed with minimal user bias. Automation forms the core for the wider application of Raman in ecological screening [50].

7.4 *Surface-Enhanced Raman Spectroscopy (SERS) for Microplastic Detection*

SERS achieves ultra-sensitive detection using amplification of Raman signals with metallic nanostructures like silver or gold nanoparticles. SERS has been studied to detect nano plastics in environmental matrices below the sensitivity limits of conventional Raman systems. SERS holds immense potential in the development of standard fieldwork for tracking ultra-tiny plastic particles and additives [58]. SERS employs plasmonic and chemical amplification to enhance Raman signals. The Co_3O_4/Co_3S_4/AgNPs sensor enables ultra-sensitive detection of nano plastics in water. Detection was stable to 1 nm particle sizes, even in unfiltered lake water. This is an excellent enhancement of the sensitivity limit of portable Raman methods. SERS has potential for nano plastic monitoring in actual environmental conditions [52].

7.5 *Integration with Spectroscopy Databases and Remote Sensing*

Raman spectroscopy is increasingly associated with polymer reference libraries and worldwide spectroscopy databases to facilitate easy microplastic identification. Cloud-based integration enables researchers to remotely and collaboratively match spectra, enhancing standardization and minimizing false positives. Additionally, new research is investigating how Raman-compatible sensors might augment remote sensing platforms for large-scale environmental surveillance, such as airborne and satellite-linked platforms [50]. Li et al. used hyperspectral

Raman imaging coupled with MCR-ALS on intricate matrices such as fish feces. Without purification or separation, they could identify several polymers (PE, PS, PP, PET). The method offered spatial and spectral data in real-time for accurate visualization. This calls for the integration of Raman devices with environmental mapping networks. It enhances polymer identification and traceability of locations [59].

8 Conclusion

In the current world, excess use of plastics is a major threat to our environment. Their breakdown into microplastics and nanoplastics is riskier as it penetrates our ecological system and affects human health. Raman spectroscopy was a sensitive and non-destructive method used to identify microplastics which could overcome the limitations of traditional techniques. The fact that microplastics penetrate our ecosystems was proven to be true when Raman spectroscopy identified them in soil, water and air samples. The non-invasive nature of this technique helped us to identify microplastics in human tissues such as the circulatory system and placenta.

Excessive use of plastics has been posing a threat to our environment for ages and their breakdown into micro and nanoplastics is even more risky as it can infiltrate the human and ecological ecosystems, hence becoming a major global issue. Raman spectroscopy has successfully overcome the drawbacks of traditional approaches to identify microplastics in a variety of situations being a sensitive and non-destructive method. It was used to identify microplastics in soil, water and air thereby validating their infiltrative nature. In medical and forensic settings, Raman's label-free, non-invasive detection provides a window into the existence of microplastics in human tissues, such as the circulatory system and placenta.

Limitations such as background fluorescence, spectrum overlap hold back the broad use of this technique especially while dealing with environmental samples. The principles of SERS and confocal Raman imaging increased sensitivity and spatial resolution thereby helping in detecting smaller and rare particles. AI-ML tools are proving to be useful in analysing the Raman spectra of various plastics accurately. This approach reduces human bias, simplifies the task of spectral interpretation and increases the prediction accuracy. Raman spectroscopy is better compared to FTIR, NMR and pyrolysis-GC/MS in terms of spatial resolution and particle-level mapping. There is still a need for multimodal approaches as no single method is sufficient on its own. Standardizing Raman-based protocols, incorporating AI tools, and extending reference libraries are essential for the future of microplastic research to guarantee that Raman spectroscopy remains a fundamental method for addressing the worldwide microplastic problem.

References

1. Arikan EB, Ozsoy HD (2015) A review: investigation of bioplastics. JCEA 9. https://doi.org/10.17265/1934-7359/2015.02.007
2. Schmidt J, Haave M, Underhaug J, Wang W (2024) Unlocking the potential of NMR spectroscopy for precise and efficient quantification of microplastics. Microplast Nanoplast 4:17. https://doi.org/10.1186/s43591-024-00095-5
3. Abina A, Korošec T, Puc U, Zidanšek A (2023) Review of bioplastics characterisation by terahertz techniques in the view of ensuring a circular economy. Photonics 10:883. https://doi.org/10.3390/photonics10080883
4. Acharjee SA, Bharali P, Gogoi B et al (2023) PHA-based bioplastic: a potential alternative to address microplastic pollution. Water Air Soil Pollut 234:21. https://doi.org/10.1007/s11270-022-06029-2
5. Fang C, Zhang X, Zhang Z, Naidu R (2024) Characterising fragmentation of compostable bioplastic: releasing microplastics or small bioplastic debris. Environ Sci Eur 36:121. https://doi.org/10.1186/s12302-024-00946-1
6. Kutralam-Muniasamy G, Shruti VC, Pérez-Guevara F, Roy PD (2023) Microplastic diagnostics in humans: "the 3Ps" Progress, problems, and prospects. Sci Total Environ 856:159164. https://doi.org/10.1016/j.scitotenv.2022.159164
7. La Nasa J, Biale G, Fabbri D, Modugno F (2020) A review on challenges and developments of analytical pyrolysis and other thermoanalytical techniques for the quali-quantitative determination of microplastics. J Anal Appl Pyrolysis 149:104841. https://doi.org/10.1016/j.jaap.2020.104841
8. Peez N, Janiska M-C, Imhof W (2019) The first application of quantitative 1H NMR spectroscopy as a simple and fast method of identification and quantification of microplastic particles (PE, PET, and PS). Anal Bioanal Chem 411:823–833. https://doi.org/10.1007/s00216-018-1510-z
9. Wang T, Li B, Zou X et al (2019) Emission of primary microplastics in mainland China: invisible but not negligible. Water Res 162:214–224. https://doi.org/10.1016/j.watres.2019.06.042
10. Fojt J, Románeková I, Procházková P et al (2022) A simple method for quantification of Polyhydroxybutyrate and Polylactic acid micro-bioplastics in soils by evolved gas analysis. Molecules 27:1898. https://doi.org/10.3390/molecules27061898
11. Nanda S, Patra BR, Patel R et al (2022) Innovations in applications and prospects of bioplastics and biopolymers: a review. Environ Chem Lett 20:379–395. https://doi.org/10.1007/s10311-021-01334-4
12. Moon J, Li M, Ramirez-Cuesta AJ, Wu Z (2023) Raman Spectroscopy. In: Wachs IE, Bañares MA (eds) Springer handbook of advanced catalyst characterization. Springer International Publishing, Cham, pp 75–110
13. Boateng D (2025) Advances in deep learning-based applications for Raman spectroscopy analysis: a mini-review of the progress and challenges. Microchem J 209:112692. https://doi.org/10.1016/j.microc.2025.112692
14. Roy D, Kanojia S, Mukhopadhyay K, Eswara Prasad N (2021) Analysis of carbon-based nanomaterials using Raman spectroscopy: principles and case studies. Bull Mater Sci 44. https://doi.org/10.1007/s12034-020-02327-9
15. Qu C, Li Y, Du S et al (2022) Raman spectroscopy for rapid fingerprint analysis of meat quality and security: principles, progress and prospects. Food Res Int 161:111805. https://doi.org/10.1016/j.foodres.2022.111805
16. Rodriguez-Saona L, Ayvaz H (2024) Infrared and Raman spectroscopy. In: Ismail BP, Nielsen SS (eds) Nielsen's food analysis. Springer International Publishing, Cham, pp 95–116
17. Saletnik A, Saletnik B, Puchalski C (2021) Overview of popular techniques of Raman spectroscopy and their potential in the study of plant tissues. Molecules 26:1537. https://doi.org/10.3390/molecules26061537

18. Nava V, Frezzotti ML, Leoni B (2021) Raman spectroscopy for the analysis of microplastics in aquatic systems. Appl Spectrosc 75:1341–1357. https://doi.org/10.1177/00037028211043119
19. Leung MM-L, Ho Y-W, Lee C-H et al (2021) Improved Raman spectroscopy-based approach to assess microplastics in seafood. Environ Pollut 289:117648. https://doi.org/10.1016/j.envpol.2021.117648
20. Terry LR, Sanders S, Potoff RH et al (2022) Applications of surface-enhanced Raman spectroscopy in environmental detection. Anal Sci Adv 3:113–145. https://doi.org/10.1002/ansa.202200003
21. Chakraborty I, Banik S, Biswas R et al (2023) Raman spectroscopy for microplastic detection in water sources: a systematic review. Int J Environ Sci Technol 20:10435–10448. https://doi.org/10.1007/s13762-022-04505-0
22. Dey T (2023) Microplastic pollutant detection by surface enhanced Raman spectroscopy (SERS): a mini-review. Nanotechnol Environ Eng 8:41–48. https://doi.org/10.1007/s41204-022-00223-7
23. Caldwell J, Rodriguez-Lorenzo L, Espiña B et al (2024) Detection of submicron- and nanoplastics spiked in environmental fresh- and saltwater with Raman spectroscopy. Mar Pollut Bull 203:116468. https://doi.org/10.1016/j.marpolbul.2024.116468
24. Käppler A, Fischer D, Oberbeckmann S et al (2016) Analysis of environmental microplastics by vibrational microspectroscopy: FTIR, Raman or both? Anal Bioanal Chem 408:8377–8391. https://doi.org/10.1007/s00216-016-9956-3
25. Liu J, Zhang X, Du Z et al (2020) Application of confocal laser Raman spectroscopy on marine sediment microplastics. J Ocean Limnol 38:1502–1516. https://doi.org/10.1007/s00343-020-0129-z
26. Munno K, De Frond H, O'Donnell B, Rochman CM (2020) Increasing the accessibility for characterizing microplastics: introducing new application-based and spectral libraries of plastic particles (SLoPP and SLoPP-E). Anal Chem 92:2443–2451. https://doi.org/10.1021/acs.analchem.9b03626
27. Perraki M, Skliros V, Mecaj P et al (2024) Identification of microplastics using μ-Raman spectroscopy in surface and groundwater bodies of SE Attica. Greece Water 16:843. https://doi.org/10.3390/w16060843
28. Li F, Liu D, Guo X et al (2024) Identification and visualization of environmental microplastics by Raman imaging based on hyperspectral unmixing coupled machine learning. J Hazard Mater 465:133336. https://doi.org/10.1016/j.jhazmat.2023.133336
29. Akbari Dana P, Gaga EO, Gedik K (2024) Analytical challenges and strategies for particle-based analysis of airborne micro(nano)plastics in size-fractionated samples using microscopy, SEM/EDX, and Raman spectroscopy. Anal Chem 96:20622–20634. https://doi.org/10.1021/acs.analchem.4c05335
30. Kong L, Zhang P, Setlow P, Li Y (2011) Multifocus confocal Raman microspectroscopy for rapid single-particle analysis. J Biomed Opt 16:120503. https://doi.org/10.1117/1.3662456
31. Mei T, Wang J, Xiao X et al (2022) Identification and evaluation of microplastics from tea filter bags based on Raman imaging. Foods 11:2871. https://doi.org/10.3390/foods11182871
32. Gambino I, Malitesta C, Bagordo F et al (2023) Characterization of microplastics in water bottled in different packaging by Raman spectroscopy. Environ Sci: Water Res Technol 9:3391–3397. https://doi.org/10.1039/D3EW00197K
33. Luo Y, Naidu R, Fang C (2023) Raman imaging to capture microplastics and nanoplastics carried by smartphones. Sci Total Environ 864:160959. https://doi.org/10.1016/j.scitotenv.2022.160959
34. Araujo CF, Nolasco MM, Ribeiro AMP, Ribeiro-Claro PJA (2018) Identification of microplastics using Raman spectroscopy: latest developments and future prospects. Water Res 142:426–440. https://doi.org/10.1016/j.watres.2018.05.060
35. Jacob O, Ramírez-Piñero A, Elsner M, Ivleva NP (2023) TUM-ParticleTyper 2: automated quantitative analysis of (microplastic) particles and fibers down to 1 m by Raman microspectroscopy. Anal Bioanal Chem 415:2947–2961. https://doi.org/10.1007/s00216-023-04712-9

36. Frère L et al (2016) A semi-automated Raman micro-spectroscopy method for morphological and chemical characterizations of microplastic litter. Mar Pollut Bull 113:461–468. https://doi.org/10.1016/j.marpolbul.2016.10.051
37. Jin N, Song Y, Ma R et al (2022) Characterization and identification of microplastics using Raman spectroscopy coupled with multivariate analysis. Anal Chim Acta 1197:339519. https://doi.org/10.1016/j.aca.2022.339519
38. Cabernard L, Roscher L, Lorenz C et al (2018) Comparison of Raman and Fourier transform infrared spectroscopy for the quantification of microplastics in the aquatic environment. Environ Sci Technol 52:13279–13288. https://doi.org/10.1021/acs.est.8b03438
39. Rytelewska S, Dąbrowska A (2022) The Raman spectroscopy approach to different freshwater microplastics and quantitative characterization of polyethylene aged in the environment. Microplastics 1:263–281. https://doi.org/10.3390/microplastics1020019
40. Sharma P, Sharma P, Abhishek K (2024) Sampling, separation, and characterization methodology for quantification of microplastic from the environment. J Hazard Mater Adv 14:100416. https://doi.org/10.1016/j.hazadv.2024.100416
41. Gómez-Mascaraque LG, Tran C, O'Callaghan T, Hogan SA (2021) Use of confocal Raman imaging to understand the microstructure of anhydrous milk fat-based oleogels. Food Struct 30:100228. https://doi.org/10.1016/j.foostr.2021.100228
42. Orlando A, Franceschini F, Muscas C et al (2021) A comprehensive review on Raman spectroscopy applications. Chem 9:262. https://doi.org/10.3390/chemosensors9090262
43. Aloia R, Annunziata L, Di Giacinto F et al (2025) Application of Raman spectroscopy for the analysis of microplastics in food and beverages: a comprehensive review. Appl Spectrosc Rev. https://doi.org/10.1080/05704928.2025.2512771
44. Luo Y, Gibson CT, Chuah C et al (2022) Raman imaging for the identification of Teflon microplastics and nanoplastics released from non-stick cookware. Sci Total Environ 851:158293. https://doi.org/10.1016/j.scitotenv.2022.158293
45. Luo Y, Gibson CT, Tang Y et al (2022) Characterising microplastics in shower wastewater with Raman imaging. Sci Total Environ 811:152409. https://doi.org/10.1016/j.scitotenv.2021.152409
46. Glöckler F, Foschum F, Kienle A (2023) Continuous sizing and identification of microplastics in water. Sensors 23:781. https://doi.org/10.3390/s23020781
47. He W, Liu R, Fei F et al (2024) In situ real-time pathway to study the polyethylene long-term degradation process by a marine fungus through confocal Raman quantitative imaging. Sci Total Environ 939:173582. https://doi.org/10.1016/j.scitotenv.2024.173582
48. Li Y, Lin X, Wang J et al (2023) Quantification of nanoplastics uptake and transport in lettuce by pyrolysis gas chromatography-mass spectrometry. Talanta 265:124837. https://doi.org/10.1016/j.talanta.2023.124837
49. Papini G, Petrella G, Cicero DO et al (2024) Identification and quantification of polystyrene microplastics in marine sediments facing a river mouth through NMR spectroscopy. Mar Pollut Bull 198:115784. https://doi.org/10.1016/j.marpolbul.2023.115784
50. Primpke S, Wirth M, Lorenz C, Gerdts G (2018) Reference database design for the automated analysis of microplastic samples based on Fourier transform infrared (FTIR) spectroscopy. Anal Bioanal Chem 410:5131–5141. https://doi.org/10.1007/s00216-018-1156-x
51. Gigault J, Halle AT, Baudrimont M et al (2018) Current opinion: what is a nanoplastic? Environ Pollut 235:1030–1034. https://doi.org/10.1016/j.envpol.2018.01.024
52. Wang C, Shi X, Bao Z et al (2023) Semiconductor heterojunction-AgNPs mediated surface-enhanced Raman spectroscopy (SERS) sensor for portable miniaturized detection platform. Chem 11:490. https://doi.org/10.3390/chemosensors11090490
53. Anger PM, Von Der Esch E, Baumann T et al (2018) Raman microspectroscopy as a tool for microplastic particle analysis. TrAC Trends Anal Chem 109:214–226. https://doi.org/10.1016/j.trac.2018.10.010
54. Win, Cowger Andy M., Booth Bonnie M., Hamilton Clara, Thaysen Sebastian, Primpke Keenan, Munno Amy L., Lusher Alexandre, Dehaut Vitor P., Vaz Max, Liboiron Lisa I.,

Devriese Ludovic, Hermabessiere Chelsea, Rochman Samantha N., Athey Jennifer M., Lynch Hannah, De Frond Andrew, Gray Oliver A.H., Jones Susanne, Brander Clare, Steele Shelly, Moore Alterra, Sanchez Holly, Nel (2020) Reporting Guidelines to Increase the Reproducibility and Comparability of Research on Microplastics Applied Spectroscopy 74(9) 1066–1077. https://doi.org/10.1177/0003702820930292

55. Das RS, Agrawal YK (2011) Raman spectroscopy: recent advancements, techniques and applications. Vib Spectrosc 57:163–176. https://doi.org/10.1016/j.vibspec.2011.08.003
56. Käppler A, Fischer M, Scholz-Böttcher BM et al (2018) Comparison of μ-ATR-FTIR spectroscopy and py-GCMS as identification tools for microplastic particles and fibers isolated from river sediments. Anal Bioanal Chem 410:5313–5327. https://doi.org/10.1007/s00216-018-1185-5
57. Qiang, Chen Jiamiao, Wang Fuqi, Yao Wei, Zhang Xiaohua, Qi Xia, Gao Yan, Liu Jiamin, Wang Mingqiang, Zou Pei, Liang (2023) A review of recent progress in the application of Raman spectroscopy and SERS detection of microplastics and derivatives Microchimica Acta 190(12). https://doi.org/10.1007/s00604-023-06044-y
58. Xu J-L, Thomas KV, Luo Z, Gowen AA (2019) FTIR and Raman imaging for microplastics analysis: state of the art, challenges and prospects. TrAC Trends Anal Chem 119:115629. https://doi.org/10.1016/j.trac.2019.115629
59. Li Y, Cheng D, Tian J et al (2025) Unlocking high lithium-ion transport in solid polymer electrolytes with dynamic hydrogen-bonded zinc-adeninate framework. Cell Rep Phys Sci 6:102624. https://doi.org/10.1016/j.xcrp.2025.102624

Microplastics and their Detection Via Fourier Transform Infrared Spectroscopy

Gouri Nair, Snigdha Adhikary, Rajib Biswas, Rajesh Kumar, and Giuseppe Sancataldo

Abstract Microplastics, typically smaller than 5 mm (and in some cases less than 1 μm), are now widespread environmental pollutants that originate from both primary sources, such as personal care products and industrial abrasives, and secondary sources formed through the degradation of larger plastic materials. This chapter examines microplastic detection methods, environmental distributions, and human health implications. Fourier transform infrared (FTIR) spectroscopy serves as the primary analytical tool for microplastic identification, with ATR-FTIR, micro-FTIR (μFTIR) imaging, transmission, and reflection modes enabling the detection of particles across various matrices. Microplastics contaminate all environmental compartments; marine systems accumulate particles in subtropical gyres; freshwater systems show variable concentrations linked to human activities; drinking water contains detectable levels in both tap and bottled sources; and agricultural soils accumulate particles through sewage sludge and plastic mulch. Microplastics have also been found in blood, placental tissue, and feces, and humans are exposed through ingestion, inhalation, and skin contact. An estimated 39,000–52,000 particles are consumed by humans annually, which raises questions regarding chemical contamination, cellular toxicity, and gastrointestinal inflammation. FTIR-based detection methods are essential for monitoring contamination patterns and human exposure, highlighting the urgent need for regulatory frameworks and mitigation strategies to address this growing environmental and public health challenge.

G. Nair · S. Adhikary
Manipal School of Life Sciences, Manipal Academy of Higher Education, Manipal, Karnataka, India

R. Biswas
Department of Physics, Tezpur University, Tezpur, Assam, India

R. Kumar
Department of Physics, Indian Institute of Technology, Rooper, India

G. Sancataldo (✉)
Department of Physics and Chemistry, University of Palermo, Palermo, Italy
e-mail: giuseppe.sancataldo@unipa.it

N. Mazumder et al. (eds.), *Vibrational Spectroscopy for Biopolymer Characterization*, Biological and Medical Physics, Biomedical Engineering,
https://doi.org/10.1007/978-3-032-15306-7_9

Keywords Microplastic · FTIR spectroscopy · Spectroscopy · Detection

1 Introduction

One of the most prevalent types of environmental contamination in the modern era is microplastics (MP). The term "microplastics" was first used in 2004, which laid the groundwork for the quickly growing field of environmental research. The size classification of microplastics encompasses a broad spectrum of these particles, ranging from 5 mm to 1 µm, bridging the gap between visible debris and molecular-scale contaminants. This range of sizes is especially important because it includes particles that are both large enough to endure in environmental systems for extended periods of time and small enough to pass through biological barriers and have become a serious threat to both human health and the integrity of ecosystems [1, 2]. Microplastics originate from both primary and secondary sources (Fig. 1). The primary sources of microplastics are intentionally manufactured particles found in personal care products, cosmetics, and industrial abrasives, whereas secondary microplastics result from the degradation of larger plastic items through weathering, mechanical abrasion, and photodegradation processes [3]. Tire wear particles represent one of the major sources contributing significantly to microplastic pollution in the environment, with substantial quantities released globally each year through road transportation [4]. Synthetic textiles released during washing cycles have long been a significant contributor, along with the breakdown of marine plastic debris and paint particles [5]. The prevalence of microplastics in a variety of environments has been well documented; ocean systems serve as major repositories for microplastic accumulation, with concentrations varying widely across different marine regions [6]. Freshwater systems exhibit widespread contamination, with relatively high abundances found in areas with intense human economic activities [7]. Terrestrial environments are also increasingly recognized as significant reservoirs, with agricultural soils receiving inputs through sewage sludge and atmospheric deposition [1, 8]. The transport of microplastics through the atmosphere enables long-range dispersal, contributing to the contamination of remote environments [9].

To analyze the presence and types of microplastics, Fourier Transform Infrared Spectroscopy (FTIR) is considered a gold standard, as it allows the detection of most particles larger than about 20 µm due to its optical resolution. Following proper sample preparation processes, the images acquired from FTIR analysis might be utilized to quantify the microplastic particles present in the sample [10]. Furthermore, the FTIR approach is frequently employed in the construction of optical sensors that can continually detect microplastics. Moreover, compared with alternative methods, the micro-FTIR technique is more reliable, as it typically preserves the integrity of the sample and is recognized for its efficiency in the characterization of polymers [11]. FTIR spectroscopy measures the infrared (IR) radiation absorbed by the MP sample, allowing for the analysis of its molecular composition. An infrared spectrum is a sample's fingerprint (MP), with absorption peaks

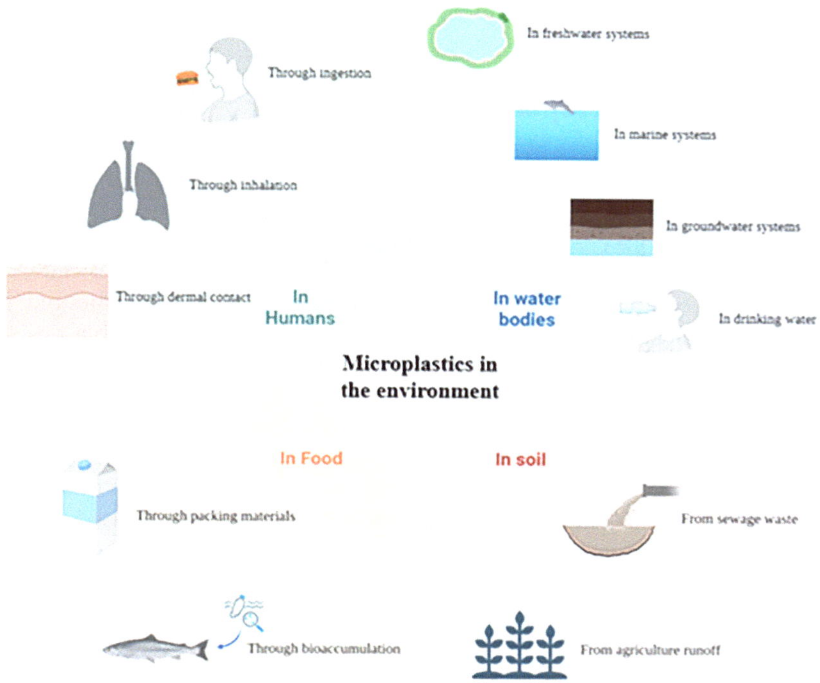

Fig. 1 Microplastics in the environment: schematic representation of microplastic sources and ingestion pathways

representing the frequencies of vibration between the bonds of the atoms that make up the substance. Thus, FTIR can be used to determine the chemical structure of a polymer molecule [12]. The IR region of the electromagnetic spectrum is divided into three regions: the higher-energy near-infrared (NIR) region, with wavenumbers of 14,000–4000 cm^{-1} (0.78–2.5 μm wavelength) range, which is sensitive to vibration combinations and overtones; the mid-infrared (MIR) region, with wavenumbers of 4000–400 cm^{-1} (2.5–30 μm wavelength) for fundamental vibration studies; and the far-infrared (FIR) region, with wavenumbers of 400–10 cm^{-1} (30–1000 μm wavelength). MIR is commonly used in the characterization of MPs among the three IR spectral areas. This technique can be used to analyze gases, solids and liquids [11]. Molecules are continually moving because the atoms vibrate, and these vibrations are quantized. When the energy of infrared radiation matches that required to induce a molecular vibration, the molecule absorbs it. As each molecule has its own infrared spectrum, this method is very useful for distinguishing various materials in samples. Typical infrared spectra feature identifiable peaks that correspond to the molecules in the sample's vibrational modes [13].

2 Detection of Microplastics Via FTIR

FTIR analysis is a popular method for detecting microplastics since it is reliable, simple, and noninvasive. FTIR spectroscopy is an effective analytical approach for determining the unique chemical bonding of particles. This approach generates distinct band patterns for diverse plastic types, utilizing specific infrared spectra, and thus makes it very efficient [14]. The quick frequency scanning capabilities of FTIR spectrometers can also be used to generate high-quality spectra. Microplastics may be identified by infrared spectroscopy because of the unique infrared spectra of several polymers commonly used in plastic production, such as polyvinyl chloride (PVC), polyester (PET), polyethylene (PE) and polypropylene (PP) [15].

Environmental samples must be adequately processed to remove microplastics prior to testing. Microplastics may be collected from environmental materials via a variety of physical and chemical processes. Several commonly used physical approaches for microplastic segregation include density separation, electrostatic separation, sieve or mesh filtering, and optical sorting. Common chemical methods for separation include digestion, density gradient separation, flotation, oxidative treatments, selective solvents, and surface modification [16]. Microplastics are typically placed on a suitable substrate, such as a potassium bromide (KBr) pellet, to facilitate infrared analysis after separation. In FTIR analysis, an infrared beam is commonly utilized to pass through a sample and record its absorbance spectrum. The resulting spectrum can subsequently be compared to reference spectra of known materials to determine the microplastic constituents [17]. When subjected to infrared light with a wavenumber between 400 and 4000 cm^{-1} for mid-IR, the molecular composition of microplastics influences whether they absorb and measure in transmission or reflection mode [16]. Polar carbonyl groups can be observed because IR absorption alters the chemical bonds in the sample's dipole. FTIR can more correctly identify microplastic polymer types and offer information on their physiochemical weathering by evaluating the degree of oxidation. The availability of a reliable polymer spectrum database facilitates the identification of specific polymer types as well as plastic validation [18]. Nonetheless, the micro-FTIR approach alternates between the object lens and the IR probe, allowing microscopic examination of microsized plastic-like particles on a single platform prior to spectroscopic verification. This approach is regarded as necessary for minute microplastics in particular. Micro-FTIR can characterize samples smaller than 10 μm [19]. Harrison et al. [20] assessed the possibility of spectral property-based molecular mapping approaches, as well as the efficacy of micro-FTIR and reflectance micro-FTIR techniques for detecting microplastics in marine sediment. Because of their variable composition, wear, and biofouling, environmental samples are difficult to match with the commercially available spectrum library [20].

2.1 Sample Preparation

Accurate detection and identification of microplastics (MPs) in complex environmental and biological matrices necessitate matrix specific, stringent sample preparation techniques. The goal of these approaches is to extract and purify MPs without changing their physical structure or chemical composition, allowing accurate spectroscopic characterization. Sample preparation procedures differ depending on the source matrix: soil, water, food, atmospheric fallout, or human biological samples, each providing distinct problems in terms of particle interference, organic content, and sample volume [21]. The soil and sediment samples are normally dried at 40–60 °C and then sieved (<5 mm) to eliminate macrodebris. The organic material in soil is digested with 30% hydrogen peroxide (H_2O_2) or Fenton's reagent (a mixture of H_2O_2 and $FeSO_4$) and incubated at 50 °C for 24–72 h under moderate agitation ([22].). After digestion, density separation is performed via saturated zinc chloride ($ZnCl_2$) or sodium iodide (NaI) solutions with densities ranging from 1.6 to 1.8 g/cm^3. Lighter polymer particles float, whereas denser mineral particles sink. The supernatant is filtered via glass fiber or polycarbonate membranes (usually 0.45 μm), washed with deionized water to eliminate residual salts, and dried before FTIR analysis [21]. Water samples, particularly those collected from seas, rivers, or drinking water sources, often have lower organic contents but have to be processed in large quantities. The samples are first passed through a sieve (e.g., 5 mm, 300 μm, 50 μm) before vacuum filtering through 0.45 μm membranes to remove large debris [22]. To remove biofouling or algae, oxidative digestion was performed with 30% H_2O_2 at 50 °C for 24 h. After mixing the sediments with water, $ZnCl_2$ was used for density separation. MPs are then concentrated for spectroscopic examination by filtration and drying [23]. Protocols for food samples must be customized to a variety of food matrices, including salt and sugar, shellfish, and processed meat. To guarantee consistent particle dispersion, the food matrix is first homogenized via a blender or homogenizer. Protein or lipid-rich samples (e.g., seafood, meats) are enzymatically digested for 24–48 h at 37–50 °C via specialized enzymes such as proteinase K, lipase, and cellulase. Additional oxidative digestion with H_2O_2 or Fenton's reagent is carried out for matrices with relatively high organic contents. Next, the MPs were extracted via $ZnCl_2$ density separation, filtered through 0.45 μm polycarbonate membranes, rinsed, and dried for FTIR imaging [23]. Human samples such as products of excretion, placental tissue, or blood are handled with considerable caution to minimize contamination while preserving polymer integrity. After sample homogenization, enzymatic digestion with proteinase K is followed by oxidative digestion at 50 °C using 30% H_2O_2. Filtration uses IR-transparent membranes such as aluminum oxide with 0.45 μm pore diameters. Filters and reagents must be precleaned, and blank controls are provided to avoid false positives due to airborne or procedural contamination. The filters are then air dried or placed in a desiccator to prepare them for FTIR analysis [24]. Passive deposition traps or active high-volume air samplers are used to capture atmospheric microplastics, which are then deposited on precleaned quartz or aluminum filters. Following collection, the material was washed and transferred to IR-compatible membranes. Low organic content requires minimal digestion, although 30% H_2O_2 can be used if necessary. Filters are dried under dust-free conditions to prevent airborne contamination before being used directly for FTIR analysis [25].

2.2 Types of FTIR Methods

2.2.1 Attenuated Total Reflectance FTIR (ATR-FTIR) (Fig. 2)

Attenuated total reflectance FTIR (ATR-FTIR) is a popular method for identifying microplastics, especially those larger than 500 μm. This method generates an evanescent wave that penetrates only 1–2 μm into the particle, whereas an infrared beam undergoes total internal reflection within a high-refractive-index crystal, such as germanium (Ge), diamond or zinc selenide (ZnSe) [26]. Microplastic particles are placed on the crystal and pushed down with an anvil for proper contact. The evanescent wave interacts with the sample surface, causing vibrational absorption corresponding to the functional groups of the polymer. Spectral data are typically collected over the range of 4000–650 cm^{-1}, with a resolution of 4 cm^{-1} and 16—64 scans per particle [27]. ATR-FTIR requires minimal sample preparation and provides quick analysis, making it excellent for laboratory-based confirmation of known microplastics in usually clean matrices [28]. However, it is surface sensitive and inefficient for particles smaller than 500 μm or uneven surfaces that fail to gain appropriate contact with the crystal. This results in spectral noise or incomplete absorption bands. Furthermore, worn microplastics may undergo surface oxidation or biofouling, which might obscure the initial polymer signatures. Despite these

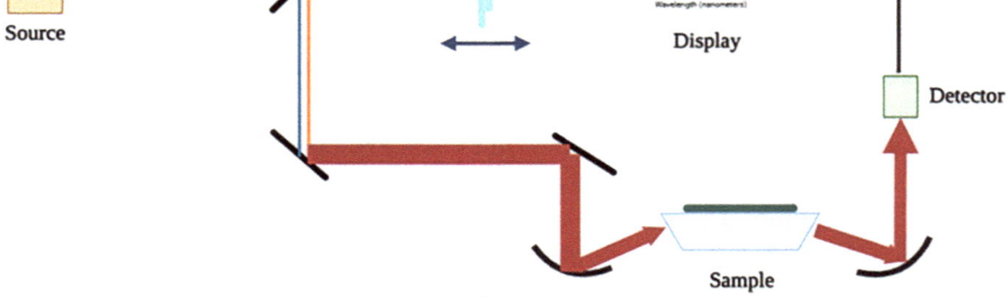

Fig. 2 Schematic representation of attenuated total reflectance FTIR (ATR-FTIR)

limitations, ATR-FTIR is a popular choice in food and freshwater microplastic research because of its ease of use and high dependability [29].

2.2.2 Micro-FTIR Imaging (μFTIR)

μFTIR combines infrared spectroscopy and microscopy, allowing the automated detection of microplastics ranging from 10 to 500 μm in size. This method uses a Fourier transform infrared spectrometer coupled with a microscope and a focal plane array (FPA) detector to raster-scan large sample regions and acquire thousands of individual spectra at the same time [28]. Dried samples are placed on IR-transparent filters, such as aluminum oxide or silicon, and the device captures spectra pixel by pixel. The detector matrix consists of 64 × 64 or 128 × 128 pixels, producing chemical maps that can identify different polymer types in diverse samples [30]. μFTIR software may automate spectrum interpretation by matching acquired spectra to polymer libraries via correlation scores or machine learning classifiers [31]. Although very precise, μFTIR is time- and resource-intensive, requiring several hours per sample frequently, and the equipment is expensive, necessitating specialized skills [27]. The diffraction limit of infrared light prevents the detection of particles smaller than 10 μm. Moreover, overlapping or clustered particles might complicate analysis [32]. μFTIR is the preferred method for quantifying microplastics in environmental and biological materials because it is automated, reproducible, and has high spatial resolution [33].

2.2.3 Transmission FTIR (Fig. 3)

Transmission FTIR is a method used to detect microplastics, mostly thin and flat particles less than 500 μm in size. This approach detects the attenuation of infrared light passing through a particle put on an IR-transparent substrate such as potassium

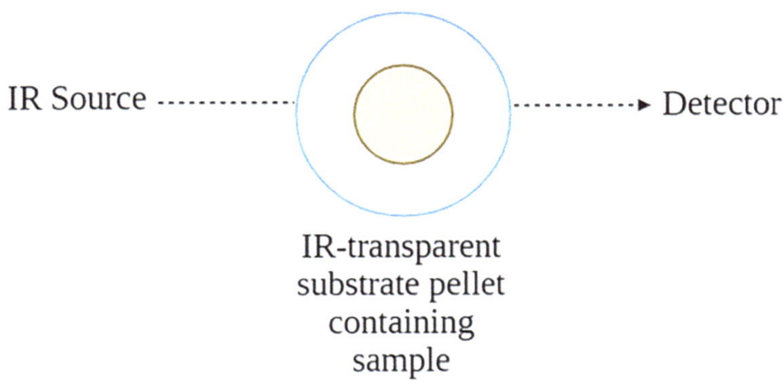

Fig. 3 Schematic representation of transmission FTIR

bromide (KBr), calcium fluoride (CaF_2), or barium fluoride (BaF_2) [34]. The sample should be thin (preferably <100 μm) to avoid scattering or absorption. Microplastics may be crushed or sectioned with a microtome to achieve these dimensional constraints. After being correctly placed, the IR beam travels straight through the sample and collects the transmission spectrum with typical characteristics of 4 cm^{-1} resolution and 16–64 coadded scans. Transmission FTIR has outstanding spectrum quality, with crisp, well-resolved peaks due to the direct contact of the IR beam with the sample bulk [27]. However, this approach is labor intensive, requires precise manipulation of individual particles onto delicate IR windows, and is prone to contamination or breakage during preparation. It is ideally suited for isolating particles in controlled laboratory conditions or confirming the results of other approaches [33].

2.2.4 Reflection FTIR (Fig. 4)

Reflection FTIR includes both specular and diffuse reflection modes and is used when microplastics are adhered to or embedded in reflective substrates such as aluminum-coated filters or food packaging materials. An infrared beam is aimed at the sample surface, and the reflected radiation is collected to provide a reflection spectrum. Specular reflection works best on smooth, mirror-like surfaces, whereas diffuse reflection is better on uneven or particulate-laden substrates. Although reflection spectra are useful for analyzing particles that cannot be removed or modified, they frequently exhibit aberrations such as derivative-shaped bands, baseline tilts, and inverted peaks [32]. To solve these difficulties, spectral correction methods such as Fourier self-deconvolution or Kramers–Kronig transformation are used to turn reflection spectra into absorbance-like curves [27]. The technique is less sensitive to sample thickness but more susceptible to surface roughness and noise interference. Its main advantage is the nondestructive examination of intact materials on their original substrate, making it excellent for forensic applications, packaging analysis, and onsite monitoring of airborne microplastics. However, compared with

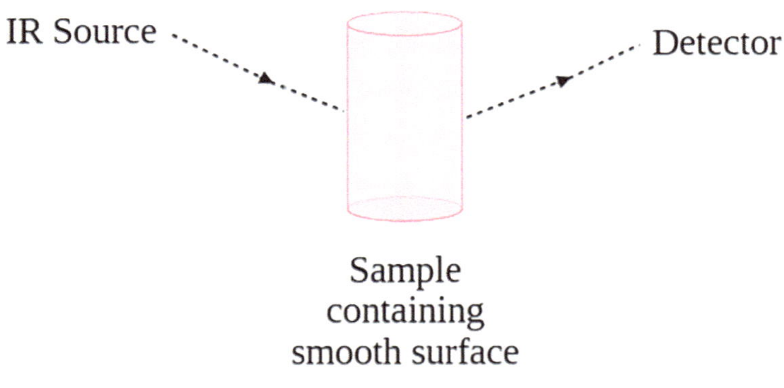

Fig. 4 Schematic representation of reflection FTIR

ATR or μFTIR, the spectrum quality is often inferior, making interpretation difficult without significant software support or operator skill [29].

3 Microplastics in Water Systems

A major environmental and public health concern is the pervasiveness of microplastics in different water systems. Assessing the environmental impact of microplastics and developing effective detection and mitigation strategies require an understanding of their distribution, concentration, and characteristics in various aquatic environments. This section examines microplastic contamination across major water systems, including freshwater environments, marine systems, drinking water supplies, and groundwater resources.

3.1 Freshwater Systems

Freshwater environments, including rivers, lakes, streams, and reservoirs, serve as critical pathways for microplastic transport and significant repositories of plastic contamination. The proximity of these systems to cities, industrial operations, and agricultural land use make them especially susceptible to microplastic pollution. In terms of population density, industrial activity, and proximity to urban centers, microplastic concentrations in freshwater systems vary significantly, with reported concentrations ranging widely across different water bodies globally [35]. Rivers function as major channels for microplastic transport from terrestrial sources to marine environments, whereas lakes and reservoirs serve as accumulation zones where particles can settle and concentrate in sediments. Fibers and fragments make up the majority of microplastic morphological features in freshwater systems, with fibers typically comprising the dominant fraction of particles found in most studies [7]. The most commonly recognized polymers are polyethylene (PE), polypropylene (PP), and polyethylene terephthalate (PET), which reflects their extensive usage in consumer goods and packaging materials [36].

For the analysis of freshwater microplastics, FTIR spectroscopy has shown great efficacy. Automated methods that use focal plane array FTIR microscopy offer trustworthy identification capabilities [37]. To accurately identify the predominant PE, PP, and PET polymers present in freshwater systems, the analytical method usually involves filtration through suitable membranes followed by FTIR analysis. This method is especially useful for separating natural organic materials, which are frequently found in freshwater environments, from synthetic fibers. Seasonal variations in microplastic concentrations are also evident across freshwater systems, with higher loads typically observed during wet seasons because of increased surface runoff and stormwater discharge [36]. The Great Lakes system exemplifies extensive freshwater contamination, with significant surface water concentrations documented across the basin [38]. The predominance of small particles in many freshwater systems suggests ongoing fragmentation processes and the potential for long-range atmospheric transport.

3.2 Marine Systems

Marine environments represent the ultimate sink for a significant portion of global microplastic pollution, with oceanic systems receiving inputs from terrestrial sources through riverine discharge, direct coastal inputs, and atmospheric deposition. There is significant variability in the distribution of microplastics in marine waters, with concentrations typically declining as one moves away from coastal regions and population centers. Surface water concentrations vary widely globally, with subtropical gyres serving as major accumulation zones [6]. The North Pacific Subtropical Gyre, also referred to as the "Great Pacific Garbage Patch", contains some of the highest concentrations of microplastics ever observed in marine systems [39]. The microplastic characteristics of marine surface waters are dominated by fragments and pellets, with weathered PE and PP particles being most prevalent due to their buoyancy characteristics [40].

ATR-FTIR spectroscopy has emerged as a primary method for identifying microplastics in marine environments and is particularly effective for analyzing larger particles extracted from marine sediments and water samples [32]. Vibrational microspectroscopy of marine samples necessitates careful consideration of particle characteristics; for thorough characterization of weathered marine particles, both FTIR and Raman techniques are frequently used. Advanced FTIR techniques enable automated detection across samples, which is essential for processing the extensive samples required in marine microplastic research. Recent studies have demonstrated the widespread distribution of microplastics throughout the oceanic water column, challenging previous assumptions about their surface-bound nature. Although concentrations of microplastics generally decline with depth, notable amounts have been found in deep-sea environments [41]. Particle density, biofouling, marine snow formation, and thermohaline circulation patterns are some of the variables that affect the vertical distribution. Deep-sea sediments serve as important repositories for the accumulation of microplastics, with concentrations varying significantly across different oceanic regions [42].

3.3 Groundwater Systems

Groundwater contamination by microplastics represents an emerging area of concern, with recent research revealing its widespread presence in aquifer systems globally. Groundwater serves as a primary drinking water source for more than 2 billion people worldwide, making microplastic contamination a significant public health issue. Studies have documented microplastic concentrations in groundwater systems, with urban areas showing higher contamination levels than rural regions do [43]. Fractured limestone aquifers are particularly vulnerable to contamination because of their high permeability and close proximity to surface waters, with contamination patterns that mirror surface inputs through recharge processes. The most common polymer types found in groundwater are PE and polyvinyl chloride (PVC), reflecting both surface contamination sources and potential leaching from subsurface infrastructure [43]. The morphological characteristics indicate a dominance of fragments and fibers, with particles generally smaller than those found in surface waters, indicating filtration effects during subsurface transport. FTIR analysis of groundwater microplastics presents unique challenges because of lower particle concentrations and smaller size distributions. This technique requires careful sample preparation to remove mineral precipitates and organic matter that interfere with spectral analysis. Advanced FTIR approaches have been developed for microplastic detection in various water matrices, with standardized methodologies being established to enable comparisons across studies [31]. Groundwater microplastic contamination occurs through various pathways, including soil infiltration, surface runoff, agricultural activities, domestic waste, landfills, and industrial processes [44].

3.4 *Drinking Water Systems*

Microplastic contamination has been detected in tap and bottled water supplies across the globe, raising serious public health concerns. The contamination patterns reflect various sources, including source water quality, treatment processes, distribution infrastructure, and packaging materials. Microplastic concentrations in tap water systems vary significantly by region and range widely worldwide, with notable differences between developed and developing countries [45]. European tap water systems typically show lower concentrations than other regions do, whereas developing countries often exhibit higher contamination levels. The predominant microplastic types in tap water are fibers and fragments, with sizes typically ranging from 10 to 1000 μm. Polymer identification reveals PE, PP, PET, and polystyrene (PS) as the most common types, likely originating from source water contamination and infrastructure degradation [46]. Bottled water systems also exhibit widespread microplastic contamination, often exceeding concentrations found in tap water. Studies have documented variable concentrations across different brands and packaging types, with single-use plastic bottles typically showing higher contamination levels than glass containers do [38]. Recent research has revealed significant numbers of nanoscale plastic particles in bottled water, highlighting the complexity of plastic contamination in packaged beverages [47]. Conventional water treatment methods such as coagulation, sedimentation, and filtration can reduce microplastic loads significantly, although smaller particles may evade treatment barriers [45].

Owing to its accurate quantitative and qualitative analysis capabilities, FTIR spectroscopy is widely used to monitor microplastics in drinking water. Research has effectively used FTIR imaging to identify microplastics, such as polyester, polyethylene, polyamide, and polyvinyl chloride, which range in size from 50–150 μm in drinking water from groundwater sources [48]. The method works particularly well for separating natural cellulose fibers, which are frequently found in treated water systems, from synthetic polymer fibers. The distribution of microplastics across different water systems reveals distinct contamination patterns and characteristics. Marine surface waters generally exhibit variable particle concentrations with notable accumulation in convergence zones, whereas freshwater systems show high spatial and temporal variability. Drinking water systems, particularly bottled water, frequently contain detectable microplastic concentrations, although these are predominantly smaller particles. The polymer composition varies across systems, with PE and PP dominating marine environments owing to their buoyancy characteristics, whereas PET and synthetic fibers are more prevalent in freshwater and drinking water systems. The size distribution also varies; groundwater systems contain smaller particles due to filtration during subsurface transport, whereas marine systems contain a broader size range reflecting various input sources and environmental processes (Fig. 5).

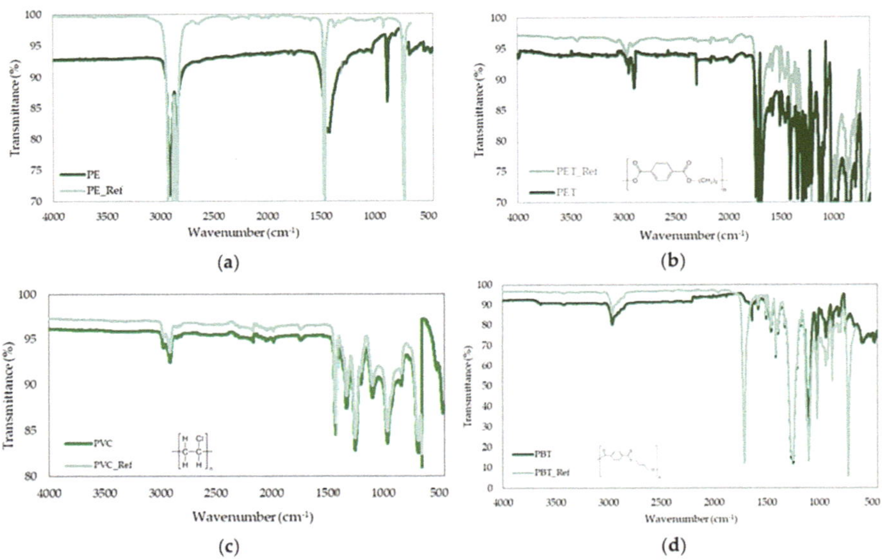

Fig. 5 FTIR spectra of (**a**) PE, (**b**) PET, (**c**) PVC (**d**) PBT found in water [49]

4 Microplastics in Soil Systems

Soil contamination by microplastics has emerged as a critical environmental concern, representing a rapidly expanding area of research within the broader microplastic pollution paradigm. In contrast to aquatic systems, where studies on microplastic contamination have been conducted for more than two decades, soil microplastic contamination research has only recently gained momentum, revealing extensive contamination in a variety of terrestrial environments. The persistence and accumulation of microplastics in soil systems have significant implications for ecosystem sustainability, food security, and agricultural productivity [1].

4.1 Sources and Pathways of Soil Contamination

The contamination of soil systems by microplastics originates from multiple anthropogenic sources, with agricultural practices representing a significant contributor. Plastic agricultural films, including mulching films, greenhouse covers, and tunneling materials, constitute a primary source of microplastic contamination in agricultural soils. Studies have identified plastic mulch as a major source of microplastic contamination, highlighting the widespread impact of modern agricultural practices on soil quality [50]. The application of sewage sludge represents another significant pathway for microplastics to enter agricultural soils. Research has revealed substantial microplastic contents in sewage sludge, with concentrations varying significantly across different treatment facilities [51]. The application of contaminated sewage sludge to agricultural fields results in direct microplastic inputs, with contamination persisting for extended periods following application. Atmospheric deposition and compost application also contribute to microplastic contamination, particularly in areas distant from direct sources [52].

FTIR spectroscopy has become essential for identifying microplastics in agricultural soils, with studies consistently detecting polyethylene as the dominant polymer type in mulched systems [53]. The density flotation method using saturated NaCl solution combined with H_2O_2 predigestion followed by FTIR identification represents the standard analytical approach for soil microplastic analysis [54]. Agricultural plastic waste, including deteriorated plastic sheeting and irrigation equipment, represents additional sources of soil contamination (Fig. 6). The breakdown of these materials through weathering processes contributes to the accumulation of microplastic particles in agricultural soils over time [56]. Furthermore, tire wear particles from agricultural machinery and transportation contribute to microplastic loads in rural soils, although these loads are typically lower than those in urban environments [4].

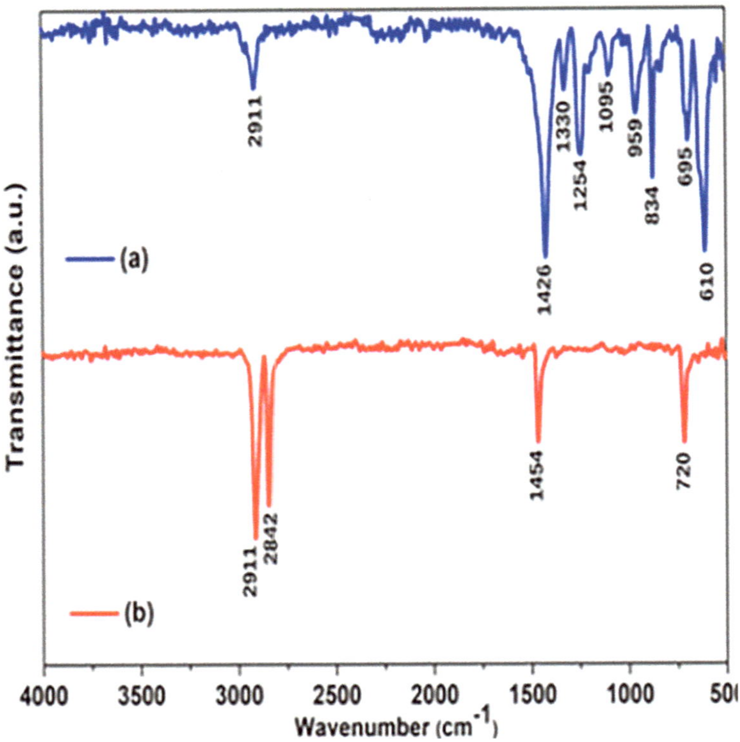

Fig. 6 FTIR spectra of (a) PVC and (b) low-density polyethylene (LDPE) found in soil [55]

5 Microplastics in Food

The widespread prevalence of microplastics in the environment has resulted in their unavoidable penetration into the human food chain. Their presence in food stems from a variety of factors, including farming methods, marine and freshwater pollution, packing materials, and food processing equipment. Marine species such as fish, shellfish, and sea salt are among the best-documented sources of microplastic ingestion in aquatic ecosystems because of their bioaccumulation and trophic transmission [57]. Furthermore, MPs are found in bottled and tap water, beer, honey, sugar, and even fruits and vegetables and are frequently linked to atmospheric deposition or plastic-containing irrigation systems [58, 59]. The contamination is worsened by food packing and processing. The plastic containers, utensils, and films used in food storage can leak MPs into the food, especially when subjected to heat or mechanical stress [60]. Substantial MP concentrations in tea bags made of nylon and PET when submerged in boiling water have been reported, indicating direct consumer exposure from processed and packaged items [61].

The consumption of microplastics presents serious concerns about human health; however, the full toxicological effects are still being investigated. They can be harmful either intrinsically or by the adsorption of environmental contaminants such as heavy metals, persistent organic pollutants (POPs), and antibiotics, which can desorb when consumed [62]. According to experimental studies, microplastics can disrupt the gut microbiota, impair nutrient absorption, cause oxidative stress, and trigger inflammatory responses in mammalian models [63]. Furthermore, nanoplastics have the ability to bypass epithelial barriers, possibly accumulating in important organs and causing genotoxic or neurotoxic consequences [64]. Given the worldwide concern and possible health consequences, the identification and quantification of microplastics in food matrices has become a grave issue in analytical research. Among the different techniques currently available, FTIR is regarded as one of the most reliable and extensively used methods for microplastic detection. FTIR works by detecting the infrared absorption spectra of polymer particles, which reflect their particular molecular vibrations, allowing for accurate identification of different kinds of plastic [65]. Using ATR-FTIR and micro-FTIR imaging, tiny particles as small as 10 μm may be analyzed in complicated food matrices without requiring considerable sample preparation [48]. When combined with filtering and enzymatic digestion techniques to remove organic matter, FTIR provides a nondestructive, repeatable method for identifying MPs in a variety of food samples, including shellfish, salts, and even drinks. Furthermore, FTIR spectrum libraries help to differentiate synthetic polymers (such as polyethylene, polypropylene, and polystyrene) from natural materials, reducing the number of false positives [32]. Zhou et al. [66] used focal-plane-array (FPA) micro-FTIR imaging in takeaway food containers to detect microplastics (mostly polyamide, polyurethane, and polystyrene) at abundances of 29–552 particles per container across seven container types in China. They used an automated spectral database matching over 4000–700 cm^{-1} with 25 μm resolution, suggesting that humans are exposed to thousands of MPs per month (pmc) [66]. Dang et al. [67] developed a TGA-FTIR chemometric approach for detecting polystyrene microplastics in milk and simulants. This method combines mass-based detection with spectrum identification in complicated matrices, resulting in enhanced sensitivity and specificity for polymer quantification [67]. Parvin et al. [68] analyzed commercial sea salts from Bangladesh via FTIR to confirm the presence of polymers such as PET, HDPE, nylon (PA), and EVA. PET revealed characteristic absorption bands (~720, 1029/1210, 1713 cm^{-1}) and nylon at ~1538 (NH bend/C-N), 1654 (C=O), and 3376 cm^{-1} (N–H), quantifying average contamination at ~2676 MPs/kg, indicating dietary exposure via salt [68]. Kuttykattil et al. [69] used ATR-FTIR (plus SEM–EDS) to analyze Australian commercial salts (Fig. 7). They identified 78.8% of the particles as microplastics between 23.2 μm and 3.9 mm, with the dominant polymers being polyamide (~34%) and polyurethane (~31%). They estimated a typical adult annual salt intake of ~155 MPs/year [69].

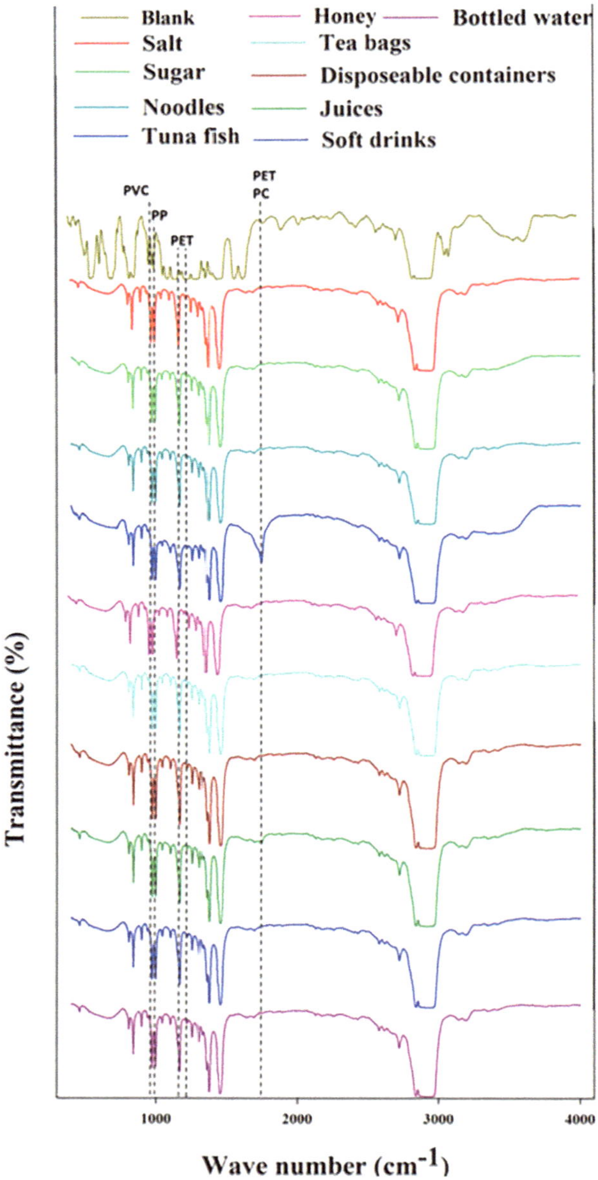

Fig. 7 FTIR spectra of PP, PE, PVC and PC found in salt, sugar, noodles, tuna fish, honey, tea bags, disposable plastic containers, juices, soft drinks, and bottled drinking water [70]

6 Microplastics in Humans

Microplastics are ubiquitous in the environment, causing them to infiltrate the human body via multifarious exposure pathways, increasing concerns about their possible health implications. Humans are exposed to microplastics primarily through ingestion, inhalation, and dermal contact, with ingestion being the most documented route. Microplastics have been found in a variety of food products, including seafood, drinking water, sea salt, honey, fruits, and vegetables [23, 59]. Recent studies predict that individuals may consume 39,000 to 52,000 microplastic particles per year, with considerably greater exposure when inhaling is included [58].

Airborne microplastics from synthetic fabrics, urban dust, and tire wear can be absorbed and deposited in the lungs, thereby causing respiratory irritation and inflammation [71]. Dermal exposure, albeit less investigated, is conceivable through the use of microbead-containing cosmetics and contact with contaminated water sources, but transdermal penetration is thought to be limited [72]. Given the absence of longitudinal human epidemiological data, the biological effects of microplastics on human health are complex and are mostly derived from in vitro and animal research. Microplastics can cause physical stresses in the gastrointestinal system, resulting in intestinal barrier failure, inflammation, and altered microbiota composition [63]. Ingested MPs have been demonstrated to pass through epithelial barriers via endocytosis or paracellular routes, possibly entering the circulation and accumulating in secondary organs such as the liver, kidneys, spleen, and even the brain [73]. Nanoplastics (particles <1 μm) are highly absorptive and cytotoxic, causing oxidative stress, mitochondrial dysfunction, apoptosis, and genotoxicity [64]. Chemical contaminants such as persistent organic pollutants (POPs), heavy metals, phthalates, and bisphenols adsorb to microplastic surfaces and may be desorbed in biological environments, resulting in toxicity through endocrine disruption, neurotoxicity, and carcinogenic potential [62].

Emerging research has begun to support these worries. Leslie et al. [73] found microplastics in human blood, including polymers such as PET, polystyrene, and polymethylmethacrylate (PMMA), indicating systemic exposure and the possibility of tissue buildup [73]. Furthermore, microplastics have been found in human stool samples from many nations, indicating extensive gastrointestinal exposure [72]. The analytical identification and characterization of microplastics in biological matrices is methodologically difficult because of the complexity of human tissues and the potential for sample contamination. FTIR allows for the qualitative and semiquantitative identification of microplastics on the basis of their distinct vibrational absorption spectra, which represent the molecular structure of the polymer chains [32]. In biological situations, microplastics must first be separated by digesting organic materials with agents such as hydrogen peroxide, nitric acid, or enzymatic cocktails and then filtered and mounted on reflecting substrates appropriate for FTIR measurement [29]. Lee et al. [74] used μ-FTIR spectroscopy to detect microplastics in the blood of 36 healthy adults. They reported polystyrene and polypropylene particles at an average concentration of 4.2 MPs/ml. A relatively high MP

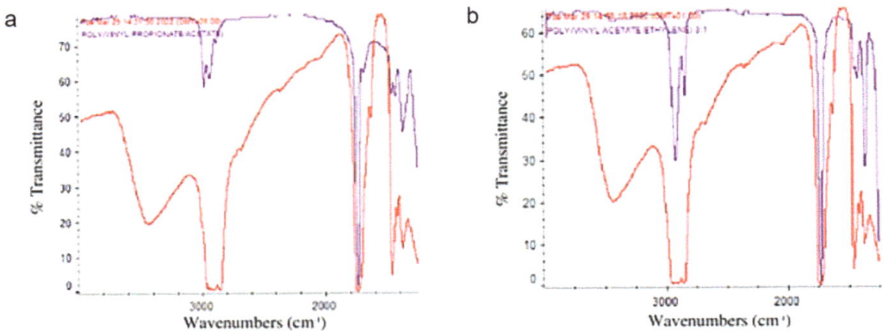

Fig. 8 FTIR spectra of (**a**) PVAC fragments and (**b**) PVAE fragments found in human veins and tissues [76]

load (≥3 MPs/mL) was associated with elevated coagulation markers, such as aPTT, C-reactive protein, and fibrinogen, suggesting potential cardiovascular effects [74]. Leonard et al. [75] used micro-FTIR (m-FTIR) to analyze the blood of 20 volunteers, measuring concentrations ranging from 1.84 to 4.65 μg/mL. They detected 24 polymer types, including polyethylene (~32%), ethylene propylene diene (~14%), and EVA/alcohol (~12%), highlighting systemic translocation and polymer diversity in circulation [75]. Rotchell et al. [76] employed μ-FTIR spectroscopy to analyze digested human saphenous vein tissue (n = 5), finding microplastics as tiny as 5 μm, characterizing pieces and fibers within vascular tissue, and highlighting the existence of MPs in human arterial systems [76]. Jenner et al. [77] used μ-FTIR to confirm microplastics in human lung tissue (n = 13), detecting different polymer types and proving that inhalation is a realistic route of exposure and accumulation [77] (Fig. 8).

7 Conclusion

The entry of microplastics into the human body via numerous exposure pathways constitutes a public health danger, particularly considering their durability, bioaccumulation potential, and propensity to function as chemical transporters. While the long-term health effects are unknown, preliminary evidence suggests that precautionary regulation, exposure reduction strategies, and ongoing methodological refinements in MP detection, particularly the use of FTIR-based techniques in biological matrices, are necessary to better understand and mitigate the human health burden of plastic pollution. Nonetheless, micro-FTIR imaging and ATR-FTIR can identify and map particles as small as 10 μm, enabling polymer identification in complicated tissue or fluid samples. Furthermore, FTIR spectrum libraries may be utilized to compare misidentified particles to known polymer spectra, ensuring accurate identification. Despite its lower resolution than that of Raman spectroscopy, FTIR remains a cornerstone in MP analysis because of its robustness, accessibility, and nondestructive nature and has been used successfully in studies detecting MPs in human placenta, stool, and blood.

References

1. de Souza Machado AA, Kloas W, Zarfl C, Hempel S, Rillig MC (2018) Microplastics as an emerging threat to terrestrial ecosystems. Glob Chang Biol 24:1405–1416. https://doi.org/10.1111/gcb.14020
2. Thompson RC, Olsen Y, Mitchell RP, Davis A, Rowland SJ, John AWG, McGonigle D, Russell AE (2004) Lost at sea: where is all the plastic? Science 304:838–838. https://doi.org/10.1126/science.1094559
3. Dehghani S, Moore F, Akhbarizadeh R (2017) Microplastic pollution in deposited urban dust, Tehran metropolis, Iran. Environ Sci Pollut Res 24:20360–20371. https://doi.org/10.1007/s11356-017-9674-1
4. Kole PJ, Löhr AJ, Van Belleghem FGAJ, Ragas AMJ (2017) Wear and tear of tyres: A stealthy source of microplastics in the environment. Int J Environ Res Public Health 14:1265. https://doi.org/10.3390/ijerph14101265
5. Browne MA, Crump P, Niven SJ, Teuten E, Tonkin A, Galloway T, Thompson R (2011) Accumulation of microplastic onshorelines woldwide: sources and sinks. Environ Sci Technol 45:9175–9179. https://doi.org/10.1021/es201811s
6. Law KL, Starr N, Siegler TR, Jambeck JR, Mallos NJ, Leonard GH (2020) The United States' contribution of plastic waste to land and ocean. Sci Adv 6. https://doi.org/10.1126/sciadv.abd0288
7. Windsor FM, Tilley RM, Tyler CR, Ormerod SJ (2019) Microplastic ingestion by riverine macroinvertebrates. Sci Total Environ 646:68–74. https://doi.org/10.1016/j.scitotenv.2018.07.271
8. Horton AA, Walton A, Spurgeon DJ, Lahive E, Svendsen C (2017) Microplastics in freshwater and terrestrial environments: evaluating the current understanding to identify the knowledge gaps and future research priorities. Sci Total Environ 586:127–141. https://doi.org/10.1016/j.scitotenv.2017.01.190
9. Allen S, Allen D, Phoenix VR, Le Roux G, Durántez Jiménez P, Simonneau A, Binet S, Galop D (2019) Atmospheric transport and deposition of microplastics in a remote mountain catchment. Nat Geosci 12:339–344. https://doi.org/10.1038/s41561-019-0335-5
10. Olesen KB, Stephansen DA, Van Alst N, Vollertsen J (2019) Microplastics in a Stormwater pond. Water 11:1466. https://doi.org/10.3390/w11071466
11. Cunsolo S, Williams J, Hale M, Read DS, Couceiro F (2021) Optimizing sample preparation for FTIR-based microplastic analysis in wastewater and sludge samples: multiple digestions. Anal Bioanal Chem 413:3789–3799. https://doi.org/10.1007/s00216-021-03331-6
12. Gong Y, Chen X, Wu W (2024) Application of Fourier transform infrared (FTIR) spectroscopy in sample preparation: material characterization and mechanism investigation. Adv Sample Prep 11:100122. https://doi.org/10.1016/j.sampre.2024.100122
13. Samson JJ (2024) Characterization of microplastics using Fourier infrared spectroscopy. In: Sivasankar V, Sunitha TG (eds) Microplastics and pollutants: interactions, degradations and mechanisms. Springer, Cham, pp 129–148. https://doi.org/10.1007/978-3-031-54565-8_6
14. Dela-Torre G, Mendoza-Castilla L, Pilar R (2019) Microplastic contamination in market bivalve Argopecten purpuratus from Lima, Peru. Man 16:85–89. https://doi.org/10.17268/manglar.2019.012
15. Davidson K, Dudas SE (2016) Microplastic ingestion by wild and cultured Manila clams (*Venerupis philippinarum)* from Baynes sound, British Columbia. Arch Environ Contam Toxicol 71:147–156. https://doi.org/10.1007/s00244-016-0286-4

16. Willans M, Szczecinski E, Roocke C, Williams S, Timalsina S, Vongsvivut J, McIlwain J, Naderi G, Linge L, K., & J. Hackett, M. (2023) Development of a rapid detection protocol for microplastics using reflectance-FTIR spectroscopic imaging and multivariate classification. Environ Sci Adv 2:663–674. https://doi.org/10.1039/D2VA00313A
17. Xu J-L, Thomas KV, Luo Z, Gowen AA (2019) FTIR and Raman imaging for microplastics analysis: state of the art, challenges and prospects. TrAC Trends Anal Chem 119:115629. https://doi.org/10.1016/j.trac.2019.115629
18. Mobaraki N, Hemmateenejad B (2011) Structural characterization of carbonyl compounds by IR spectroscopy and chemometrics data analysis. Chemom Intell Lab Syst 109:171–177. https://doi.org/10.1016/j.chemolab.2011.08.011
19. Löder MGJ, Gerdts G (2015) Methodology used for the detection and identification of microplastics—A critical appraisal. In: Bergmann M, Gutow L, Klages M (eds) Marine anthropogenic litter. Springer International Publishing, Cham, pp 201–227. https://doi.org/10.1007/978-3-319-16510-3_8
20. Harrison JP, Ojeda JJ, Romero-González ME (2012) The applicability of reflectance micro-Fourier transform infrared spectroscopy for the detection of synthetic microplastics in marine sediments. Sci Total Environ 416:455–463. https://doi.org/10.1016/j.scitotenv.2011.11.078
21. Imhof HK, Schmid J, Niessner R, Ivleva NP, Laforsch C (2012) A novel, highly efficient method for the separation and quantification of plastic particles in sediments of aquatic environments. Limnol Oceanogr Methods 10:524–537. https://doi.org/10.4319/lom.2012.10.524
22. Masura J, Baker JE, Foster GD, Gregory D, Arthur C, Herring C. (2015). Laboratory methods for the analysis of microplastics in the marine environment: Recommendations for quantifying synthetic particles in waters and sediments. https://repository.library.noaa.gov/view/noaa/10296
23. Cole M, Webb H, Lindeque PK, Fileman ES, Halsband C, Galloway TS (2014) Isolation of microplastics in biota-rich seawater samples and marine organisms. Sci Rep 4:4528. https://doi.org/10.1038/srep04528
24. Ragusa A, Svelato A, Santacroce C, Catalano P, Notarstefano V, Carnevali O, Papa F, Rongioletti MCA, Baiocco F, Draghi S, D'Amore E, Rinaldo D, Matta M, Giorgini E (2021) Plasticenta: first evidence of microplastics in human placenta. Environ Int 146:106274. https://doi.org/10.1016/j.envint.2020.106274
25. Dris R, Gasperi J, Rocher V, Saad M, Renault N, Tassin B (2015) Microplastic contamination in an urban area: A case study in greater Paris. Environ Chem 12:592. https://doi.org/10.1071/EN14167
26. Araujo CF, Nolasco MM, Ribeiro AMP, Ribeiro-Claro PJA (2018) Identification of microplastics using Raman spectroscopy: latest developments and future prospects. Water Res 142:426–440. https://doi.org/10.1016/j.watres.2018.05.060
27. Renner G, Schmidt TC, Schram J (2018) Analytical methodologies for monitoring micro(nano) plastics: which are fit for purpose? Curr Opin Environ Sci Health 1:55–61. https://doi.org/10.1016/j.coesh.2017.11.001
28. Primpke S, Cross RK, Mintenig SM, Simon M, Vianello A, Gerdts G, Vollertsen J (2020) Toward the systematic identification of microplastics in the environment: evaluation of a new independent software tool (siMPle) for spectroscopic analysis. Appl Spectrosc 74:1127–1138. https://doi.org/10.1177/0003702820917760
29. Munno K, De Frond H, O'Donnell B, Rochman CM (2020) Increasing the accessibility for characterizing microplastics: introducing new application-based and spectral libraries of plastic particles (SLoPP and SLoPP-E). Anal Chem 92:2443–2451. https://doi.org/10.1021/acs.analchem.9b03626
30. Mintenig SM, Int-Veen I, Löder MGJ, Primpke S, Gerdts G (2017) Identification of microplastic in effluents of waste water treatment plants using focal plane array-based micro-Fourier transform infrared imaging. Water Res 108:365–372. https://doi.org/10.1016/j.watres.2016.11.015

31. Primpke S, Wirth M, Lorenz C, Gerdts G (2018) Reference database design for the automated analysis of microplastic samples based on Fourier transform infrared (FTIR) spectroscopy. Anal Bioanal Chem 410:5131–5141. https://doi.org/10.1007/s00216-018-1156-x
32. Käppler A, Fischer D, Oberbeckmann S, Schernewski G, Labrenz M, Eichhorn K-J, Voit B (2016) Analysis of environmental microplastics by vibrational microspectroscopy: FTIR, Raman or both? Anal Bioanal Chem 408:8377–8391. https://doi.org/10.1007/s00216-016-9956-3
33. Frère L, Paul-Pont I, Moreau J, Soudant P, Lambert C, Huvet A, Rinnert E (2016) A semi-automated Raman microspectroscopy method for morphological and chemical characterizations of microplastic litter. Mar Pollut Bull 113:461–468. https://doi.org/10.1016/j.marpolbul.2016.10.051
34. Käppler A, Windrich F, Löder MGJ, Malanin M, Fischer D, Labrenz M, Eichhorn K-J, Voit B (2015) Identification of microplastics by FTIR and Raman microscopy: A novel silicon filter substrate opens the important spectral range below 1300 cm(−1) for FTIR transmission measurements. Anal Bioanal Chem 407:6791–6801. https://doi.org/10.1007/s00216-015-8850-8
35. Koelmans AA, Mohamed Nor NH, Hermsen E, Kooi M, Mintenig SM, De France J (2019) Microplastics in freshwaters and drinking water: critical review and assessment of data quality. Water Res 155:410–422. https://doi.org/10.1016/j.watres.2019.02.054
36. Ding L, Mao R, Fan, Guo X, Yang X, Zhang Q, Yang C (2019) Microplastics in surface waters and sediments of the Wei River, in the northwest of China. Sci Total Environ 667:427–434. https://doi.org/10.1016/j.scitotenv.2019.02.332
37. Primpke S, Lorenz C, Rascher-Friesenhausen R, Gerdts G (2017) An automated approach for microplastics analysis using focal plane array (FPA) FTIR microscopy and image analysis. Anal Methods 9:1499–1511. https://doi.org/10.1039/C6AY02476A
38. Mason SA, Welch VG, Neratko J (2018) Synthetic polymer contamination in bottled water. Front Chem 6:407. https://doi.org/10.3389/fchem.2018.00407
39. Lebreton L, Slat B, Ferrari F, Sainte-Rose B, Aitken J, Marthouse R, Hajbane S, Cunsolo S, Schwarz A, Levivier A, Noble K, Debeljak P, Maral H, Schoeneich-Argent R, Brambini R, Reisser J (2018) Evidence that the great pacific garbage patch is rapidly accumulating plastic. Sci Rep 8:4666. https://doi.org/10.1038/s41598-018-22939-w
40. Erni-Cassola G, Zadjelovic V, Gibson MI, Christie-Oleza JA (2019) Distribution of plastic polymer types in the marine environment; A meta-analysis. J Hazard Mater 369:691–698. https://doi.org/10.1016/j.jhazmat.2019.02.067
41. Choy CA, Robison BH, Gagne TO, Erwin B, Firl E, Halden RU, Hamilton JA, Katija K, Lisin SE, Rolsky C, Van Houtan S, K. (2019) The vertical distribution and biological transport of marine microplastics across the epipelagic and mesopelagic water column. Sci Rep 9:7843. https://doi.org/10.1038/s41598-019-44117-2
42. Woodall LC, Sanchez-Vidal A, Canals M, Paterson GLJ, Coppock R, Sleight V, Calafat A, Rogers AD, Narayanaswamy BE, Thompson RC (2014) The deep sea is a major sink for microplastic debris. R Soc Open Sci 1:140317. https://doi.org/10.1098/rsos.140317
43. Panno SV, Kelly WR, Scott J, Zheng W, McNeish RE, Holm N, Hoellein TJ, Baranski EL (2019) Microplastic contamination in karst groundwater systems. Groundwater 57:189–196. https://doi.org/10.1111/gwat.12862
44. Xu B, Liu F, Cryder Z, Huang D, Lu Z, He Y, Wang H, Lu Z, Brookes PC, Tang C, Gan J, Xu J (2020) Microplastics in the soil environment: occurrence, risks, interactions and fate—A review. Crit Rev Environ Sci Technol 50:2175–2222. https://doi.org/10.1080/10643389.2019.1694822
45. Pivokonsky M, Cermakova L, Novotna K, Peer P, Cajthaml T, Janda V (2018) Occurrence of microplastics in raw and treated drinking water. Sci Total Environ 643:1644–1651. https://doi.org/10.1016/j.scitotenv.2018.08.102
46. Kosuth M, Mason SA, Wattenberg EV (2018) Anthropogenic contamination of tap water, beer, and sea salt. PLoS One 13. https://doi.org/10.1371/journal.pone.0194970

47. Schymanski D, Goldbeck C, Humpf H-U, Fürst P (2018) Analysis of microplastics in water by micro-Raman spectroscopy: release of plastic particles from different packaging into mineral water. Water Res 129:154–162. https://doi.org/10.1016/j.watres.2017.11.011
48. Mintenig SM, Löder MG, Primpke S, Gerdts G (2019) Low numbers of microplastics detected in drinking water from ground water sources. Sci Total Environ 648:631–635. https://doi.org/10.1016/j.scitotenv.2018.08.178
49. Hernández-Morales G, López-Mendez MC, Rico-Barragán AA, Pérez-Moreno J, Peña-Montes C, Peralta-Pelaez LA, González-Moreno HR (2025) Microplastic pollution in Tropical River: Fourier transform infrared spectroscopy-based characterization of abundance and polymer composition in water and sediments from Filobobos River, Mexico. Hydrology 12:124. https://doi.org/10.3390/hydrology12050124
50. Huang Y, Liu Q, Jia W, Yan C, Wang J (2020) Agricultural plastic mulching as a source of microplastics in the terrestrial environment. Environ Pollut 260:114096. https://doi.org/10.1016/j.envpol.2020.114096
51. Li X, Chen L, Mei Q, Dong B, Dai X, Ding G, Zeng EY (2018) Microplastics in sewage sludge from the wastewater treatment plants in China. Water Res 142:75–85. https://doi.org/10.1016/j.watres.2018.05.034
52. Liu M, Lu S, Song Y, Lei L, Hu J, Lv W, Zhou W, Cao C, Shi H, Yang X, He D (2018) Microplastic and mesoplastic pollution in farmland soils in suburbs of Shanghai, China. Environ Pollut 242:855–862. https://doi.org/10.1016/j.envpol.2018.07.051
53. Miao H, Zhang S, Gao W, Zhou J, Cai H, Wu L, Liu J, Wang Z, Liu T (2024) Microplastics occurrence and distribution characteristics in mulched agricultural soils of Guizhou province. Sci Rep 14:21505. https://doi.org/10.1038/s41598-024-72829-7
54. Thakur B, Singh J, Singh J, Angmo D, Vig AP (2023) Identification and characterization of extracted microplastics from agricultural soil near industrial area: FTIR and X-ray diffraction method. Environ Qual Manag 33:173–181. https://doi.org/10.1002/tqem.22035
55. Arevalo-Hernandez JJ, Barrera de Brito AD, Curi N, Avanzi JC, Silva MLN (2025) A method for the extraction and analysis of microplastics from tropical agricultural soils in Southeastern Brazil. Soil Syst 9:34. https://doi.org/10.3390/soilsystems9020034
56. Steinmetz Z, Wollmann C, Schaefer M, Buchmann C, David J, Tröger J, Muñoz K, Frör O, Schaumann GE (2016) Plastic mulching in agriculture. Trading short-term agronomic benefits for long-term soil degradation? Sci Total Environ 550:690–705. https://doi.org/10.1016/j.scitotenv.2016.01.153
57. Barboza LGA, Dick Vethaak A, Lavorante BRBO, Lundebye A-K, Guilhermino L (2018) Marine microplastic debris: an emerging issue for food security, food safety and human health. Mar Pollut Bull 133:336–348. https://doi.org/10.1016/j.marpolbul.2018.05.047
58. Senathirajah K, Attwood S, Bhagwat G, Carbery M, Wilson S, Palanisami T (2021) Estimation of the mass of microplastics ingested—A pivotal first step toward human health risk assessment. J Hazard Mater 404:124004. https://doi.org/10.1016/j.jhazmat.2020.124004
59. Zhang Y, Kang S, Allen S, Allen D, Gao T, Sillanpää M (2020) Atmospheric microplastics: A review on current status and perspectives. Earth Sci Rev 203:103118. https://doi.org/10.1016/j.earscirev.2020.103118
60. Smith M, Love DC, Rochman CM, Neff RA (2018) Microplastics in seafood and the implications for human health. Curr Environ Health Rep 5:375–386. https://doi.org/10.1007/s40572-018-0206-z
61. Hernandez LM, Yousefi N, Tufenkji N (2019) Plastic teabags release billions of microparticles and nanoparticles into tea. Environ Sci Technol 53:12300–12310. https://doi.org/10.1021/acs.est.9b02540
62. Wright SL, Kelly FJ (2017) Plastic and human health: A micro issue? Environ Sci Technol 51:6634–6647. https://doi.org/10.1021/acs.est.7b00423
63. Jin Y, Lu L, Tu W, Luo T, Fu Z (2019) Impacts of polystyrene microplastic on the gut barrier, microbiota and metabolism of mice. Sci Total Environ 649:308–317. https://doi.org/10.1016/j.scitotenv.2018.08.353

64. Prüst M, Meijer J, Westerink RHS (2020) The plastic brain: neurotoxicity of micro- and nanoplastics. Part Fibre Toxicol 17:24. https://doi.org/10.1186/s12989-020-00358-y
65. Shim WJ, Song YK, Hong SH, Jang M (2016) Identification and quantification of microplastics using Nile red staining. Mar Pollut Bull 113:469–476. https://doi.org/10.1016/j.marpolbul.2016.10.049
66. Zhou X, Wang J, Ren J (2022) Analysis of microplastics in takeaway food containers in China using fpa-ftir whole filter analysis. Molecules 27:9. https://doi.org/10.3390/molecules27092646
67. Dang TT, Sogut E, Uysal-Unalan I, Corredig M (2024) Quantification of polystyrene microplastics in water, milk, and coffee using thermogravimetry coupled with Fourier transform infrared spectroscopy (TGA-FTIR). Chemosphere 368:143777. https://doi.org/10.1016/j.chemosphere.2024.143777
68. Parvin F, Nath J, Hannan T, Tareq SM (2022) Proliferation of microplastics in commercial sea salts from the world longest sea beach of Bangladesh. Environ Adv 7:100173. https://doi.org/10.1016/j.envadv.2022.100173
69. Kuttykattil A, Raju S, Vanka KS, Bhagwat G, Carbery M, Vincent SGT, Raja S, Palanisami T (2022) Consuming microplastics? Investigation of commercial salts as a source of microplastics (MPs) in diet. Environ Sci Pollut Res Int 30:930. https://doi.org/10.1007/s11356-022-22101-0
70. Ahmad M, Ahmad J, Usama M, Al-Swadi HA, Mousa MA, Rafique MI, Al-Wabel MI, Al-Farraj ASF (2025) Microplastic contamination in commercial food and drink products and associated risk of potential human intake in Riyadh, Saudi Arabia. Environ Monit Assess 197:246. https://doi.org/10.1007/s10661-025-13680-9
71. Prata JC (2018) Airborne microplastics: consequences to human health? Environ Pollut 234:115–126. https://doi.org/10.1016/j.envpol.2017.11.043
72. Schwabl P, Köppel S, Königshofer P, Bucsics T, Trauner M, Reiberger T, Liebmann B (2019) Detection of various microplastics in human stool: A prospective case series. Ann Intern Med 171:453–457. https://doi.org/10.7326/M19-0618
73. Leslie HA, van Velzen MJM, Brandsma SH, Vethaak AD, Garcia-Vallejo JJ, Lamoree MH (2022) Discovery and quantification of plastic particle pollution in human blood. Environ Int 163:107199. https://doi.org/10.1016/j.envint.2022.107199
74. Lee D-W, Jung J, Park S, Lee Y, Kim J, Han C, Kim H-C, Lee JH, Hong Y-C (2024) Microplastic particles in human blood and their association with coagulation markers. Sci Rep 14:30419. https://doi.org/10.1038/s41598-024-81931-9
75. Leonard S, Liddle CR, Atherall CA, Chapman E, Watkins M, Calaminus S, Rotchell JM (2024) Microplastics in human blood: polymer types, concentrations and characterization using μFTIR. Environ Int 188:108751. https://doi.org/10.1016/j.envint.2024.108751
76. Rotchell JM, Jenner LC, Chapman E, Bennett RT, Bolanle IO, Loubani M, Sadofsky L, Hobkirk J, Palmer TM (2023) Detection of microplastics in human saphenous vein tissue using μFTIR: A pilot study. PLoS One 18:e0280594. https://doi.org/10.1371/journal.pone.0280594
77. Jenner LC, Rotchell JM, Bennett RT, Cowen M, Tentzeris V, Sadofsky LR (2022) Detection of microplastics in human lung tissue using μFTIR spectroscopy. Sci Total Environ 831:154907. https://doi.org/10.1016/j.scitotenv.2022.154907

FTIR Spectroscopy for Microplastic Analysis

Mihir Nakul, Samruddhi Kothawade, Sanket Dinesh Rao, Budheswar Dehury, Nirmal Mazumder, and Rupak Kumar Sarma

Abstract Microplastic contamination has emerged as a critical environmental and public health concern, necessitating robust analytical methods for detection and characterisation. This chapter comprehensively reviews the application of Fourier Transform Infrared (FTIR) spectroscopy for microplastic analysis across environmental, food safety, and biomedical research domains. FTIR spectroscopy enables precise identification of microplastic polymers through their unique spectral fingerprints, with characteristic absorption bands distinguishing common plastics including polyethylene (PE), polypropylene (PP), polystyrene (PS), polyethylene terephthalate (PET), and polyvinyl chloride (PVC). The chapter examines various FTIR spectroscopy modes, including Attenuated Total Reflectance (ATR-FTIR), transmission, and reflectance FTIR spectroscopy, analysing their advantages and limitations for different sample types. Advanced techniques such as micro-FTIR (μFTIR) spectroscopy enable the detection of particles as small as 10–20 μm, while FTIR imaging provides spatial mapping capabilities for high-throughput analysis. Integration with artificial intelligence (AI) and machine learning approaches, including Convolutional Neural Networks (CNNs) and Principal Component Analysis (PCA), has enhanced identification accuracy and enabled automated classification processes. Applications span from riverine and marine microplastic monitoring to detection in seafood, drinking water, and human biological samples, demonstrating concentrations ranging from 1 to 302 particles per individual in various matrices. This establishes FTIR spectroscopy as an indispensable tool for microplastic research, providing both qualitative and quantitative analytical capa-

M. Nakul · S. Kothawade · S. D. Rao · B. Dehury
Department of Bioinformatics, Manipal School of Life Sciences, Manipal Academy of Higher Education, Manipal, India

N. Mazumder
Department of Biophysics, Manipal School of Life Sciences, Manipal Academy of Higher Education, Manipal, India

R. K. Sarma (✉)
Department of Botany, Nalbari College, Nalbari, Assam, India

N. Mazumder et al. (eds.), *Vibrational Spectroscopy for Biopolymer Characterization*, Biological and Medical Physics, Biomedical Engineering,
https://doi.org/10.1007/978-3-032-15306-7_10

bilities that support environmental monitoring, assessment of food safety, and human health evaluation in the context of global plastic pollution.

1 Introduction

Plastic pollution is driven by widespread human activities observed throughout the terrestrial and remote marine ecosystems. These plastics include PET, HDPE (High-density Polyethylene), PVC, LDPE (Low-density Polyethylene), PP, and PS which are widely used in industries because of their properties like durability, flexibility and resilience [1]. However, very few proportions of these plastics are effectively recycled or degraded. Exposure of these plastics to diverse natural calamities promotes their fragmentation into smaller particles, fostering their distribution and long-term stability within the environment. The degradation of macroplastics (MAPs) results in the formation of microplastics (MPs) and nanoplastics (NPs). Microplastics, particles smaller than 5 mm, are persistent environmental pollutants that accumulate widely in oceans, sediments, and aquatic organisms, particularly within the gastrointestinal tracts of fish destined for human consumption. Their presence poses multiple risks, including inflammation, gastrointestinal blockages, and exposure to adsorbed contaminants such as pesticides and heavy metals. Furthermore, MPs provide surfaces favourable to the infection by pathogenic microorganisms, amplifying their ecological and health-related impacts [1]. The persistence and aggregation of MPs in the ecosystem present significant environmental integrity and human health challenges, underscoring the urgent need for effective waste management strategies and innovative remediation technologies.

Advanced analytical methods are employed to investigate and identify MPs and related microparticles. Among the most prominent are pyrolysis–gas chromatography/mass spectrometry (Py-GC/MS) and vibrational spectroscopy. Vibrational spectroscopy, including FTIR and Raman spectroscopy, is particularly valuable as it enables detailed chemical characterisation of microparticles in a non-destructive manner. FTIR spectroscopy is widely used among vibrational spectroscopy techniques. It measures the interactions of light with the particles and creates a unique molecular fingerprint. This fingerprint is compared against reference libraries to identify materials [2]. Instead of tedious and time-consuming manual comparisons, automated matching algorithms and reference libraries are preferred. It calculates a Hit Quality Index (HQI), which measures the sample's close resemblance to a known material. These HQIs can be affected by noise or other disturbances in the data. Thus, to achieve maximum accuracy, HQI is combined with visual checks of top matches, filtering out the results by subsampling and accepting the matches above a particular threshold. FTIR spectroscopy is highly dependent on the quality of reference libraries created [2]. FTIR spectroscopy analysis cycle follows sequential steps from spectral fingerprint generation to material identification, validation, and library updating for improved accuracy. A detailed workflow of FTIR spectroscopy has been represented in Fig. 1.

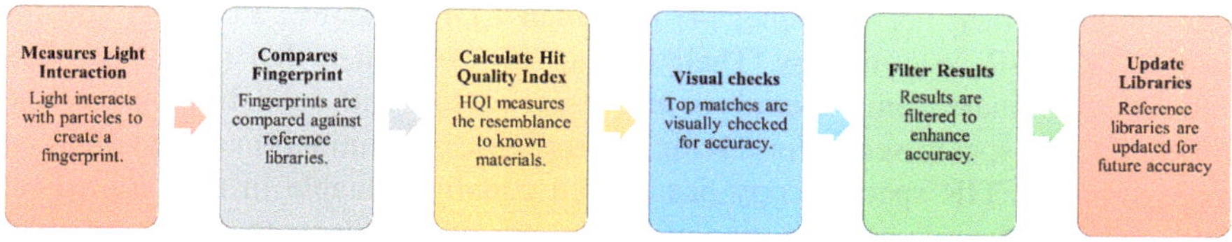

Fig. 1 FTIR spectroscopy analysis cycle

FTIR spectroscopy-based techniques further include μFTIR and ATR-FTIR due to their capability of generating polymer-specific vibrational spectra, which allows reliable classification of commonly identified polymers like PE, PP, PS, PET, and PVC [3]. FTIR provides high identification accuracy of approximately 91% across the particles of various sizes in microns [4].

Raman spectroscopy offers high spatial resolution and is helpful for analysis of <20 μm microplastic particles. Ultimately, specific assessments depicted that the detection efficiency of Raman spectroscopy is lower than that of μFTIR spectroscopy for particular polymers. It does not work well with plastics containing pigments or carbon due to fluorescence interference, leading to weaker signals. It limits the ability of Raman spectroscopy to be used to identify specific polymers. On a similar note, Py-GC/MS is considered a destructive but highly sensitive method that confirms microplastics' chemical composition. It performs thermal decomposition of polymers and analyses their breakdown into their by-products. It permits precise polymer quantification, achieving mass recovery rates of approximately 86% [3]. It is used to determine fibrous microplastics and degraded/mixed polymer particles. It persists with a drawback, its inability to retain information on shape or size, which makes it less suitable for morphological analysis or rapid screening. NIR (Near-Infrared spectroscopy) is a fast, non-destructive, reflectance-based method. It is a simple technique for real-time screening, suitable for larger microplastics >100 μm and detects diverse polymers, demonstrating correlation values close to ATR-FTIR spectroscopy [4]. On comparing four polymers (PE, PP, PS, PET, and PVC), similar average correlation coefficients were found ($R^2 \approx 0.77$–0.80) using this technique. Observations stated that NIR performs well for specific polymers, PP and PET, but failed to detect PS, which was successfully identified by ATR-FTIR spectroscopy. Studies depicted a lack of standardised NIR spectral libraries, which hinders accuracy in classification across different polymers [4].

Compared to other techniques, FTIR spectroscopy extends its scope beyond comparative analytical experiments when applied to environmental, food-related and biomedical research domains. In ecological studies, μFTIR spectroscopy is extensively used to classify the diverse range of microplastics. It identified common polymers with higher accuracy and minimal spectral interference, making them a preferred choice for analysing complex and variable environmental samples [3, 4]. Although the food sample analysis is not the only focus of cited studies, the microplastics detected using the FTIR spectroscopy are associated with food-related resources involved in packaging materials, consumer plastics and fibres from household utilities. For example, [4] identified LDPE, HDPE, PET, PP, and PS using ATR-FTIR spectroscopy. Their findings demonstrated that these polymers frequently originate from materials in prolonged contact with food, highlighting the utility of FTIR spectroscopy for tracing microplastic contamination across the food supply chain. FTIR spectroscopy has proven equally valuable in the biomedical field, enabling the reliable detection of minute plastic fragments that pose potential health hazards. When coupled with computational approaches such as CNNs and multilayer perceptrons (MLPs), FTIR-based analysis can be automated, achieving near-perfect identification accuracy of up to 100% [1]. These advances establish FTIR spectroscopy as a robust identification tool and a promising technique for medical screening.

Recent progress in FTIR spectroscopy methodology has further demonstrated its scalability for high-throughput monitoring. Incorporating preprocessing techniques, including first-derivative correction, smoothing, and baseline normalisation, has significantly enhanced spectral match quality and reduced misclassification rates [2]. These developments underscore the expanding role of FTIR spectroscopy in addressing complex analytical challenges across environmental, food and biomedical research, where accurate microplastic detection is increasingly critical. The flowchart in Figure 2 illustrates the pathway from plastic pollution to microplastic identification using analytical techniques such as Py-GC/MS and vibrational spectroscopy, with applications spanning environmental studies, food analysis, and biomedical fields.

2 Spectral Properties of Microplastics

The categorisation of different plastic types relies on their distinct physicochemical properties, which can be accurately characterised using FTIR spectroscopy. Each type of plastic has a unique spectral fingerprint of vibrations, which depicts absorption bands at defined wavenumbers (measured in cm^{-1}), demonstrating an FTIR spectrum. The spectral fingerprint enables precise identification and differentiation of various plastics based on their molecular structure, as shown in Table 1. The FTIR spectra of multiple microplastics in the environment are shown in Fig. 3.

The summaries of the functional group and key FTIR spectroscopy absorption peaks for commonly encountered polymers in microplastic analysis have been described in Table 1. These spectral features form the foundation for FTIR spectroscopy's polymer identification, as evidenced in recent analytical studies [8, 9].

These spectral fingerprints' clarity and accuracy also rely on the method by which the spectra are recorded. Techniques like ATR-FTIR, Transmission-FTIR and Reflectance-FTIR spectroscopy influence the interaction between the infrared light and the sample, affecting the detection efficiency and practical application. A comparative overview of the advantages and limitations of different FTIR spectroscopy modes applied to microplastic analysis has been discussed in Table 2. This comparison offers valuable insights derived from recent methodological studies,

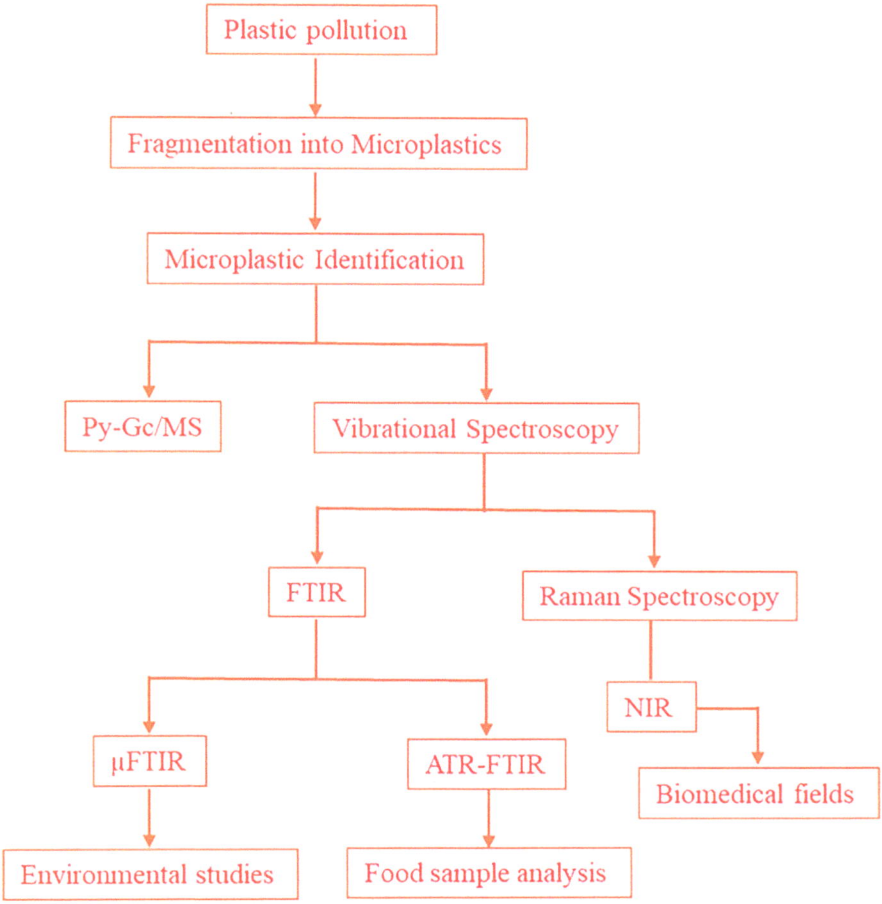

Fig. 2 Approaches for microplastic identification

highlighting how various infrared spectroscopy techniques contribute to the analysis of both microplastics and nanoplastics.

In microplastic research, polymer identification relies on single-point FTIR measurements such as ATR and transmission modes. While effective, these approaches require the analysis of individual particles, making them time-intensive and impractical for large-scale environmental monitoring. The advent of FTIR imaging, particularly with Focal Plane Arrays (FPA) detectors, has significantly advanced analytical capabilities. This technique enables the acquisition of spatially resolved spectra across entire sample areas, eliminating the need for manual particle selection. FTIR imaging accelerates analysis and enhances objectivity by simultaneously covering multiple particles [11]. Integrating reflectance-FTIR imaging further expands its utility by enabling contactless analysis, which is particularly

Table 1 Characteristic FTIR spectroscopy peaks and Functional groups of common microplastic polymers

Polymer	Functional groups	Key FTIR peaks (cm^{-1})	Spectral Significance	References
PE (polyethylene)	Methylene groups (–CH_2–)	2915, 2845 (C–H stretch); 1472, 1462 (CH_2 bending); 730, 717 (CH_2 rocking)	Long, simple carbon chains typical of PE; the ~719 cm^{-1} peak is a marker for crystalline structure	Jung et al. [5]
PP (polypropylene)	Methyl (–CH_3–) and methylene (–CH_2–) groups	2950, 2915, 2838 (C–H stretch); 1455 (CH_2 bend); 1377 (CH_3 symmetric deformation); 997, 972, 840, 808 (CH_2 rock, C–C stretch, C–CH stretch)	The 1377 cm^{-1} peak is characteristic of PP and helps differentiate it from PE	Jung et al. [5]
PS (polystyrene)	Aromatic ring (C_6H_5)	3024 (aromatic C–H stretch); 1601, 1492 (aromatic ring stretch); 1452, 1027, 753, 700, 694 (ring vibrations, aromatic CH bend); 2000–1650 ("benzene fingers")	Strong benzene ring signals make PS easy to identify; the "benzene fingers" region is unique to aromatic rings	Jung et al. [5]
PET (polyethylene terephthalate)	Ester (–COO–), aromatic ring	1715 (C=O stretch); 1241–1094 (C–O stretch); 1400–1600 (ring stretch); 720 (aromatic CH out-of-plane bend)	The strong 1715 cm^{-1} C=O stretch is a signature for PET and other polyesters; additional peaks confirm aromatic and ester groups	Jung et al. [5]
PVC (polyvinyl chloride)	C–Cl, –CH_2– Groups	616 (C–Cl stretch); 610–690 (C–Cl stretch range); 1427 (CH_2 bend); 1331, 1255 (CH bend); 966 (CH_2 rock); 2910, 2845 (C–H stretch)	Peaks at 610–690 cm^{-1} confirm chlorine, unique to PVC; CH_2 bending and rocking further verify polymer backbone	Jung et al. [5]

advantageous for heterogeneous or fragile samples [8, 10]. Beyond improving efficiency, this approach addresses the pressing need for reproducible and scalable methodologies in microplastic studies. For instance, mapping particles directly on filter substrates using reflectance-FTIR imaging has proven effective for rapidly screening large sample areas, removing the reliance on infrared-transparent supports [10]. Additionally, coupling spectral imaging with statistical models such as Partial Least Squares Discriminant Analysis (PLS-DA) has enhanced polymer identification accuracy in complex matrices.

These methodological advancements represent a pivotal shift in microplastic analysis—from labour-intensive, low-throughput techniques to semi-automated,

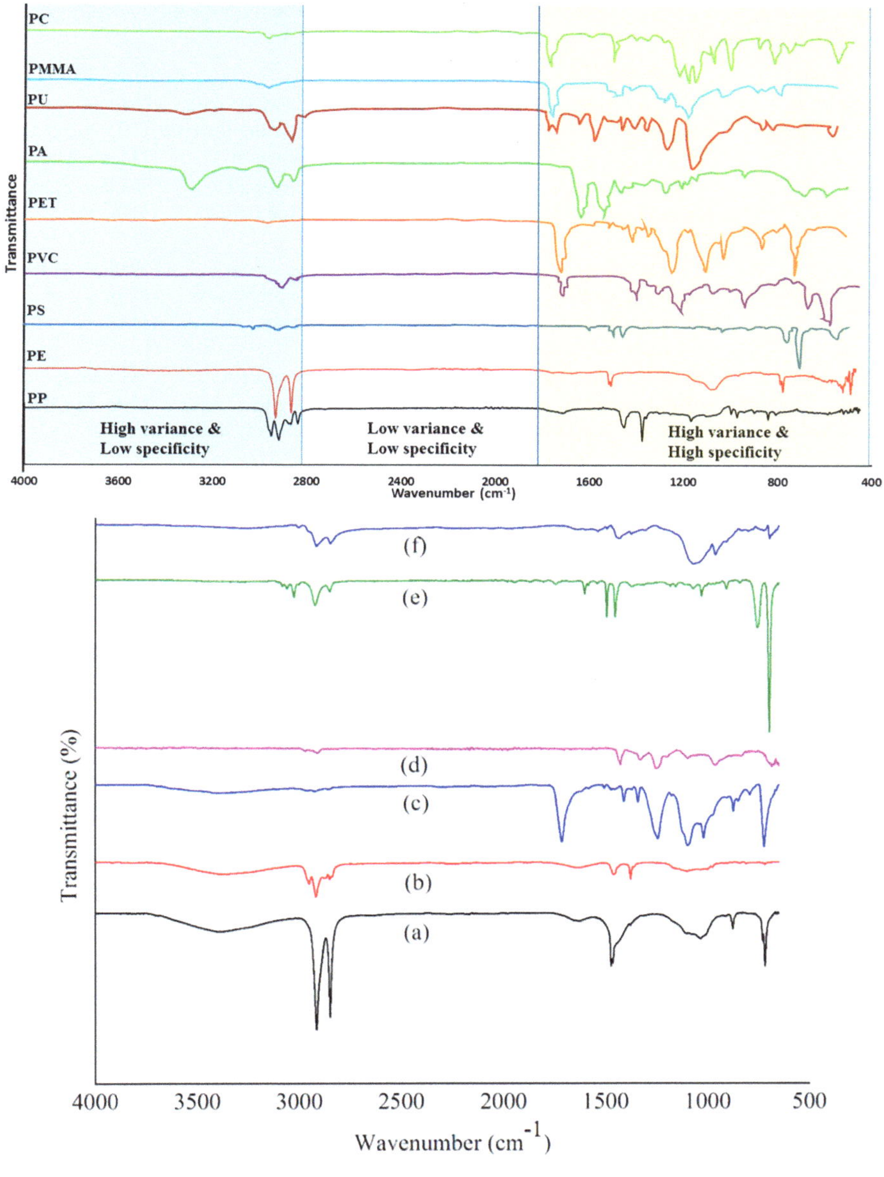

Fig. 3 The spectral features of FTIR spectroscopy of various microplastics found in the environment. Reproduced from Veerasingam et al. [6], under the terms of the CC BY 4.0. 2) FTIR spectra of microplastics: (**a**) polyethylene (PE), (**b**) polypropylene (PP), (**c**) polyethylene terephthalate (PET), (**d**) polyvinyl chloride (PVC), (**e**) polystyrene (PS), and (**f**) styrene butadiene rubber (SBR). Reproduced from Islam et al. [7], under the terms of the CC BY 4.0

Table 2 Comparative overview of ATR-FTIR, transmission, and reflectance spectroscopy modes for microplastic analysis

FTIR mode	Advantages	Drawbacks	References
ATR-FTIR spectroscopy	Produces high-quality spectra	Requires direct contact with the sample, increasing the risk of cross-contamination	Willans et al. [10], Xie et al. [11]
	Suitable for thick, strongly absorbing particles	Less effective for small particles due to the limited interaction area	Willans et al. [10], Xie et al. [11]
	Commonly used for manual polymer identification in microplastic research	Time-consuming when analysing large numbers of particles effectively	Willans et al. [10], Xie et al. [11]
Transmission FTIR spectroscopy	Provides good spectral resolution if the sample thickness is optimal	Requires IR-transparent substrates, adding to procedural complexity and cost	Willans et al. [10], Xie et al. [11]
	Ideal for thin samples with weak IR absorption	Limited to small particles that permit IR transmission	Willans et al. [10], Xie et al. [11]
	Commonly used for hyperspectral imaging applications	Labor intensive	Willans et al. [10], Xie et al. [11]
Reflectance FTIR spectroscopy	Non-contact analysis allows high-throughput and semi-automated workflows	Spectral quality can be affected by refractive index variations and surface irregularities	Willans et al. [10] Xie et al. [11]
	Effective for thick or opaque particles	Careful calibration and preprocessing are required for data interpretation	Willans et al. [10], Xie et al. [11]
	Suitable for large area imaging and mapping	Limited by the availability of spectral libraries for reflectance modes	Willans et al. [10], Xie et al. [11]

high-volume systems with improved reliability. Significantly, they also mark progress toward standardising microplastic detection workflows, ensuring applicability across scientific research and environmental monitoring contexts [8, 10].

3 Sampling and Sample Preparation

Accurately identifying and quantifying microplastics in environmental matrices relies heavily on standardised sampling and pre-treatment protocols. Careful sample preparation is essential to minimise contamination and ensure data reliability. The structured steps involved in sampling and pre-treatment are illustrated in Fig. 4. The diagram outlines the procedural steps commonly applied in freshwater

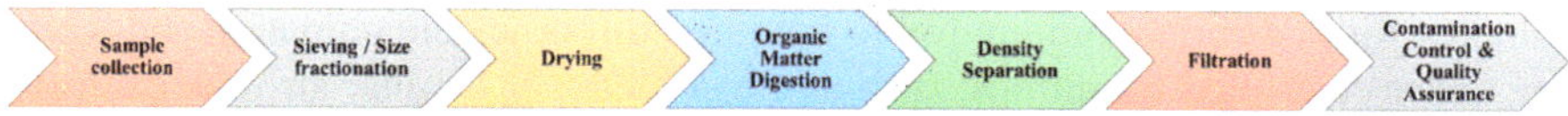

Fig. 4 Workflow for the collection, pre-treatment, and contamination control of microplastic samples

sediment, wastewater, and sludge analyses, depending on the type of environmental matrix.

Sample collection: In wastewater studies, water samples are typically obtained using grab samplers or automatic sampling devices, with volumes ranging from approximately 1 L for influent to up to 500 L for effluent. These samples are subsequently processed to ensure representativeness [12]. Soil and sediment samples are generally collected with stainless-steel or grab samplers, focusing on the upper 2–5 cm sediment layer. Using stainless-steel equipment is strongly recommended to prevent the introduction of synthetic polymers from plastic tools, thereby reducing the risk of contamination [13]. Controlled sampling conditions are essential for air and biological samples to minimise airborne contamination during collection.

Sample Pretreatment: Several pre-treatment steps are necessary to process microplastic samples effectively. Sieving or size fractioning separates particles by size using mesh sieves (0.3–5 mm), typically through wet, dry, or sequential sieving. Wet sieving reduces static cling and particle loss, while sequential sieving helps classify particles into distinct size ranges [13]. Drying is then performed to eliminate moisture via oven drying (≤60 °C), air drying, or freeze drying, with care taken to avoid overheating, which may damage polymers [13]. Organic matter digestion follows, targeting the removal of organic debris and proteins using chemical agents (e.g., hydrogen peroxide, potassium hydroxide, Fenton's reagent, HCl) or enzymatic digestion (e.g., proteinase-K, cellulase). While enzymatic digestion is gentler, it may obscure some polymers, such as HDPE. Acids and alkalis are prohibited for PET, and PS is vulnerable to spectral damage, making hydrogen peroxide preferable in such cases. Density separation is then employed, utilising salt solutions such as NaCl (~1.2 g/cm^3), $ZnCl_2$ (~1.5 g/cm^3), or NaI (~1.6–1.7 g/cm^3) to float plastics away from heavier materials. NaCl is cost-effective but insufficient for dense polymers like PET and PVC. Filtration represents the final step, isolating MPs using fine glass fibre or stainless steel mesh filters, carefully choosing pore sizes to match target size ranges. Efficiency is improved through vacuum or pressure filtration [12, 13].

Contamination control and quality assurance: The reliability of microplastic analysis critically depends on strict contamination control throughout sampling, preparation, and laboratory processing. Given the widespread presence of microplastics and the risk of laboratory contamination, rigorous protocols are essential. Key measures include the use of non-plastic tools and containers (glass, metal, or stainless steel) to prevent the introduction of synthetic polymers, prefiltering all solutions and reagents before use and incorporating "process blanks" as quality controls, wherein a blank sample is processed alongside test samples to detect contamination [12, 13]. Specific security measures can be involved, like covering samples with lids or aluminium foil to minimise airborne contamination and working under controlled conditions, such as clean rooms or laminar air-flow cabinets, during filtration and digestion steps. This collectively emphasises contamination control as the essential part of microplastic research, portraying accurate results without being distorted by laboratory-introduced artefacts. Spike recovery tests are recommended to validate recovery rates and detect potential losses or contamination during pre-treatment [12].

4 Identification and Quantification

4.1 Building and Curating Spectral Libraries

A significant step towards accurate microplastic identification using FTIR spectroscopy is the creation of high-quality spectral libraries. These databases emphasise the characteristic absorption spectra for various polymers—PET, HDPE, PVC, LDPE, PP, and PS-enabling automated and reproducible material classification. The construction of effective spectral libraries involves several key considerations, including the diverse polymer types, spectral resolution, and baseline correction techniques [1]. The study by Villegas-Camacho et al. [1] incorporated the application of diverse normalisation methods such as min-max, vector, and z-score normalisation, before adding spectra to the library. It also demonstrates that choosing the appropriate normalisation method significantly impacts the healthy-classified system, distinguishing between polymers. Moreover, researchers recommend continuously updating and validating spectral libraries using real environmental samples and known standards to improve their relevance and reliability in microplastic research. By focusing on these elements, researchers can move beyond manual, subjective visual comparisons and towards highly reproducible, automated systems for microplastic classification, paving the way for consistent and reliable results in environmental studies.

4.2 Use of Chemometric Tools for Automated Classification

As spectral libraries become increasingly refined and comprehensive, microplastic identification has advanced by integrating chemometric and machine learning tools to automate classification. Traditional manual matching of FTIR spectra is time-consuming and prone to human error. Advanced computational approaches are now widely adopted to address these limitations. Specialised spectral libraries, such as the FLOPP and FLOPP-e databases, have been developed specifically for μATR-FTIR analysis of commonly encountered polymers [14]. These databases are compatible with automated algorithms that employ multivariate techniques, including

PCA, hierarchical clustering, and supervised classification models, to rapidly and accurately identify polymers from complex spectral datasets. Transparency in algorithm design and adopting standardised protocols are emphasised to ensure reproducibility across laboratories and environmental conditions. Integrating curated spectral libraries with chemometric workflows represents a critical step in transforming FTIR spectroscopy from a semi-qualitative tool into a quantitative, high-throughput diagnostic platform for microplastic research.

4.3 Micro-FTIR Spectroscopy for Particle Size Down to 10–20 μm

Recent advances in FTIR instrumentation have significantly improved the resolution and sensitivity required to detect and classify microplastic particles at smaller scales. μFTIR spectroscopy has emerged as a key technique for analysing particles as small as 10–20 μm, a size range highly relevant to environmental and biological monitoring. Equipped with FPA detectors in transmission mode, μFTIR systems enable high-resolution spectral imaging that supports the identification and quantification of microplastics below 100 μm, with demonstrated performance approaching the 10 μm threshold [14]. By integrating infrared microscopy with spatially resolved detection, μFTIR produces comprehensive spectral maps that capture particle size, shape, and composition variability across heterogeneous samples. When paired with automated particle recognition software, μFTIR spectroscopy increases throughput and reduces operator bias. Consequently, this method is particularly effective for detecting the smaller microplastic fraction often overlooked by conventional bulk FTIR techniques.

4.4 FTIR Spectroscopy Imaging for In Situ Polymer Mapping

Beyond particle-level identification, FTIR imaging extends analytical capabilities by enabling in situ mapping of microplastic spatial distribution. Unlike point-based approaches, FTIR imaging collects thousands of spectra simultaneously across a sample area, generating chemical maps that delineate the location, size, and polymer type of individual particles. This methodology has been demonstrated through μATR-FTIR imaging in reflection mode, consistently identifying polymer particles smaller than 20 μm even within complex and heterogeneous matrices [15]. The combination of spectral data with spatial resolution provides valuable insight into deposition patterns of microplastics, particularly in sediments, biological tissues, and surface films. Integration with automated classification software further enhances reproducibility and throughput, enabling the rapid screening of large sample sets with minimal operator involvement. FTIR imaging offers a precise, non-destructive, and scalable approach to microplastic mapping, advancing our understanding of their environmental distribution, supporting risk assessments, and clarifying polymer-specific pathways within ecosystems.

5 Environmental and Biological Case Studies

5.1 Riverine and Marine Microplastic Monitoring

FTIR spectroscopy offers high accuracy in polymer identification and quantification. Its versatility is exemplified by initiatives ranging from local river surveys to investigations in remote oceanic regions, enabling assessments of spatial distribution, sources, and types of plastic pollution. One notable example is the Microplastic Detectives citizen science project along the German coast. By engaging volunteers in collecting 1139 samples from 71 sites, the study employed ATR-FTIR spectroscopy to analyse suspected particles larger than 1 mm. Plastics were detected in 15.5% of samples (177 out of 1139), demonstrating both spatial variability in contamination and the effectiveness of combining citizen-driven sampling with rigorous analytical methods [16]. In freshwater systems, FTIR spectroscopy has shed light on sources and transport pathways of microplastics from land to sea. For example, in Indonesia's Martapura River, FTIR analysis identified PS and polycarbonate as dominant polymers in both water and sediment. Particles were found in diverse morphologies, including fragments, fibres, and films, with fibres strongly associated with aquaculture activities [17]. Expanding to the global marine context, FTIR imaging has enabled detection in remote environments such as the Ross Sea, Antarctica, where concentrations ranged from 0.0032 to 1.18 particles per cubic meter of seawater. PE and PP were the most common polymers, with fragments (71.9%) and fibres (12.7%) dominating the morphologies [18]. Together, these findings highlight both the ubiquity of plastic pollution and the ability of FTIR spectroscopy to characterise its composition and physical forms across diverse aquatic systems.

5.2 FTIR Detection in Seafood, Drinking Water, and Human Samples

FTIR spectroscopy has also been widely applied to samples with direct implications for human health, including seafood, drinking water, and human biological material. Its sensitivity and specificity in distinguishing polymer types make it indispensable for tracing microplastic contamination along the environmental to human exposure continuum. FTIR spectroscopy has become a standard tool for microplastic detection and polymer identification in seafood. For instance, anchovies collected from Indonesian harbours showed substantial contamination: individuals from Alor contained an average of 302 ± 1 particles, predominantly microfibers (95%) measuring 50–500 μm, while those from Balikpapan averaged 130 ± 1.73 particles, mostly microfilms (43.83%) in the 20–50 μm range [19]. Similarly, studies on green mussels identified a variety of polymers, including cellulose, PA, PET, PP, PVC, and PE, through the combined use of micro-Raman, ATR-FTIR, and scanning electron microscopy. This integrative approach enabled detailed chemical and morphological characterisation of microplastics in aquaculture systems [20]. Employing a combination of micro-Raman spectroscopy, ATR-FTIR spectroscopy, and scanning electron microscopy allows a detailed, multi-faceted understanding of microplastics in aquaculture systems, providing chemical and morphological characterisation. Drinking water has also been a focus of FTIR-based research. Imaging approaches have consistently detected microplastics in bottled water samples, with concentrations ranging from 172 to 944 particles per 500 mL. The predominant polymers included cellulose (70.4%) and polyamide (17.8%), alongside PET, PP, PVC, and PE [21]. In response to such findings, regulatory bodies such as the European Union have adopted standardised protocols based on micro-FTIR spectroscopy to ensure reliable and reproducible detection of particles within the 20 μm–5 mm size range [21]. Most striking are studies demonstrating microplastic presence in human biological systems. Analyses of faecal samples consistently identified 1–20 particles per sample, with PE being the most prevalent polymer (57%), followed by PP (24%) and PS (19%) [22]. Lifestyle factors also appear to influence exposure: frequent use of plastic packaging and higher consumption of processed foods were correlated with elevated microplastic loads in stool samples [23].

5.3 *Degradation Studies Involving Microplastics*

Understanding the degradation pathways of microplastics in the environment is crucial for assessing their long-term fate and ecological impacts. FTIR spectroscopy studies have proven invaluable in characterising these transformation processes, particularly in revealing molecular-level structural changes induced by multiple environmental factors.

Ultraviolet (UV) radiation is a major driver of microplastic degradation [24]. FTIR spectroscopy analysis reveals the accumulation of functional groups such as carbonyl and hydroxyl, which serve as indicators of oxidation and polymer chain scission [25]. Accelerated weathering experiments confirm that elevated temperature and humidity amplify degradation rates, leading to increased fragmentation and the generation of smaller particles. The carbonyl index, derived from FTIR spectra, is commonly used to quantify the extent of oxidative degradation [26, 27]. Biological processes also play a role. FTIR-based studies have shown that biofilm formation can induce significant physicochemical modifications within weeks. For example, PE, PP, and PS beads exposed to natural bacterioplankton in the Baltic Sea exhibited altered crystallinity, stiffness, and compression properties [28]. Shifts in surface chemistry parallelled these changes, readily detected using ATR-FTIR spectroscopy, highlighting the interplay between microbial colonisation and polymer structure. Moreover, FTIR techniques have enhanced quantification of microplastic fragmentation and overall breakdown. Using adapted NanoRelease protocols, researchers demonstrated that UV exposure fragments polymers into micro- and nanoplastic particles and drives substantial mass loss by releasing dissolved organic compounds [29]. This dual process of particle generation and chemical dissolution highlights the multifaceted nature of microplastic degradation and underscores the importance of quantifying both particulate and dissolved fractions when modelling environmental fate. Together, these findings reinforce the essential role of FTIR spectroscopy in various fields of microplastic research.

6 Advanced Techniques and AI Integration

The growing complexity of microplastic pollution analysis has created a pressing need for approaches beyond traditional laboratory methods. As the diversity of polymer types, environmental matrices, and demand for high-throughput analysis increases, advanced techniques have emerged to enhance accuracy, automation, and scalability. Notably, AI, machine learning, and portable instrumentation innovations have transformed FTIR spectroscopy into a more powerful and accessible analytical platform.

One of the most transformative developments is the application of deep learning for spectral classification. CNNs, particularly one-dimensional variants (1D-CNNs), have shown strong proficiency in processing FTIR spectral data [30]. Unlike conventional approaches that rely on manual feature extraction or basic chemometric methods, 1D-CNNs can automatically learn and extract relevant spectral features, enabling highly accurate classification across multiple polymer types. Importantly, their robustness persists even when spectra are complicated by noise or variability in real-world samples. Architectures inspired by LeNet5 outperform traditional machine learning models such as decision trees and random forests on small and large datasets. Beyond CNNs, specialised architectures further advance the field: autoencoders can reconstruct and correct distorted spectra, semi-supervised autoencoders integrate unsupervised feature learning with classification to boost robustness [31], and dual-path CNNs simultaneously perform classification and regression, expanding their applicability for detailed microplastic characterisation [9, 32].

Although deep learning has gained prominence, classical chemometric techniques remain essential. PCA is widely used for dimensionality reduction, condensing high-dimensional FTIR data into fewer variables without significant information loss. This facilitates classification and visualisation of clustering patterns, enabling discrimination of polymer types and particle sizes in complex mixtures [33]. Hierarchical Cluster Analysis (HCA), an unsupervised method, allows data to self-organise into meaningful clusters, making it valuable for exploratory studies, identifying unknown polymers, and detecting anomalies [30]. These approaches also enhance spectral library organisation, improving subsequent identification workflows. While less complex, decision tree algorithms offer interpretability and transparency, making them particularly valuable in regulatory contexts and standard operating procedures.

Parallel to advances in computational methods, portable FTIR systems have increased accessibility and efficiency in microplastic analysis. Compact and robust, these instruments enable in situ measurements across diverse environments, from aquatic ecosystems to industrial sites, reducing the need for extensive sample transport [34]. Near-infrared spectrometers operating in the 900–1700 nm range exemplify this trend, delivering much of the analytical capability of benchtop systems while offering cost, speed, and usability advantages. The democratisation of such technologies has broadened participation in microplastic monitoring, empowering research institutions, smaller organisations, and citizen science initiatives.

Large-scale projects such as Microplastic Detectives illustrate the potential of this approach. Volunteers equipped with portable FTIR units have collected and analysed thousands of samples, extending monitoring to scales previously unattainable by conventional research teams [16]. This model enhances spatial and temporal coverage while increasing public engagement and providing rich datasets for environmental policy development. Recent innovations have further accelerated efficiency by integrating automation into portable systems. For example, the MARS platform (Microplastic Analysis using Reflectance Spectroscopy) combines reflectance accessories with automated imaging and sample handling. Such systems achieve classification accuracies above 98%, particularly for degraded plastics, while reducing analysis times by several-fold compared with traditional methods [8]. The convergence of portable instrumentation, automation, and AI-driven analysis signals a new frontier in microplastic research. Features such as real-time classification, adaptive machine learning algorithms, and cloud-based spectral library integration are rapidly becoming standard. These innovations ensure that FTIR spectroscopy remains a cornerstone of environmental monitoring and microplastic research in the digital age.

7 Challenges, Trends and Future Directions

Despite FTIR spectroscopy's strengths, persistent challenges limit its accuracy and efficiency, especially when it comes to detecting tiny microplastic particles in complex matrices. Drawing insights from several recent studies, this section highlights the main obstacles researchers face and the practical workarounds they employ.

One significant difficulty while using FTIR spectroscopy for microplastic analysis is detecting particles below 10–20 μm. To put this in perspective, human hair is typically about 70–100 μm wide, so these MPs are roughly ten times smaller than a strand of hair. This size limit arises from how infrared light interacts with matter. Due to the diffraction limit, traditional μFTIR imaging tools cannot detect or map particles smaller than 10 μm reliably. As described in a study Othman et al. [15] secondary metabolites or weathered plastic fragments that break below this size threshold can easily evade detection. To overcome this, the researchers used ATR-FTIR, which can handle smaller particles more effectively. Interestingly, they observed that a perfectly shaped 6 μm sphere produced stronger absorption peaks than larger particles (10–12 μm) because it filled more of the interaction area with the light beam. However, this only holds if the particles are ideally shaped and well-packed.

Another study Othman et al. [15] expands on the technical resolution challenge. In FTIR imaging, the analysis area is divided into pixels, each representing a specific spatial resolution that depends on the microscope's objective lens. For example, with a standard 15× objective, each pixel might be about 2.7 μm wide, while a 3.5× objective yields pixels about 11.6 μm wide. If microplastic fragments are smaller than the pixel size, their spectral signals become weak and may blend into the background, reducing detection confidence. Each FTIR spectroscopy mode—whether transmission, reflection, or ATR—has its own minimum detection size, known as the limit of detection (LOD). For ATR-FTIR, the practical LOD is around 10–20 μm under optimal conditions, but anything smaller often slips through, especially if the particle does not make good contact with the crystal.

Detection limits are also influenced by how samples are prepared and handled. In surface water studies [35]. Researchers tackled the challenge of detecting particles under 20 μm using specialised filters like Ano Disc® membranes with pore sizes as small as 0.2 μm. Even then, the microscopic area analysed per image is minimal, often just 4 mm^2, only about 0.82% of the total filter area. Scanning the entire filter would take upwards of 48 h, so researchers frequently settle for subsampling, which carries the risk of missing rare or tiny particles. Moreover, particle agglomeration is another hurdle; pristine microplastics are highly hydrophobic, causing them to clump together or stick to labware. To counteract this, the study used a surfactant (0.05% NovaChem SF100) to keep particles dispersed and evenly distributed on the filter. This showcases how the limit of detection is not just about size but also particle distribution and number.

Interference from organic and inorganic contaminants presents yet another significant challenge. In complex matrices like wastewater or sludge [36]. Organic debris can mask the distinct spectral fingerprints of plastic polymers, leading to poor recovery rates and misidentification. Residual inorganic salts, such as zinc chloride ($ZnCl_2$), used during density separation, can remain on filters and distort spectral acquisition. A single-step chemical digestion often fails to remove persistent organic matter completely. To address this, researchers optimised a multistep digestion protocol using Fenton's reagent, which effectively degrades organic debris without harming polymer structures. They also added an acid-wash step with HCl to dissolve residual salts, ensuring cleaner spectra and reducing contamination-related noise.

A comprehensive review [37] Furthermore, in environmental samples, organic materials such as biofilms and natural debris tend to cling to microplastic surfaces, masking their polymer-specific IR signals. Likewise, inorganic materials such as sediments and minerals contribute overlapping absorbance bands, making distinguishing target polymer peaks from background noise challenging. Compounding this, common additives and surfactants like sodium dodecyl sulfate (SDS) or sodium stearate can stick to MPs during sample prep and mimic the IR signature of certain plastics (e.g., PE), creating false positives. The same paper notes the additional complication of heavy metals and microbial colonies that can alter spectra, further underlining the need for robust cleaning protocols. To tackle these obstacles, the authors recommend a combination of pretreatment steps: digesting organic debris with Fenton's reagent or hydrogen peroxide, treating inorganic residues with acid washes or density separation, and even employing Nile Red staining paired with fluorescence microscopy as a supportive screening step before FTIR spectroscopy. They also stress the importance of expanding spectral libraries to include possible additives and residues that can otherwise lead to misidentification.

Pigments pose their own unique problem. One study [38] discusses how certain pigments, such as titanium dioxide (TiO_2) and black carbon-based additives, cause spatial interference that can obscure polymer-specific peaks, complicate the spectra, and risk misidentification. Although FTIR spectroscopy is generally more resistant to pigment interference than Raman spectroscopy, additional steps like photobleaching and baseline corrections are often needed to resolve overlapping peaks. Environmental weathering and by-products absorbed on MP surfaces only exacerbate this issue, highlighting the necessity for thorough sample cleaning and developing more robust spectral libraries.

The combined complexity of organic matter, inorganic debris, and microplastics makes sample pretreatment indispensable, as emphasised in another study [11]. Here, tissues, cellulose, chitin, silica, and minerals are noted for masking or overlapping the target polymer signatures. To mitigate this, the authors used a suite of pretreatment methods: strong alkalis, acids, or oxidative reagents like Fenton's or hydrogen peroxide for digesting organic matter; and density separation with high-density solutions like $ZnCl_2$ for isolating MPs from inorganic debris. They also employed a Universal Enzymatic Purification Protocol (UEPP) to improve recovery rates while minimising polymer damage.

8 Conclusion

FTIR spectroscopy is the definitive method for identifying and characterising microplastics across environmental and biological samples, leveraging unique spectral fingerprints for polymer-specific detection. Advances in FPA-based FTIR imaging and μFTIR spectroscopy now enable high-resolution mapping and analysis of particles down to 10 μm, dramatically improving throughput and reducing operator bias. Integrating machine learning with curated spectral libraries has transformed FTIR spectroscopy into a quantitative, automated platform, achieving near-perfect classification accuracy. Despite its strengths, challenges remain in standardising protocols, expanding libraries to include weathered polymers, and enhancing sensitivity for particles below 10 µm. Future efforts should focus on harmonising sample preparation and analysis workflows, developing comprehensive databases for aged plastics, and combining FTIR with complementary techniques, such as Raman spectroscopy, to extend detection limits and validate results. As microplastic pollution intensifies, FTIR spectroscopy's evolving capabilities will be essential for robust monitoring, risk assessment, and informing effective mitigation strategies.

References

1. Villegas-Camacho O, Francisco-Valencia I, Alejo-Eleuterio R, Granda-Gutiérrez EE, Martínez-Gallegos S, Villanueva-Vásquez D (2025) FTIR-based microplastic classification: a comprehensive study on normalisation and ML techniques. Recycling 10(2):46
2. Kozloski R, Cowger W, Arienzo MM (2024) Moving toward automated µFTIR spectra matching for microplastic identification: addressing false identifications and improving accuracy. Microplast Nanoplast 4(1):27
3. Wang Z, Wang X, Hu C, Ge T, Wang L, Xing J, He X, Zhao Y (2024) Comparative evaluation of analytical techniques for quantifying and characterizing polyethylene microplastics in farmland soil samples. Agriculture 14(4):554
4. Circelli L, Cheng Z, Garwood E, Yuksel K, Di Iorio E, Angelico R, Colombo C (2024) Comparison of ATR-FTIR and NIR spectroscopy for identification of microplastics in biosolids. Sci Total Environ 916:170215
5. Jung MR, Horgen FD, Orski SV, Rodriguez V, Beers KL, Balazs GH, Jones TT, Work TM, Brignac KC, Royer S-J (2018) Validation of ATR FT-IR to identify polymers of plastic marine debris, including those ingested by aquatic organisms. Mar Pollut Bull 127:704–716
6. Veerasingam S, Ranjani M, Venkatachalapathy R, Bagaev A, Mukhanov V, Litvinyuk D, Mugilarasan M, Gurumoorthi K, Guganathan L, Aboobacker VM (2021) Contributions of Fourier transform infrared spectroscopy in microplastic pollution research: a review. Crit Rev Environ Sci Technol 51(22):2681–2743
7. Islam MS, Islam Z, Islam D (2024) Abundance, source apportionment, and surface characteristics of microplastics in agricultural soil in a flood-prone area of Central Bangladesh. Water Air Soil Pollut 235(3):164
8. Nakajima R, Sawada H, Hayashi S, Nara A, Hattori M (2025) Development of a novel semi-automated analytical system of microplastics using reflectance-FTIR spectrometry: designed for the analysis of large microplastics. Environ Sci Adv 4:901–911
9. Thar AS, Laitrakun S, Deepaisarn S, Opaprakasit P, Somnuake P, Athikulwongse K (2023). A comparative study of machine learning and deep learning models for microplastic classification using FTIR spectra. In 2023 18th international joint symposium on artificial intelligence and natural language processing (ISAI-NLP), pp 1–6
10. Willans M, Szczecinski E, Roocke C, Williams S, Timalsina S, Vongsvivut J, McIlwain J, Naderi G, Linge KL, Hackett MJ (2023) Developing a rapid microplastic detection protocol using reflectance-FTIR spectroscopic imaging and multivariate classification. Environ Sci Adv 2(4):663–674
11. Xie J, Gowen A, Xu W, Xu J (2024) Analysing micro-and nanoplastics with cutting-edge infrared spectroscopy techniques: a critical review. Anal Methods 16(15):2177–2197
12. Cavazzoli S, Ferrentino R, Scopetani C, Monperrus M, Andreottola G (2023) Analysis of micro-and nanoplastics in wastewater treatment plants: key steps and environmental risk considerations. Environ Monit Assess 195(12):1483

13. Lakchani DT, Jayasinghe A, Maithreepala RA, Ulrich U, Collard F (2025) Microplastics in freshwater sediment in the indo-Sri Lankan region: a review of methodologies. Microplast Nanoplast 5(1):16
14. Da Silva VH, Murphy F, Amigo JM, Stedmon C, Strand J (2020) Classification and quantification of microplastics (< 100 μm) using a focal plane array–Fourier transform infrared imaging system and machine learning. Anal Chem 92(20):13724–13733
15. Othman AM, Elsayed AA, Sabry YM, Khalil D, Bourouina T (2023) Detection of sub-20 μm microplastic particles by attenuated total reflection Fourier transform infrared spectroscopy and comparison with Raman spectroscopy. ACS Omega 8(11):10335–10341
16. Walther BA, Pasolini F, Korez Lupše Š, Bergmann M (2024) Microplastic detectives: a citizen-science project reveals significant variation in meso-and microplastic pollution along German coastlines. Front Environ Sci 12:1458565
17. Annisa N, Mahmud M, Fatimah A, Khotidjah N, Riduan R, Mahyudin RP, Nirtha I, Firdausy MA, Sumantri I, Prasetia H (2024) Identifying microplastic types in the Martapura River's water, sediment, and fish using FTIR (case study: Loktangga Village and Teluk Muara Kelayan), South Kalimantan. E3S Web Conf 503:01001
18. Conterosito E, Roncoli M, Ivaldi C, Ferretti M, De Felice B, Parolini M, Gazzotti S, Ortenzi MA, Gianotti V (2025) μ-FTIR reflectance spectroscopy coupled with multivariate analysis: a rapid and robust method for identifying the extent of photodegradation on microplastics. Anal Chem 97(6):3263–3273
19. Ningrum EW, Patria MP (2019). Microplastics and mercury detection on anchovies from Alor and Balikpapan harbours, Indonesia. In 2019 IEEE R10 humanitarian technology conference (R10-HTC)(47129), pp 254–257
20. Miserli K, Lykos C, Kalampounias AG, Konstantinou I (2023) Screening of microplastics in aquaculture systems (fish, mussel, and water samples) by FTIR, scanning electron microscopy–energy dispersive spectroscopy and micro-Raman spectroscopies. Appl Sci 13(17):9705
21. Zhou X, Wang Q, Wang J, Li H, Ren J, Tang S (2023) Quantification of microplastics in plastic-bottled Chinese baijiu using micro-FTIR in imaging mode. Appl Sci 13(20):11142
22. Daniel DB, Daniel DB, Refosco A, Mastad OCL, Kögel T, Dankel S, Gomiero A, Dierkes J (2024) Microplastics in human faeces and their correlation with dietary behaviour: a pilot study. Norsk Tidsskrift Ernæring 22(5):169
23. Hartmann C, Lomako I, Schachner C, El Said E, Abert J, Satrapa V, Kaiser A-M, Walch H, Köppel S (2024) Assessment of microplastics in human stool: a pilot study investigating the potential impact of diet-associated scenarios on oral microplastics exposure. Sci Total Environ 951:175825
24. Ćurčić M, Kolarž P, Hadžić B, Stajčić I, Lazarević Z, Maletić D, Aničić UM, Ilić A (2025) Plasma, UV radiation and ozone for microplastics degradation: optical characterisation of polystyrene, polyethene and polypropylene degradation using FTIR and Raman spectroscopy. Sci Sinter 00:25
25. Martinho SD, Fernandes VC, Figueiredo SA, Vilarinho R, Moreira JA, Delerue-Matos C (2023) Laboratory studies about microplastic ageing and its effects on the adsorption of chlorpyrifos. Polymers 15(16):3468
26. Benjak P, Radetić L, Presečki I, Brnardić I, Sakač N, Grčić I (2024) Microplastic-related leachate from recycled rubber tiles: the role of TiO2 protective coating. Surfaces 7(3):786–800
27. Rostampour S, Jhang SS, Hsu J-K, Cook R, Li Y, Fan C, Sung L-P (2025) Evaluating the environmental factors on microplastic generation: an accelerated weathering study. Microplastics 4(1):13
28. McGivney E, Cederholm L, Barth A, Hakkarainen M, Hamacher-Barth E, Ogonowski M, Gorokhova E (2020) Rapid physicochemical changes in microplastics induced by biofilm formation. Front Bioeng Biotechnol 8:205
29. Pfohl P, Wagner M, Meyer L, Domercq P, Praetorius A, Hüffer T, Hofmann T, Wohlleben W (2022) Environmental degradation of microplastics: how to measure fragmentation rates to secondary micro-and nanoplastic fragments and dissociation into dissolved organics. Environ Sci Technol 56(16):11323–11334

30. Liu Y, Yao W, Qin F, Zhou L, Zheng Y (2023) Spectral classification of large-scale blended (micro) plastics using FT-IR raw spectra and image-based machine learning. Environ Sci Technol 57(16):6656–6663
31. Jang H-D, Kwon S, Nam H, Chang DE (2024) Semi-supervised autoencoder for chemical gas classification with FTIR Spectrum. Sensors 24(11):3601
32. Brandt J, Mattsson K, Hassellov M (2021) Deep learning for reconstructing low-quality FTIR and Raman spectra–a case study in microplastic analyses. Anal Chem 93(49):16360–16368
33. Choobbari ML, Ciaccheri L, Chalyan T, Adinolfi B, Meulebroeck W, Ottevaere H (2022) Analysis of size and concentration of microplastics in water using static light scattering combined with PCA and LDA. EPJ Web Conf 266:12004
34. Rani M, Ducoli S, Federici S, Depero LE (2023) Influx of near-infrared technology in microplastic community: a bibliometric analysis. Microplastics 2(1):107–121
35. Huppertsberg S, Knepper TP (2020) Validation of an FT-IR microscopy method for determining microplastic particles in surface waters. MethodsX 7:100874
36. Cunsolo S, Williams J, Hale M, Read DS, Couceiro F (2021) Optimising sample preparation for FTIR-based microplastic analysis in wastewater and sludge samples: multiple digestions. Anal Bioanal Chem 413(14):3789–3799
37. Randhawa JS (2023) Advanced analytical techniques for microplastics in the environment: a review. Bull Natl Res Cent 47(1):174
38. Prata JC, Padrão J, Khan MT, Walker TR (2024) Do's and don'ts of microplastic research: a comprehensive guide. OAE Publishing, Alhambra

Vibrational Spectroscopic Characterization of Collagen

W. T. Malsha Dulanjani and Ching Yern Chee

Abstract Collagen can be identified as the most abundant fibrous protein in the kingdom of Animalia. Approximately one-third of the total protein content in the human body is collagen. Being an important biopolymer, collagen consists of a unique repeating unit of tripeptide domain. This tripeptide domain is made from three specific amino acids called glycine, proline and hydroxyproline. The repeating units polymerized together to make the polypeptide chain of collagen. Collagen has certain structural integrity which is necessary to maintain proper functioning of body tissues. Characterization of these structural features and specific functionalities is important to evaluate performances of collagen in tissues like skin, cartilage, and tendons. Several conventional characterization techniques like X-ray diffraction (XRD), thermal analysis, electron microscopy and spectroscopic methods are utilized to analyze the structural and physiochemical properties of collagen. Many of these characterization techniques pose certain challenges due to the complex structure and sensitivity of collagen towards many external factors. However, vibrational spectroscopic characterization techniques like Fourier Transform Infrared (FTIR) spectroscopy and Raman spectroscopy can overcome most of these challenges and characterize collagen effectively. Both FTIR and Raman spectroscopic techniques aid to obtain comprehensive understanding about native and modified collagen states. These techniques collectively provide detailed insights into the collagen's chemical structure, bonding types, molecular interactions and conformational changes. Most importantly, these techniques facilitate rapid, highly sensitive and nondestructive approaches to characterize collagen. These specific advantages make FTIR and Raman spectroscopic techniques preferable over many other conventional

W. T. M. Dulanjani
Faculty of Applied Sciences, Department of Polymer Science, University of Sri Jayewardenepura, Nugegoda, Sri Lanka

C. Y. Chee (✉)
Faculty of Engineering, Department of Chemical Engineering, University of Malaya, Kuala Lumpur, Malaysia
e-mail: chingyc@um.edu.my

N. Mazumder et al. (eds.), *Vibrational Spectroscopy for Biopolymer Characterization*, Biological and Medical Physics, Biomedical Engineering,
https://doi.org/10.1007/978-3-032-15306-7_11

characterization methods. In this chapter, characterization of collagen using Fourier Transform Infrared (FTIR) spectroscopy and Raman spectroscopy is discussed. The major characteristic peaks observed in both techniques are mentioned while explaining their molecular origins. The sections are further clarifying the different spectral shifts and changes in intensity ratios due to structural changes in collagen-based systems. This chapter emphasizes the importance of employing FTIR and Raman spectroscopic methods to characterize collagen's structure, denaturation, and chemical modifications.

Keywords FTIR spectroscopy · Raman spectroscopy · Collagen · Triple helical structure

1 Introduction

Extracellular matrix (ECM) is a major part in the animal connective tissues which mainly consist of different types of secreted proteins, which ensure the proper maintenance of the surrounding resident cells, which are the other major part of the animal connective tissues. These secreted proteins in ECM must be responsible for mechanical and biochemical properties of connective tissues. Collagen is the highly abundant structural protein among them. Owing to its structural integrity and functionality, collagen exhibits high mechanical strength with elasticity, aiding for proper functioning of animal body tissues [1, 2].

As a protein, collagen is made from long chains of amino acids. The unique feature here is that the collagen contains a repetitive sequence pattern of tripeptide units. These tripeptide units have the general formulation of Gly-X-Y. Here, Gly means glycine, which is the smallest of the amino acids. X and Y are representing proline and hydroxyproline amino acids respectively. This unique tripeptide units polymerize and form the characteristic triple helical structure of collagen. This triple helical structure is the fundamental structural unit of collagen, which is aligned and stabilized by covalent crosslinking and inter-chain hydrogen bonds [3]. These triple helical units are packed together to build fibrils, which are ultimately bound together to build collagen fibers. These triple helical configurations ultimately result in a rigid rod-like structure of collagen. The hierarchical structure of collagen, from triple-helical polypeptide chains to very structured fibrils, gives the skin, cartilage, tendons, and bones mechanical qualities like strength, elasticity, and structural integrity [4, 5].

2 Collagen Structure and Biochemistry

Collagen is the most abundant protein found in animals mainly inside the skin, bones and tissues. It can be mainly classified into fibrillar and non-fibrillar categories [6]. Type I, II, III, and V/XI are the key types of fibrillar collagen which has a triple helix structure with long, uninterrupted Gly-X-Y amino acid repeats enabling self-assemble fibril formation through covalent cross-linking. In this Gly-X-Y repeating tripeptide unit, Glycine can be found at every third position while X and Y positions are represented by proline and hydroxyproline respectively. Here glycine, proline, and hydroxyproline together are responsible for around 57% of the total amino acids of collagen, forming the triple helix formation [7]. The strong mechanical properties of collagen are arising from these unique structural features, which help to maintain efficient functioning in our body tissues [1, 8]. In contrast, the non-fibrillar collagens mainly consist of network forming (Type IV), anchoring (Type VII), and FACITs (Type IX, XII), which contain interrupted helices structure with short triple helical domains. Due to this structural discontinuity, the non-fibrillar types of collagens typically form mesh-like membranes (Type IV), beaded filaments (Type VI), or act as adaptors, regulating the cellular microenvironment through barrier formation and fibril modulation [9, 10].

All these collagen types contain peptide bonds (–CO-NH–) in their main chain, resulting in amide I, II, and III regions in both FTIR and Raman spectra. The uniqueness of the triple helical structure of collagen comes down to its stabilization by extensive hydrogen bonding, which influences changes in these amide peak positions. During the process of protein hydroxylation, –OH groups are introduced to the collagen structure, resulting in signature peaks in each spectrum. Certain chemical modifications can significantly vary the functional groups in collagen structure. Vibrational spectroscopic techniques, including Fourier-transform infrared (FTIR) spectroscopy and Raman spectroscopy, provide a powerful nondestructive approach to characterize all these structural changes and secondary interactions within the collagen matrix. By analyzing the peak positioning and intensity variations of particular bands, these modifications and functionality changes can be effectively characterized [11, 12].

The polar functional groups in collagen can be detected by FTIR spectroscopy to provide valuable insights into the secondary structure of collagen. Many significant peaks are observable in IR range which can be utilized to evaluate different collagen samples using FTIR [13, 14]. In contrast, Raman spectroscopy is also widely employed to characterize collagen samples. Raman is useful to detect non-polar bonding and gives detailed information about the molecular conformations of collagen [2, 15]. Both FTIR and Raman spectroscopic methods are useful to obtain a comprehensive understanding about the collagen's characteristics (Fig. 1).

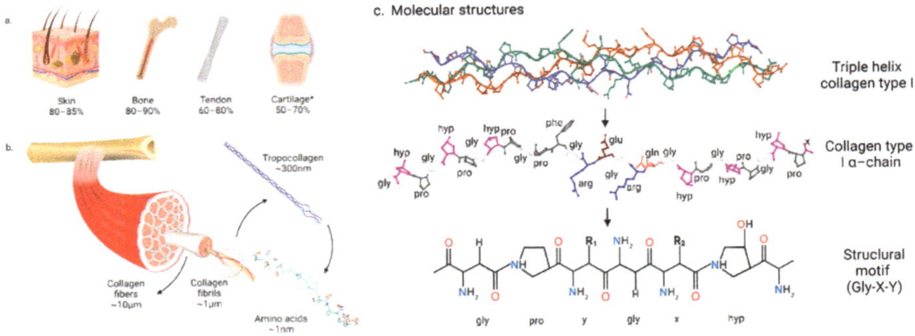

Fig. 1 Collagen type I content in various body tissues (skin, bone, tendon, cartilage) and collagen's triple helix structure. Adapted from Ma et al. [16]

3 FTIR Spectroscopy of Collagen

Collagen samples tend to give many characteristic absorption peaks within the IR range. Fourier-transform infrared (FTIR) spectra of collagen is a useful analytical tool to characterize collagen's structural features effectively. Primarily, FTIR spectroscopy visualizes the vibrational modes arising from polar functional groups in the collagen structure.

The amide bands can be recognized as the most prominent absorption bands in FTIR spectra of collagen, which help to analyze the secondary structure of collagen thoroughly [1, 17]. Following that the amide I peak is strongly influenced by the secondary structure of the collagen, which results from C=O stretching vibrations of the peptide bonds. This peak can be detected within the range of 1600–1700 cm^{-1} [18]. By analyzing the peak positioning and changes in the sharpness of this peak, several information can be gathered including, the presence of the triple helix and β-sheet structures of collagen. A sharp amide I peak at 1650 cm^{-1} suggests the presence of intact triple-helical features in the tested collagen sample. A peak shift toward 1630–1640 cm^{-1} wave numbers suggests possible denaturation of collagen or formation of β-sheet structures. If the peak broadening or splitting happens, that's an indication of random coil regions and a mix of triple helix in the observed collagen sample [17, 19].

As another prominent amide band, the amide II peak arises from the N–H bending and C–N stretching of the peptide backbone of collagen. It is visible around the range of 1546–1551 cm^{-1} and exhibits lower sensitivity towards water than the amide I band. Hence, in hydrated collagen samples, the amide II band is more stable compared to the amide I band. When hydrated collagen samples are analyzed using FTIR, amide II band is more effective to obtain structural information about collagen [20]. Different fluctuations in the peak positioning and intensity variations of this peak often explain the stability of the triple helical structure, possible degradations, and alterations in the intermolecular packing system. Hydrogen bond strength and peptide bond environment also influence the N-H bending and C–N stretching in collagen. Stronger hydrogen bond formations found in more ordered collagen fibrils, result sharper amide II band centered around 1545–1550 cm^{-1}. But the less ordered collagen fibrils yield weaker hydrogen bond formation within the collagen matrix. Under swelling or mechanical stress, these collagen fibrils tend to loosen up causing peak broadening. Such kinds of changes in collagen samples can be effectively identified by analyzing the broadening of amide II peak often with a slight shift towards lower wave numbers [12]. Denaturation of collagen causes the triple helix structure to unwind into random coils. Random coils of collagen have weaker hydrogen bonding, causing peak shift towards lower wavenumbers (around 1535–1540 cm^{-1}) while reducing the peak intensity. Since both of these amide I and amide II bands provide complementary information about collagen's structural features, they are employed together for these interpretations of the collagen [21, 22].

The amide III band, which is arising in the range of 1200–1350 cm^{-1}, is also attributed to the N–H bending and C–N stretching vibrations of peptide backbone of collagen. But this vibration strongly interacts with the CH_2 wagging motions, especially arising from glycine and proline residues in the side chains of collagen [23]. This amide III band can also act as a characteristic peak indicator for the triple helical integrity or denaturation of collagen. If the amide III band shows reduced intensity or band broadening, that attributes for partial denaturation or helix disruption of collagen. A well-defined, strong amide III band attribute for the stabilized triple helical structure of collagen [24]. The intensity ratio between this amide III and amide I or amide II bands, is often used as an index to quantify the triple helix structure content in a collagen based sample for several biomedical applications [25].

In addition to the above-mentioned amide I, II, and III signature bands, the amide A and amide B bands also provide valuable information about the collagen's structural features. These amide A and amide B bands can be found in the IR range of 3070–3300 cm^{-1}. The amide A peak is arising mainly due to the N–H stretching vibrations of peptide bond, can be observed at ~3300 cm^{-1}. This peak is influenced by the hydrogen bonds within the collagen structure, which are typically generating during hydration or triple helix formation. Down shift of this peak can be observed when the hydrogen bonds are increasing, because of the N–H bond weakening from hydrogen bonding. These weaker N–H bonds resulting require less energy to stretch and shifts towards lower wave number, indicating stronger hydrogen bonding network. The amide B peak, which is arising from the asymmetric =C–H stretching vibrations, can be observed typically in the 3070–3100 cm^{-1} region. This peak is less distinct in FTIR spectra of collagen due to the peak overlapping with the 2920 and 2850 cm^{-1} aliphatic peaks. However, this peak is still valuable to attribute several conformational changes of collagen arising due to crosslinking, denaturation or modifications [26, 27].

As another insightful functional group, CH_2 groups found in the side chains of amino acids in collagen, give a vibrational peak centered around 1450 cm^{-1}. The peak intensity of this peak typically depends upon the concentration of CH_2 groups and their change of dipole moment during the vibration. The sharpness of this peak in an FTIR spectrum indicates restricted molecular motions of CH_2 groups, resulting from the steric hindrance and intermolecular interactions within the molecules

[28]. When this peak appears as a sharp and intense peak in an FTIR spectrum of a collagen sample, it suggests that the CH_2 groups in amino acid side chains within the collagen are tightly packed, resulting in a well-organized collagen packing while restricting the motion of CH_2 groups [29]. If this peak shows broadening or shifting, that might suggest that the triple helical structure of collagen has unwound, or the packing has become disordered due to the denaturation of collagen. In addition, the intensity of this peak may provide more details about the hydration effects on collagen. The reduction in peak intensity or even a slight red shift is attributed to the disruption of hydrogen bonds between collagen helices due to the hydration effect. Using the degree of this peak change, the level of hydration disruption can be evaluated. Crosslinking also affects the intensity of this CH_2 peak, and narrowing, shifting, or increased intensity of the peak depends upon the extent of crosslinking. Thus, the intensity of this peak is helpful to confirm successful crosslinking in a collagen matrix, in applications such as collagen scaffolds, hydrogels, and films [30, 31].

In the FTIR spectrum of hydrated collagen, a broad overlapped peak can be found in the region between 3200 and 3500 cm^{-1}. Mostly, it indicates overlap of the O-H stretching band significantly with N-H stretches and weakly with amide I and II peaks [1]. Since these peak overlaps, the direct interpretation of this peak is quite difficult. Moisture absorption during the sample preparation process might also cause some interference (Fig. 2). Due to that, distinguishing between the OH groups arising from hydrogen bonds in collagen structure and OH groups from bound water are also difficult [33] But, using advanced spectroscopic techniques like Fourier self-deconvolution, these overlapped peaks can be separated to get more meaningful insights, enhancing the resolution without changing the original features in the spectrum [33, 34].

As discussed, the FTIR spectroscopic technique plays a central role in collagen characterization with many distinct advantages. This analytical approach is fast and

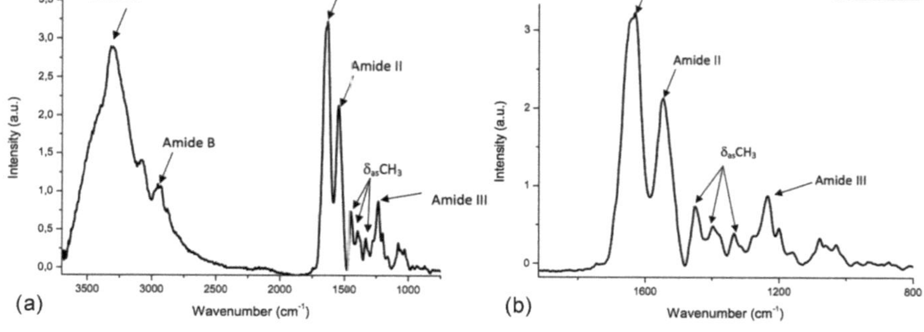

Fig. 2 (**a**) FTIR spectrum of type I collagen from pure (dry) membranes from bovine Achilles deep tendon in the region of 3700 to 800 cm^{-1}. (**b**) FTIR spectrum of type I collagen in the region of 1800 to 800 cm^{-1}. Reproduced from Júnior, Z. S. S. et al., Scientific Reports, 5, 11,448 (2015) [32]

can analyze collagen's structural features in both natural and modified states compared to many conventional methods. This method also offers a non-destructive approach while providing detailed information about conformational changes and functionalities of collagen. Overall, FTIR can be identified as a crucial analytical technique in characterizing collagen-based systems like scaffolds, drug delivery systems, biomaterials, and other polymeric applications.

4 Raman Spectroscopy of Collagen

Raman spectroscopy, the other major vibrational spectroscopic technique, is also widely used to characterize collagen's structural features. This method primarily serves detailed analysis of the backbone conformations of collagen. While being particularly sensitive to the changes in polarizability of molecules, Raman spectroscopy gives information on collagen's structure, bond types and conformational changes. Both natural and modified collagen systems are characterized using this complementary technique. Raman provides molecular-level information while analyzing non-polar or weakly polar functional groups [35]. Here, the inelastically scattered light emitted after laser excitation of the collagen samples are detected. The laser interacts with the vibrational modes in the collagen sample and alters its energy by exciting molecular vibrations, which provide a characteristic Raman spectrum by using the resulting shifts in wavelength [36]. Unlike FTIR, Raman spectroscopy can directly analyze hydrated biological samples like tissues, gels, or living tissues without strong interference because of its tolerance to water. This is one of the major advantages of Raman spectroscopy as a powerful non-destructive technique to investigate collagen [37].

Collagen contains non-polar amino acids such as glycine, proline, and hydroxyproline and typically produce water-laden matrices in tissues. Due to these specific amino acid side chains and the peptide backbone, collagen gives several characteristic peaks in Raman spectroscopy including amide I, amide III, CH deformation band and C-C/C-N skeletal vibrations. All these peaks are prominent for identifying specific structural states of collagen while accessing conformational changes due to crosslinking, denaturation, or mechanical stress [35].

The amide I band is arising due to the C=O stretching and N–H bending modes of peptide chain, which is centered between ~1655–1670 cm^{-1} [38]. Since both C=O and N–H groups are containing in the peptide backbone, they are strongly affected by the protein's hydrogen bonding behaviors. In well-preserved native collagen, this peak can be identified as a sharp peak near the range of ~1655–1665 cm^{-1} attributing to the triple helical structure of native collagen. If denaturation or degradation happens due to enzymes, this peak typically shifts downwards suggesting that the triple helical structure has being converted into a random coil or α-helical structures [39]. Even broadening or splitting may also occur if the triple helical structure has being converted into mixture of structural states [40]. Likewise, the position and the shape of this amide I peak in Raman spectrum, depends on the secondary structure of collagen, and the main amide I peak is attributed to the collagen I fibrils while the shoulders at ~1630 cm^{-1} and ~1675 cm^{-1} are corresponding to the 3_{10}-helix and β-sheet features of collagen respectively [38]. Since this peak is arising mainly from C=O stretching, the changes in hydrogen bonding can also be attributed using the shift of this peak. If the peak has shifted to lower wavenumbers, it typically suggests that the hydrogen bonds are strong, whereas an upward shift of this peak or r2eduction in intensity might suggests dehydration [41].

The amide III peak can be found in the ~1240–1270 cm^{-1} region which is arising due to N–H bending and C–N stretching vibrations within the collagen backbone. The position and intensity of this peak vary depending upon the structural changes of the collagen backbone. If the collagen sample contains ordered triple helical structures, the N–H in-plane bending produce the ~1270 cm^{-1} sub-band whereas the ~1240 cm^{-1} sub-band originated from the disordered or random-coil configuration of collagen [42]. The intensity ratio between these two sub-bands act as a quantitative indicator of the stability of triple helical structure of collagen. A higher ratio of this sub-bands' intensities indicate more ordered triple helical content and well-ordered packing within the collagen sample while denoting the "maturity" in collagen tissue studies. The lower intensity ratio between these sub-bands indicates the presence of more random coils or partially unfolded chains in the tested sample. Because of biological diseases like osteoarthritis, aging, mechanical degradation, or thermal denaturation, the collagen network becomes more disrupted and provides a reduced intensity ratio of these two sub-groups. Therefore, this ratio differences can be analyzed to predict such conditions in collagen samples. In some studies, this ratio has been used to distinguish between diseased bone with a disorganized collagen network from a healthy bone that maintains a well-organized collagen network [42].

Another characteristic peak in the Raman spectra of collagen appears near 1450 cm^{-1}. This peak is resulting from CH_2 scissoring mode and CH_3 asymmetric bending vibrations in collagen. The aliphatic side chains and backbone C–H groups in collagen are primarily responsible for this peak [43]. Compared to the amide I and III bands, this 1450 cm^{-1} band is not that sensitive under variations such as hydration, pH, or temperature. Therefore, researchers use this peak as a normalization reference to normalize the intensities of other bands. For example, the Mineral-to-Matrix ratio in a bone sample is typically evaluated by comparing the intensities of mineral-specific bands normalized to the 1450 cm^{-1} band [44]. However, this peak can be utilized to attribute several changes in side-chain environments and packing in the collagen network, which might be influenced by crosslinking, hydration state or tissue density.

The polypeptide skeletal motions in Raman spectroscopy, are originated specially from C–C stretching vibrations in proline (~855 cm^{-1}), ring vibrations in hydroxyproline (~877 cm^{-1}), and C–C backbone stretching vibrations arises from the proline/hydroxyproline linkage (~920 cm^{-1}). Since collagen contains a high amount of these amino acids, these vibrational modes are significant in Raman spectroscopy of collagen [45]. These peaks are visible in native collagen as well defined and sharp peaks suggesting a well-ordered triple helical structure and rigid molecular environment. Peak broadening or reduction in intensities might suggest denaturation, which results in disordered configurations of collagen [46]. Using the intensity ratio between the peaks arising from the stretching vibrations in proline (~855 cm^{-1}) and the proline/hydroxyproline linkage (~920 cm^{-1}), the hydroxylation extent can be estimated (Fig. 3). Since the hydroxyproline is essential for the stability of the triple helical structure of the collagen, this ratio correlates to the stability of the collagen structure [47].

During bone glycation, this ratio is the only Raman spectral parameter which changes indicating the glycation-specific modification of collagen. Because of this specific sensitivity, this intensity ratio is used to identify the collagen glycation in Raman spectroscopy [48]. If this ratio increased, that suggests the level of hydroxylated proline has increased with time, which enhances the stability of collagen structure. Usually in biomedical applications like tissue engineering, the increment of this ratio is used as a key indicator to identify the maturation of ECM of collagen. According to the literature, the collagen maturing in osteoblast cultures can be identified by monitoring this intensity ratio (~920/855 cm^{-1}), and it helps to facilitate real-time observations of ECM assembly of collagen [49].

CH_3 symmetric stretching is also an important vibrational mode visible in a Raman spectrum of collagen. This peak mainly appears due to the methyl groups in amino acids and can be noticed at ~2930 cm^{-1}. The intensity and position of this peak is strongly influenced by the side chain packing in collagen's backbone. Closely packed side chains often reduce their movements resulting sharper and

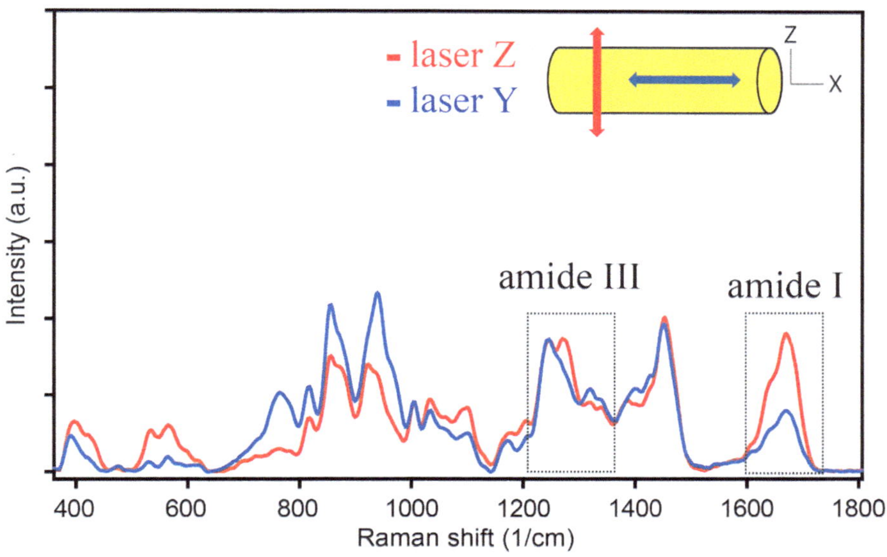

Fig. 3 Raman spectra of collagen tendon under two different laser polarization orientations. Reproduced from Galvis, L., Dunlop, J. W., Duda, G., Fratzl, P., & Masic, A. (2013). Polarized Raman anisotropic response of collagen in tendon: towards 3D orientation mapping of collagen in tissues. PLOS ONE, 8(5), e63518 [40]

more intense peak around ~2930 cm^{-1}. Hydration also has a pronounced effect on this peak, often resulting peak broadening or intensity reduction because of interactions between the side chains and water molecules [50]. In many ECM studies, CH_3 peak has been studied under different hydration conditions to understand the factors like tissue swelling, degradation of extracellular matrix, and denaturation of proteins [51].

Around 1003 cm^{-1} in a Raman spectrum of collagen, the symmetric phenylalanine ring breathing mode can be found as another notable peak. The aromatic amino acid residues are responsible for this particular peak, and it is typically visible as an isolated, sharp peak. Compared to the other peaks in Raman spectroscopy, this peak exhibit lower sensitivity towards hydration and conformational changes of collagen. Hence, this peak is typically used for normalization in Raman studies of collagen [11, 52]. The amount of phenylalanine present is responsible for the intensity of this peak. Thus, the intensity of this peak has been analyzed in many studies to detect the protein density in biochemical samples. Likewise, this peak can also be utilized to estimate collagen content in different body tissues. This peak is also helpful to monitor protein denaturation and alterations in the ECM when comparing diseased and healthy connective tissues [51].

Accordingly, Raman spectroscopy can also provide a systematic pathway to characterize collagen. As a powerful analytical tool, Raman offers valuable insights into the structural features of collagen. It is also a label-free and nondestructive vibrational spectroscopic approach. The water tolerance ability of Raman spectroscopy makes it more suitable for the analysis of hydrated biological samples and living systems, enhancing its capabilities of characterizing the collagen structure. To analyze complex tissue matrices, Raman mapping can be utilized because of its improved sensitivity and resolution. All these capabilities of Raman spectroscopy make it a highly effective vibrational spectroscopic technique to characterize collagen.

5 Conclusion

Both FTIR and Raman spectroscopic techniques provide unique approaches to investigating the collagen structure, enabling the assessment of external effects on the secondary structure of collagen. As a fibrous structural protein in the extracellular matrix, collagen provides mechanical resilience and other biological functions, utilizing its unique triple helical architecture. Vibrational spectroscopic techniques aid in analyzing the behavior of this triple helical structure upon degradation, hydration, and post-translational modifications of collagen. As a nondestructive method to characterize structural features of collagen at the molecular level, FTIR spectroscopy mainly provides valuable insights on amide I, II, III, A, and B bands along with C–H and O–H vibrations. It reveals molecular details on the secondary structure of collagen, emphasizing the hydration effects, hydrogen bonding, and denaturation mechanism. As the other major non-destructive, label-free vibrational spectroscopic technique, Raman spectroscopy characterizes collagen, providing complementary insight into the secondary structure and triple helical architecture, through the amide I and III bands. Using some specific peak ratios and peak positioning, Raman spectroscopy can also access the molecular packing and structural stability of collagen. Peaks like C–C/C–N stretches (850–950 cm^{-1}) and CH_3 stretching (2930 cm^{-1}) in a Raman spectrum of collagen further helpful to analyze the backbone conformations, local hydration, and molecular orientations. FTIR and Raman spectroscopic techniques together offer a fast, non-destructive, and highly informative analytical approach to characterize collagen. The insights gaining from these vibrational spectroscopic methods can be employed for advanced collagen research and a wide range of biochemical applications.

References

1. Shoulders MD, Raines RT (2009) Collagen structure and stability. Annu Rev Biochem 78(1):929–958
2. Martinez MG et al (2019) Characterisation of structural changes in collagen with Raman spectroscopy. Appl Spectrosc Rev 54(6):509–542
3. Shoulders MD, Raines RT (2009) Collagen structure and stability. Annu Rev Biochem 78:929–958
4. Wick G et al (2001) Really old-palaeoimmunology: immunohistochemical analysis of extracellular matrix proteins in historic and pre-historic material. Exp Gerontol 36(9):1565–1579
5. Exposito J-Y et al (2010) The fibrillar collagen family. Int J Mol Sci 11(2):407–426
6. Bella J, Hulmes DJ (2017) Fibrillar collagens. In: Fibrous proteins: structures and mechanisms. Springer, Cham, pp 457–490
7. Li P, Wu G (2018) Roles of dietary glycine, proline, and hydroxyproline in collagen synthesis and animal growth. Amino Acids 50(1):29–38
8. Brodsky B, Persikov AV (2005) Molecular structure of the collagen triple helix. Adv Protein Chem 70:301–339
9. Ricard-Blum S (2011) The collagen family. Cold Spring Harb Perspect Biol 3(1):a004978
10. Kadler KE et al (2007) Collagens at a glance. J Cell Sci 120(12):1955–1958
11. Movasaghi Z, Rehman S, Rehman IU (2007) Raman spectroscopy of biological tissues. Appl Spectrosc Rev 42(5):493–541
12. Barth A (2007) Infrared spectroscopy of proteins. Biochim Biophy Acta 1767(9):1073–1101
13. de Campos Vidal B, Mello MLS (2011) Collagen type I amide I band infrared spectroscopy. Micron 42(3):283–289
14. Kristoffersen KA et al (2023) FTIR-based prediction of collagen content in hydrolyzed protein samples. Spectrochim Acta A Mol Biomol Spectrosc 301:122919
15. Querido W et al (2017) Vibrational spectroscopy and imaging: applications for tissue engineering. Analyst 142(21):4005–4017
16. Amirrah IN et al (2022) A comprehensive review on collagen type I development of biomaterials for tissue engineering: from biosynthesis to bioscaffold. Biomedicine 10(9):2307
17. Ramshaw JA, Shah NK, Brodsky B (1998) Gly-XY tripeptide frequencies in collagen: a context for host–guest triple-helical peptides. J Struct Biol 122(1–2):86–91
18. Fratzl P (2008) Collagen: structure and mechanics, an introduction. In: Collagen: structure and mechanics. Springer, Cham, pp 1–13

19. Acuña-Nelson S-M et al (2024) Effect of UHT thermal treatment on the secondary structures of Milk proteins: insights from FTIR analysis and potential allergenic activity. Int J Food Sci 2024(1):1880779
20. Stani C et al (2020) FTIR investigation of the secondary structure of type I collagen: new insight into the amide III band. Spectrochim Acta A Mol Biomol Spectrosc 229:118006
21. Kittiphattanabawon P et al (2010) Isolation and characterisation of collagen from the skin of brownbanded bamboo shark (Chiloscyllium punctatum). Food Chem 119(4):1519–1526
22. Perez-Puyana V et al (2019) Assessment of the denaturation of collagen protein concentrates using different techniques. Biol Chem 400(12):1583–1591
23. Doyle BB, Bendit E, Blout ER (1975) Infrared spectroscopy of collagen and collagen-like polypeptides. Biopolymers 14(5):937–957
24. Zhang Z, Li G, Shi B (2006) Physicochemical properties of collagen, gelatin and collagen hydrolysate derived from bovine limed split wastes. J Soc Leather Technol Chemi 90(1):23
25. Grewal MK, Huppertz T, Vasiljevic T (2018) FTIR fingerprinting of structural changes of milk proteins induced by heat treatment, deamidation and dephosphorylation. Food Hydrocoll 80:160–167
26. Kong J, Yu S (2007) Fourier transform infrared spectroscopic analysis of protein secondary structures. Acta Biochim Biophys Sin 39(8):549–559
27. Sadat A, Joye IJ (2020) Peak fitting applied to Fourier transform infrared and Raman spectroscopic analysis of proteins. Appl Sci 10(17):5918
28. Bakir G et al (2020) Orientation matters: polarization dependent IR spectroscopy of collagen from intact tendon down to the single fibril level. Molecules 25(18):4295
29. Riaz T et al (2018) FTIR analysis of natural and synthetic collagen. Appl Spectrosc Rev 53(9):703–746
30. Silva Júnior ZS et al (2015) Effect of papain-based gel on type I collagen–spectroscopy applied for microstructural analysis. Sci Rep 5:11448. https://doi.org/10.1038/srep11448
31. Movasaghi Z, Rehman S, Rehman DI (2008) Fourier transform infrared (FTIR) spectroscopy of biological tissues. Appl Spectrosc Rev 43(2):134–179
32. Júnior ZSS et al (2015) Effect of papain-based gel on type I collagen - spectroscopy applied for microstructural analysis. Sci Rep 5(1):11448
33. Kudo S et al (2017) Adsorption of water to collagen as studied using infrared (IR) microspectroscopy combined with relative humidity control system and quartz crystal microbalance. Appl Spectrosc 71(7):1621–1632
34. Surewicz WK, Mantsch HH (1988) New insight into protein secondary structure from resolution-enhanced infrared spectra. Biochim Biophys Acta 952:115–130
35. Esmonde-White K (2014) Raman spectroscopy of soft musculoskeletal tissues. Appl Spectrosc 68(11):1203–1218
36. Bergholt MS, Serio A, Albro MB (2019) Raman spectroscopy: guiding light for the extracellular matrix. Front Bioeng Biotechnol 7:303
37. Leikin S et al (1997) Raman spectral evidence for hydration forces between collagen triple helices. Proc Natl Acad Sci USA 94(21):11312–11317
38. Gullekson C et al (2011) Surface-sensitive Raman spectroscopy of collagen I fibrils. Biophys J 100(7):1837–1845
39. Połomska M et al (2021) Effects of temperature on the FT NIR Raman spectra of fish skin collagen. Appl Sci 11(18):8358
40. Galvis L et al (2013) Polarized Raman anisotropic response of collagen in tendon: towards 3D orientation mapping of collagen in tissues. PLoS One 8(5):e63518
41. Gonapinuwala ST et al (2025) Collagen with native triple-helical structure from representative cold-water and warm-water fish skins using a modified protocol. Food Bioprocess Technol 18:7993–8007
42. Pezzotti G et al (2022) Raman spectroscopic insight into osteoarthritic cartilage regeneration by mRNA therapeutics encoding cartilage-anabolic transcription factor Runx1. Mater Today Bio 13:100210

43. Khalid M et al (2018) Raman spectroscopy detects changes in bone mineral quality and collagen cross-linkage in staphylococcus infected human bone. Sci Rep 8(1):9417
44. Nieuwoudt MK et al (2020) Raman spectroscopy reveals age- and sex-related differences in cortical bone from people with osteoarthritis. Sci Rep 10(1):19443
45. Crisford A et al (2024) Harnessing Raman spectroscopy and multimodal imaging of cartilage for osteoarthritis diagnosis. Sci Rep 14(1):31466
46. Połomska M et al (2010) Fourier transform near infrared Raman spectroscopy in studies on connective tissue. Acta Phys Pol A 118:1
47. Jenkins CL et al (2003) Effect of 3-Hydroxyproline residues on collagen stability. J Am Chem Soc 125(21):6422–6427
48. Unal M et al (2018) Assessing glycation-mediated changes in human cortical bone with Raman spectroscopy. J Biophotonics 11(8):e201700352
49. Sharma A et al (2020) Raman spectroscopy links differentiating osteoblast matrix signatures to pro-angiogenic potential. Matrix Biol Plus 5:100018
50. Paschou A et al (2017) High pressure Raman study of type-I collagen. FEBS J 285:2641–2653
51. Buckley K et al (2012) Raman spectroscopy reveals differences in collagen secondary structure which relate to the levels of mineralisation in bones that have evolved for different functions. J Raman Spectrosc 43:1237–1243
52. Dehring K et al (2006) Correlating changes in collagen secondary structure with aging and defective type II collagen by Raman spectroscopy. Appl Spectrosc 60:366–372

Integrating Artificial Intelligence Aided Vibrational Spectroscopy Applied to Medical Biopolymer Applications

Arathi Panicker, Kiran Manoharan, Srimegha Thambi, Sujadevi Vijaya Gangadharan, Prabaharan Poornachandran, and Manu Krishnan Nambudiri

Abstract Biopolymers, natural polymers derived from biological sources, have gained significant importance in cutting-edge medical applications. Traits such as environmental friendliness, biodegradability, biocompatibility, and its non-toxic qualities make them essential for advancing sustainable technologies, reducing environmental harm, and offering customisable physical, chemical, and biological properties. Vibrational spectroscopic modalities like Infra-Red, Raman spectroscopies are used to extract intricate details about the bio-polymer composition and its suitability for any application. Artificial Intelligence (AI) methodologies play a large role in assisting these spectroscopic methods decipher subtle spectral features that are otherwise elusive. This chapter discusses major biopolymers that are used in medical aspects, major medical applications that they are used in and major advancements in AI methodologies that can effectively process the spectroscopic datasets.

Keywords Artificial intelligence · Biopolymer · Spectroscopy

1 Introduction

Biopolymers have emerged to be a material of extreme importance in the current scenario. Fundamentally they are of biological origin and are large macromolecules that constitute many repeated units. The constituting repeated units are linked to

Arathi Panicker and Kiran Manoharan contributed equally with all other contributors.

A. Panicker · K. Manoharan · S. Thambi · S. V. Gangadharan · P. Poornachandran · M. K. Nambudiri (✉)
Centre for Internet Studies and Artificial Intelligence, Amrita Vishwa Vidyapeetham, Kollam, Kerala, India
e-mail: manukk@am.amrita.edu

N. Mazumder et al. (eds.), *Vibrational Spectroscopy for Biopolymer Characterization*, Biological and Medical Physics, Biomedical Engineering,
https://doi.org/10.1007/978-3-032-15306-7_12

each other via covalent bonds to form larger molecules. These biopolymers are especially useful in the medical field due to its inherent nature of biocompatibility, biodegradability and functional versatility. As the health care field continues to evolve towards more personalised and regenerative solutions, the demand for advanced biomaterials that can integrate seamlessly with biological systems is also growing proportionally. Among the various material classes available, biopolymers are uniquely positioned to meet these demands, offering tuneable physical, chemical and biological properties.

Vibrational spectroscopic techniques including Raman and Infrared (IR) spectroscopy play a crucial role in advancing the application of biopolymers in targeted medical use. These methodologies are highly effective in providing detailed insights into molecular bonds, functional groups, and conformational changes within biopolymeric systems. Despite their advantages, interpreting the vibrational spectra from complex biological matrices remains a significant challenge. The hurdles arise from various factors inherent to biopolymer structures, such as overlapping spectral signals, the presence of secondary structures, baseline noise, and the intrinsic heterogeneity of naturally derived biopolymers. To overcome these analytical challenges and extract meaningful features, researchers are increasingly turning to advanced computational tools.

AI and Machine Learning (ML) are increasingly aiding in addressing the complexities of biopolymer usage. AI /ML methodologies decipher subtle features in spectral data that often remain elusive to traditional methods and often attain better precision for automated classification, regression and feature extraction tasks, thereby facilitating speedy and accurate profiling of biopolymers. This chapter assesses the range of biopolymers, with a focus on polysaccharides and proteins and their utility in different medical applications and selected AI methodologies that are capable of precisely analysing the spectroscopic profile of the biopolymers.

2 Major Polysaccharides/Protein Biopolymers Used for Medical Applications

There is a diverse array of natural biopolymers, classified into various categories based on their chemical composition, biological origin, molecular structure, and functional properties. These divisions include polysaccharides, proteins, nucleic acids and polyesters, each having its own roles in biological systems. However, polysaccharide- and protein-based biopolymers are particularly significant from a medical perspective and generalised applications and major ones are listed in Fig. 1.

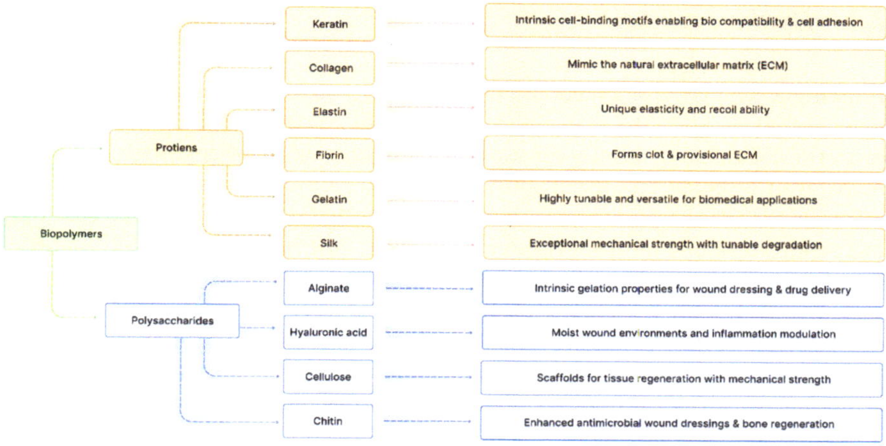

Fig. 1 Key polysaccharide and protein-based biopolymers utilised in medical applications, highlighting their primary functions and specialisations

2.1 *Alginate*

Alginate is a naturally occurring, negatively charged polymer obtained from brown algae. It has been widely investigated for medical applications because of its biocompatibility, affordability, non-toxic properties, and its capacity to readily form gels when combined with agents such as calcium.

2.2 *Keratin*

Keratin is a fibrous structural protein obtained from animal or human sources, including epithelial cells, hair, skin, nails, wool, and feathers. It contains a high proportion of cysteine, enabling the formation of disulfide cross-links, and possesses distinct peptide motifs such as leucine–aspartic acid–valine (LDV) and arginine–glycine–aspartic acid (RGD), which facilitate cell adhesion and interaction.

Structurally, keratin exhibits a fibrous architecture capable of self-assembly into hierarchical structures, including three-dimensional (3D) microstructures with tuneable surface chemistry and mechanical properties. It stems from its natural origin, and it can form hydrogen bonds with substances, enabling functionalisation for material applications. Keratin's regenerative nature and adaptability to modifications, such as combination with synthetic or natural polymers, promote its potential as a sustainable biomaterial.

2.3 Hyaluronic Acid

Hyaluronic acid (HA) is a naturally occurring glycosaminoglycan and an important structural element of the extracellular matrix (ECM), recognised for viscoelastic properties, and strong affinity for water. As a significant ECM component, HA participates in essential biological activities such as cell migration, tissue repair, and cell growth. These features make it an attractive material for developing scaffolds that replicate natural tissue environments and promote regeneration. With advancements in biomaterials research, HA has been applied in multiple functional formats, including injectable hydrogels, electro spun nanofibers, bioactive coatings, and composite scaffolds. Chemical modifications of HA—such as oxidation (Oxi-HA), thiolation (HA-DTPH), adenosine diphosphate conjugation (HA-ADP), and methacrylation (HAMA)—allow for controlled degradation, adjustable mechanical properties, and responsiveness to environmental factors like pH or redox potential. Such characteristics support the creation of in situ-gelling, self-healing, and stimuli--responsive hydrogels capable of releasing therapeutic agents in a targeted and regulated fashion.

2.4 Cellulose

Cellulose is the most copiously available natural biopolymer which is primarily obtained from plant cell walls and is familiar for its non-toxicity, and renewability. Numerous forms of cellulose—especially cellulose acetate (CA), bacterial cellulose (BC), and functionalised derivatives like tricarboxylic cellulose (TCC), ethyl cellulose (EC), carboxymethyl cellulose (CMC) and dialdehyde bacterial cellulose (DABC) are widely investigated. These forms are processed into nanofibers, cryogels, hydrogels, 3D printed scaffolds and electro spun mats customised for various applications. While native cellulose itself lacks antimicrobial properties, it is often functionalised with bioactive agents like silver nanoparticles, graphene oxide, chitosan, or plant-based extracts to impart antibacterial, anti-inflammatory and antioxidant capabilities.

2.5 Chitin

Chitin is a naturally occurring and widely available polysaccharide found in the hard outer shells of crustaceans, the protective coverings of insects, and the cell walls of fungi. It is composed of N-acetylglucosamine units and is known for its high crystallinity and chemical stability. It consists of N-acetylglucosamine units and features a high degree of crystallinity and chemical stability. Chitin exhibits low immunogenicity, enabling it to interact safely and effectively with biological systems. Its structural similarity to glycosaminoglycans further promotes favourable cell-material interactions. Although water and most organic solvents cannot dissolve chitin in its native form, researchers often apply chemical or physical modifications to convert it into more usable forms such as nanofibers, whiskers and hydrogels. These modified forms display enhanced mechanical strength, porosity and functional surface properties. Chitin also demonstrates inherent antimicrobial activity and supports haemostasis and tissue integration. Since it comes from renewable sources, chitin provides both ecological and biomedical benefits, making it a promising material for advanced healthcare and biomaterial applications.

2.6 *Collagen*

Collagen is a naturally occurring structural protein and an essential constituent of the ECM, considered one of the most important biopolymers in living systems. Its distinctive triple-helical configuration, enriched with glycine, proline, and hydroxyproline, imparts remarkable mechanical strength and resistance to heat. With excellent biocompatibility, biodegradability, and minimal immunogenicity, collagen integrates effectively with host tissues without eliciting unwanted immune reactions. Its architecture closely replicates that of native ECM, thereby supporting cell attachment, growth, and differentiation. Owing to its adaptability, collagen can be processed into nanofibers, hydrogels, films, and sponges with adjustable porosity, mechanical characteristics, and degradation profiles, making it advantageous for a wide range of biomedical uses. Its capacity to retain moisture aids in preserving a hydrated environment conducive to healing. Additionally, collagen displays inherent bioactivity by engaging with cell surface receptors, which enhances tissue elasticity, integrity, and tensile strength. Type I and III collagen, commonly found in connective tissues, are particularly important for skin regeneration and repair. Cross-linking techniques further improve its durability and allow control over degradation rates.

2.7 *Elastin*

Elastin is a structural protein prevalent in the ECM of elastic tissues such as the skin, lungs, arteries, and ligaments, where it imparts flexibility, durability, and structural stability.

Formed from crosslinked tropoelastin, its unique architecture enables repeated stretch and recoil without deformation, vital for withstanding physiological stresses. Biocompatible, biodegradable, and minimally cytotoxic, elastin integrates well into biological systems and regulates cell adhesion, proliferation, migration, and differentiation via bioactive sequences and receptor-mediated pathways. Its thermoresponsive behaviour, especially in elastin-like polypeptides (ELPs), allows reversible phase transitions with temperature, enhancing physiological adaptability. High hydrophilicity and moisture retention support hydrated tissue environments. Elastin is processable into sponges, films, gels, and nanofibers, and combines well with materials like collagen, gelatin, alginate, chitosan, PCL (Polycaprolactone), PVA (polyvinyl alcohol), and PLA (Polylactic Acid). It is often crosslinked using genipin or modified through molecular imprinting for enhanced bioactivity. Its composite fibrous–amorphous structure grants tuneable viscoelasticity ideal for scaffolds. Stability is improved using crosslinkers like genipin or tannic acid. Elastin also exhibits distinctive Raman and Fourier Transform Infrared Spectroscopy (FTIR) spectral features, aiding structural and degradation studies. Its elastic recoil, environmental responsiveness, mechanical tunability, and bioactivity make it a multifunctional biopolymer.

2.8 Fibrin

Fibrin is a natural fibrous protein biopolymer produced during blood clotting. It plays an essential role in haemostasis by forming clots at injury sites, halting bleeding, and initiating wound stabilisation. Beyond its clotting function, fibrin serves as a temporary ECM, creating a 3D network that supports the attachment, migration, and multiplication of fibroblasts, keratinocytes, and endothelial cells. This matrix promotes tissue regeneration through mechanisms such as angiogenesis, collagen synthesis, and re-epithelialisation. However, its rapid degradation and relatively low mechanical strength necessitate modifications to improve functionality. Thermal treatments such as hyperthermia or photothermal therapy can influence fibrin's stability and biological activity. Despite these limitations, fibrin remains an important biomaterial due to its biodegradability, compatibility with living tissues, and multifunctional role in clot formation and tissue repair.

2.9 Gelatin

Gelatin is a biopolymer obtained from denatured collagen, valued for its hydrophilicity, compatibility with biological systems, biodegradability, and lack of toxicity. Preserving the triple-helical domains and RGD motifs of collagen, it facilitates cell attachment, growth, and specialisation by closely resembling the ECM.

Its amino acid composition features positively charged groups that interact with negatively charged molecules, enhancing degradability and mechanical performance. Gelatin's chemical structure allows functional modifications—such as methacrylation (gelatin methacryloyl, GelMA), thiolation (thiolated gelatin, TGel), or crosslinking with glutaraldehyde, 1-ethyl-3-(3-dimethylaminopropyl) carbodiimide (EDC)/N-hydroxysuccinimide (NHS), or genipin—to improve mechanical strength, stability, and stimuli-responsiveness (pH, temperature, light). Its hydrophilicity aids moisture retention, supporting biological processes, while thermal gelling enables formation of hydrogels, cryogels, and electrospun nanofibers. Though structurally ECM-like and architecturally versatile, gelatin suffers from mechanical and physiological instability, often addressed via crosslinking or polymer hybridisation. Overall, gelatin is a low-cost, highly tuneable, and bioactive material widely used in biomedical and biomaterial research.

2.10 Silk

Silk, a natural biopolymer obtained from species such as *Bombyx mori*, *Antheraea assama*, *Antheraea mylitta*, and *Philosamia ricini*, primarily consists of two proteins—silk fibroin (SF) and sericin. SF, the main structural component, features a β-sheet-rich crystalline arrangement that provides remarkable mechanical strength and thermal stability, whereas sericin acts as a hydrophilic, glue-like coating that surrounds fibroin fibres and contributes to its functional versatility.

Both components are characterised by biocompatibility, biodegradability, low immunogenicity, and processability into diverse forms such as nanofibers, hydrogels, films, and sponges. SF's molecular structure includes hydrophobic sequences that facilitate intramolecular and intermolecular interactions, enabling stimuli-responsive self-assembly and tuneable crystallinity. Its degradation rate can be regulated through structural modifications, making it adaptable for long-term or short-term applications. Sericin demonstrates cytoprotective, antioxidant, and mitogenic properties, promoting fibroblast and keratinocyte proliferation, collagen synthesis, and inflammatory modulation. The combination of SF's mechanical resilience and sericin's bioactive traits allows silk to mimic ECM characteristics, supporting cell adhesion, differentiation, and tissue integration. Silk's compatibility with functionalisation enables the incorporation of targeting moieties or bioactive molecules without compromising its integral properties.

3 Major Medical Applications

3.1 Wound Healing

The skin, being the largest organ in humans, functions as a protective shield against threats such as pathogens, environmental stressors, and dehydration. A wound refers to any disruption in the skin's normal structure or function, or that of underlying tissues, leading to compromised cellular and biological integrity and altered physiological activity. The wound healing process is a multifaceted and vital biological mechanism that reinstates tissue integrity and restores normal function. It involves sequential stages—haemostasis, inflammation, proliferation, and remodelling—each essential for tissue restoration and regeneration.

Polysaccharides are known for their ability to form gels, films, and scaffolds. Their intrinsic properties such as hydrophilicity, structural versatility, and potential for bioactivity make them ideal materials for wound healing applications. Polysaccharides like alginate, HA, cellulose, and chitin play diverse roles in supporting haemostasis, moisture retention, inflammation control, and tissue regeneration. Through chemical modifications and composite development, these materials have transformed into versatile systems that can manage a range of wounds, from short-term injuries to long-standing ones, even in cases where healing is hindered by infections or diabetes-related complications [1]. Proteins also play a vital role in wound repair, providing both structural integrity and important biological activities. Unlike polysaccharides, which mainly provide hydration and scaffolding, proteins like collagen, elastin, fibrin, gelatin, and silk directly influence cell behaviour, angiogenesis, and tissue regeneration. Used in various forms such as hydrogels, sponges, and nanofibers, these proteins enhance healing by promoting cell adhesion, reducing inflammation, and supporting ECM remodelling, making them key materials in advanced wound care [2].

Several biopolymers, including collagen, gelatin, SF, and fibrin, facilitate cell adhesion, proliferation, and migration due to their ECM-like structures that support the functions of fibroblasts and keratinocytes. Chitin and elastin also enhance fibroblast proliferation and cellular migration, while gelatin scaffolds encourage cell differentiation. A central role of these biopolymers is to support tissue regeneration and remodelling [3]. Materials such as collagen, HA, chitin, gelatin, fibrin, elastin, silk, and cellulose are applied in various forms—hydrogels, films, sponges, nanofibers, and composite scaffolds, to regenerate tissue across acute, chronic, diabetic, infected, ocular, gastric, and burn wounds. Chitin nanostructures and collagen-based composites promote remodelling of chronic wound beds, while elastin-containing hybrids contribute to ECM reorganisation and strength restoration [4–6]. Granulation tissue formation, a key milestone in chronic wound repair, is supported especially by gelatin [7]. Preserving adequate moisture at the wound site is crucial for effective recovery, a role fulfilled by biopolymers like alginate, HA, gelatin, elastin, and silk, which help retain moisture while supporting tissue hydration and flexibility. Collagen and alginate additionally manage exudate absorption and ensure pain-free dressing removal, which is crucial in sensitive or infected wounds. Crosslinked formulations with elastin and cellulose further enhance the elasticity and conformability of dressings [8].

Biopolymers such as collagen and chitosan have demonstrated significant procoagulant activity through platelet aggregation and clot stabilisation [9]. Similar mechanisms are also observed in fibrin and alginate-based systems. Fibrin forms a provisional ECM, stabilising the wound and initiating cell infiltration. Specific composites like collagen–tranexamic acid or alginate–tannic acid enhance clotting efficiency and add antibacterial properties [10, 11]. For re-epithelialisation, biopolymers such as HA, chitin, gelatin, fibrin, and silk support keratinocyte migration and epidermal regeneration. HA membranes and burn wound dressings particularly promote rapid epidermal coverage, while gelatin and fibrin matrices help close diabetic and chronic wounds. Angiogenesis, or new blood vessel formation, is stimulated by collagen, HA, gelatin, silk, elastin, and chitin. These materials either inherently promote vascularisation or deliver angiogenic agents such as ginsenoside Rg1, catechins, and quercetin [12–14]. Elastin-derived peptides (EDPs) have additional pro-angiogenic and immunomodulatory effects, while SF materials aid vascular regeneration by mitigating oxidative stress. Biopolymers also assist in collagen synthesis and deposition, essential for structural restoration of the skin. Chitin, HA, alginate, gelatin, and fibrin stimulate fibroblasts and regulate MMPs (Matrix Metalloproteinases), encouraging endogenous collagen matrix production. Advanced materials such as HA nanofibers, strontium-loaded alginate xerogels, and chitin derivatives accelerate this process. Tracking tools like Raman spectroscopy, FTIR, Second Harmonic Generation(SHG) imaging, and markers like COL1A1 and COL3A1 are used in conjunction with collagen-based systems to monitor deposition during healing [15, 16]. Many of these materials provide antimicrobial, antioxidant, and anti-inflammatory effects, vital for preventing infection and modulating wound biochemistry. HA, chitin, alginate, gelatin, silk, and cellulose are incorporated with silver nanoparticles, zinc oxide, nitric oxide, quercetin, catechins, and propolis to fight microbial growth. Sericin, a silk protein, delivers potent antioxidant and antibacterial effects, while chitin aids in immune modulation during healing [17, 18].

For drug and bioactive delivery, collagen, HA, gelatin, silk, fibrin, alginate, and cellulose act as carriers, enabling targeted and prolonged release of growth factors, vitamins, anti-inflammatory agents, antimicrobials, and plant-derived compounds [19]. These materials are often engineered into injectable, self-healing, or in situ-forming hydrogels that respond to pH, temperature, or enzymatic changes. Silk, cellulose, and HA-based systems also enable the creation of smart wound dressings with integrated biosensing and real-time monitoring capabilities. Many of these biopolymers also contribute to ECM mimicry, improving integration and healing. SF nanofibers, gelatin scaffolds, and collagen-elastin hybrids replicate the natural ECM. Crosslinking agents like tannic acid and genipin enhance mechanical integrity, moisture retention, and tissue compatibility. In addition to these commonly recognised roles, several rare and specialised functions are also performed. Elastin has diagnostic value in burn severity assessment via its unique spectral features. Gelatin supports mucoadhesive wound healing, making it useful in oral or mucosal ulcers [20]. HA contributes to gastric and ocular wound repair by forming acid-resistant, adhesive hydrogels [21]. HA films and sponges also promote wound contraction, reducing surface area [22]. Cellulose-based scaffolds, especially when combined with growth factors or stem cells, enable targeted therapy and stem cell support [23]. Silk sutures are employed for mechanical wound closure and offer inherent antimicrobial benefits. Finally, fibrin matrices serve as provisional ECMs, and gelatin-based dressings promote granulation tissue formation, both essential for regenerating difficult-to-heal wounds [24].

3.2 Skin Substitutes

Skin substitutes are advanced biomaterials or tissue-engineered constructs designed to mimic natural skin and support wound healing [25]. In cases like burns, ulcers, and traumatic injuries. Unlike traditional dressings, they actively promote tissue regeneration and address the limitations of grafts, such as immune rejection, disease risk, and limited availability. Designed to mimic the architecture and function of skin—particularly the ECM—these substitutes facilitate cell attachment, movement, growth, and tissue remodelling. Many are bilayered, combining a regenerative dermal layer with a protective epidermal layer, as seen in gelatin–chitosan cryogels or silicone–collagen scaffolds.

Skin substitutes, made from natural biopolymers (e.g., collagen, chitosan, gelatin, SF) or synthetic polymers (e.g., PCL, PLGA), offer biocompatibility, biodegradability, and bioactivity. They can be functionalised with antibacterial, anti-inflammatory, and angiogenic agents like growth factors, stem cells, graphene oxide, or plant extracts (e.g., Aloe vera). Fabrication methods such as electrospinning, cryogelation, 3D printing, and self-assembly create scaffolds with optimal porosity, strength, and nanostructure. These constructs support wound healing by retaining moisture, enabling oxygen exchange and drug delivery, and promoting re-epithelialisation, cell proliferation, angiogenesis, and even skin appendage regeneration. Acellular scaffolds mimic ECM without triggering immune responses, while cell-laden versions boost regeneration. Innovations include injectable hydrogels, smart dressings, disease-specific models, and in vitro tools like reconstructed human epidermis for drug and cosmetic testing [26, 27].

3.3 Ophthalmology: Intraocular Implants

Ophthalmology is a medical and surgical specialty focused on diagnosing, treating, and preventing conditions and disorders affecting the eyes and the visual system. In this field, intraocular implants are increasingly being developed using biopolymers to improve biocompatibility, mechanical performance, and therapeutic outcomes. A primary application of these implants is in intraocular lenses (IOLs), which are commonly implanted following cataract surgery to replace the eye's natural lens [28]. Biopolymers such as cellulose derivatives and hydrogels are employed in the fabrication of IOLs to enhance optical clarity, reduce protein adhesion, and prevent complications like posterior capsule opacification (PCO). A significant challenge with IOLs is the development of glistenings—tiny microvacuoles formed through microphase separation—that can adversely affect visual quality [29].

Studies have shown that the hydrophobicity and elasticity of the crosslinked polymer networks significantly influence glistening formation, highlighting the importance of careful material selection and design in maintaining long-term optical performance [30, 31]. The use of biopolymers in ophthalmology has expanded beyond lenses, offering new possibilities in targeted drug delivery, inflammation-responsive implants, corneal repair, and regenerative therapy.

Biopolymers are widely used in ocular drug delivery systems like ocular inserts (ocuserts) and in situ forming gels, utilising materials such as chitosan, alginate, gelatin, HPMC, and PVA. These biodegradable matrices provide sustained, targeted drug release for conditions like glaucoma, fungal keratitis, and uveitis, enhancing corneal contact time, bioavailability, and patient compliance over conventional eye drops. Smart intraocular implants made from HA and chitosan respond to inflammatory signals, releasing anti-inflammatory or antibiotic agents on demand for diseases like posterior uveitis [32]. Biopolymers also aid in corneal applications through collagen–chitosan films and porcine cornea–polyurethane composites as synthetic keratoprostheses, supporting structural and biological integration [33]. Additionally, biopolymer-based hydrogels serve as vitreous substitutes and ocular adhesives in intraocular surgeries due to their transparency, flexibility, and ocular compatibility [32]. They also function as stem cell delivery platforms for retinal or corneal regeneration. Combining biodegradability, biocompatibility, and tuneable mechanics, biopolymers enhance precision, reduce complications, and improve outcomes in ophthalmic treatments [34].

3.4 *Tissue Engineering*

Tissue engineering is a multidisciplinary field combining principles of materials science, biology, and engineering to preserve, repair, or improve tissue function, as well as to create biological substitutes for damaged or diseased tissues.

With the rising demands for organ replacements and regenerative therapies, tissue engineering has evolved as a propitious approach to address the limitations of conventional transplants and grafts, such as immune rejection and donor shortages. An essential aspect of tissue engineering is the use of scaffolds—3D frameworks that support cell attachment, growth, and specialisation while imitating the natural ECM. Among the wide range of materials available for scaffold design, biopolymers derived from natural biological sources have attracted considerable interest.

Their ingrained alikeness to native tissue constituents makes them model candidates for scaffold development. Biopolymers are used in diverse physical forms like hydrogels, nanofibers, sponges, microbeads, films, and porous scaffolds tailored through methods like freeze-drying, electrospinning and crosslinking [35]. Their application spans a wide range of tissues including bone, cartilage, skin, nerve, vascular, cardiac, liver and adipose tissues. A common strategy involves combining these biopolymers with natural or synthetic polymers such as chitosan, gelatin, collagen, PCL, PLGA, PHBV or PVA, along with inorganic fillers like hydroxyapatite, graphene oxide and bioactive glass, to enhance mechanical strength, porosity, osteoconductive and degradation profiles [36].

Many of these polymers, such as gelatin and HA, contain bioactive sequences like RGD that further aid in cellular responses while others like chitin and fibrin promote vascularisation or matrix mineralisation. Advanced scaffold design often integrates these polymers with conductive or bioactive additives like polypyrrole, rGO, MoS_2 or magnetic nanoparticles for targeted applications in neural or cardiac tissue repair [37, 38]. Another shared feature is the ability of these biopolymers to be engineered into composite scaffolds that support ECM-like environments. Alginate, HA, and gelatin frequently form polyelectrolyte complexes or blend with polymers like chitosan or PCL to produce functional hydrogels and nanofibers. Collagen and its derivative gelatin are often structured into sponges, hydrogels, or films that support tailored porosity and degradation suited to tissue-specific needs, with spectroscopic techniques like Raman and FTIR being commonly used for their characterisation. SF and keratin are frequently incorporated with natural agents (e.g., TGF-β3, GYY4137, pectin) or used with synthetic polymers for enhanced regenerative signalling and mechanical stability. Cellulose's unique water retention, porosity, and oxygen permeability properties make it suitable for wound healing and skin regeneration. Fibrin is particularly suited to applications requiring cell adhesion and migration and is often blended with other polymers or ceramics like β-TCP [37, 39].

Alginates have potential applications in nanocomposites and bioactive glasses, as well as in the microencapsulation of various cell types, including human adipose-derived mesenchymal stem cells (hAD-MSCs), with a coating of PCL nanofibers [40]. In some innovative strategies, alginate is combined with graphene oxide/PCL to create a nanofibrous scaffold for myocardial tissue regeneration, providing a supportive environment for cardiac cell growth. It is also used in conjunction with pectin and selenium nanoparticles, which improves thermal stability, tensile strength, cellular performance, and antioxidant capacity [41]. For Bone tissue engineering, some of the recent studies demonstrated the use of cytocompatible Alginate–graphene oxide–hydroxyapatite beads utilising SA's crosslinking capabilities [42], Bead-shaped scaffolds can be formed by coagulating TCC/sodium alginate loaded with hydroxyapatite (HA) and/or graphene oxide (GO) using calcium ions [43] and alginate/hydroxyapatite (Alg/HAp) porous scaffold enhanced with bioglass (BGMS10) to improve mechanical strength and bioactivity [44]. Some of the studies used creative methodologies like use of Raman spectral imaging to trace protein secretion by cells within the HA matrices [45], buildout of injectable click-crosslinked HA hydrogels loaded with cytomodulin-2 for cartilage regeneration [46] along with incorporating nanomaterials to enhance neural and bone-forming responses.

Studies focus on combining collagen with bioactive agents like genipin or strontium to fabricate bioactive composites for regenerative medicine [47, 48]. Novel studies investigated visible light cross linkable gelatin bioinks and graphene-based gelatin scaffolds, displaying gelatin's plasticity to modern bio fabrication techniques [49–51]. Among the many inventive studies, those amalgamating drug-loaded bioactive ions, nanospheres, or conductive materials stick out for their multifunctionality [52, 53]. The integration of metformin-loaded MSNs(Mesoporous Silica Nanoparticles) in GelMA hydrogels fostered osteogenic differentiation in dental stem cells [54]. Likewise, furfuryl-gelatin bioinks enabled high-fidelity bioprinting, and Ti64 alloy modified with gelatin-chitosan composites introduced both biocompatibility and antimicrobial action [55]. An especially noteworthy study utilised electrochemically extracted chitin scaffolds from marine sponges, which provided non-toxic, natural, and porous frameworks capable of supporting various cell types, including mesenchymal stem cells and neural cells [56].

Another work used injectable chitin-polycaprolactone-HA microgels for craniofacial bone repair [57]. Furthermore, chitinous matrices customised with copper/zinc oxide nanoparticles or silver molybdate display robust antimicrobial properties [58]. Studies on liver-mimicking injectable scaffolds [59], fibrin gels infused with insulin for promoting nerve regeneration [60] and magnetic fibrin nanoparticles [61] for bone tissue engineering spotlight the evolving utility of fibrin. Few studies involve the progression of TCC/SA beads encompassing GO and HA from eggshells, which display low cytotoxicity and propitious bone regeneration properties [43] and use of customised BC with polydopamine to imitate the skin ECM, amplifying nutrient diffusion and cell survival [62]. Studies also underline cellulose composites doped with fibronectin or Fe/Zn nanoparticles to strengthen osteoblast proliferation, protein adsorption or confer antibacterial activity. Using novel methods, Keratin is combined with silk in a 40:60 ratio along with 2% gelatin and 20% CPO post which the scaffold demonstrated exceptional mechanical properties, oxygen-generative ability and non-cytotoxicity suggesting its potential suitability for urinary tract reconstruction [63]. It is positioned as a sustainable biopolymer for its adaptability to various modifications, combinations with synthetic and natural polymers and sources including industrial wastes. Keratin/PCL mats combined with VEGF and heparin are found to prolong clotting time while also promoting endothelial growth [64]. In cartilage repair, composite scaffolds made of chitosan/γ-PGA and elastin/PCL with surface elastin alterations displayed enhanced ECM secretion, chondrocyte proliferation and reduced apoptosis [65, 66]. For vascular tissue engineering, elastin-mimetic triblock polymers embedded in ePTFE grafts or electrospun polyurethane matrices improved elasticity and cellular interactions. In bone regeneration, elastin was combined with chitosan membranes and BMP-2-loaded scaffolds to enhance mechanical strength, osteogenic differentiation, and bioimaging potential.

A study utilising PLA–elastin–gelatin composite fibres loaded with clindamycin demonstrated their ability to support cell adhesion, maintain fiber alignment consistent with geometric cues, and exhibit effective antimicrobial activity [67]. In cardiac tissue engineering, elastin is integrated with alginate, gelatin, and collagen to mimic heart valve leaflet architecture, with emphasis on anisotropic bending properties and cell-friendly microstructures. In hepatic models, elastin-modified surfaces on tissue culture polystyrene (TCPS) regulated rat hepatocyte morphology and function, offering a platform for liver regeneration studies and bioartificial liver design [68, 69]. Molecular imprinting technology was also used to incorporate bioactive recognition sites (e.g., fibronectin, laminin) into elastin-based sponges, pioneering a strategy for cell-specific scaffold functionalisation [70].

3.5 *Implants*

Medical implants are devices placed within the body to assist, substitute, or improve biological structures. Implants, when made from biopolymers, offer privileges like gradual degradation (eliminating the need for removal), reduced risk of rejection and enhanced compatibility with body tissues. Common uses include dental fixtures, orthopaedic implants, as well as drug delivery applications. Merging implant technology with biopolymer science introduces innovation to regenerative medicine, tissue engineering, and personalised healthcare.

3.5.1 Vascular Implants

Vascular tissue engineering is swiftly evolving with the integration of biopolymers to create scaffolds that closely resemble the function and structure of native blood vessels. The transition from conventional autologous grafts to hybrid and synthetic scaffolds is due to the limited availability of donor tissues and complications such as intimal hyperplasia or thrombosis. Current studies showcase the capability of materials such as collagen, chitosan, SF, polyurethane and bacterial nanocellulose for developing vascular implants. Silk/PLGA/calcium phosphate composites and bacterial nanocellulose-silk hybrids display controlled release of vasculogenic/angiogenic factors like VEGF and foster deeper cell infiltration [71]. One of the prime challenges in Tissue-engineered vascular grafts (TEVGs) is attaining long-term patency and averting restenosis [72]. Approaches such as integrating keratin for nitric oxide generation (e.g., PCL/keratin scaffolds with catalytic NO release and endothelialisation) [72], incorporating heparin for thrombus resistance (e.g., PCL scaffolds layered with SePEI and heparin promoting endothelialisation and inhibiting smooth muscle proliferation) [73], and using BDE or genipin crosslinking to strengthen collagen stability (e.g., genipin-crosslinked type II collagen scaffolds with enhanced biostability and cell differentiation) [47, 74] indicate significant improvements in scaffold performance. Notably, fish scale-derived collagen constitutes a functional and sustainable substitute with high vascularisation potential—as demonstrated by in vivo murine studies showing enhanced blood and lymphatic vessel ingrowth around methylated and BDE-crosslinked fish scale collagen implants [75]. Emerging designs now give importance to targeted biological signalling, multifunctionality blending mechanical robustness and long-term integration leading to considerable potential for future clinical translation.

3.5.2 Neural Implants

Neural implants based on biopolymers make remarkable advances in addressing the complex challenges of neural tissue repair, such as biocompatibility, conductivity, immune response, and integration with host tissue. Studies explore a range of materials, from injectable CS/β-GP hydrogels for CNS repair [76] to conductive PEDOT(Poly(3,4-ethylenedioxythiophene) based scaffolds for neural regeneration [77] and chitin-CNT composites that support synaptic function and electrical signalling [78]. Innovations in surface chemistry, such as dextran-peptide coatings, provide neuron-selective adhesion (e.g., dextran surfaces grafted with the IKVAV peptide selectively promote neuronal adhesion over glial and fibroblast cells) [79], while voltage-responsive drug-releasing scaffolds introduce dynamic therapeutic control (e.g., graphene-hydrogel electroactive scaffolds enabling pulsatile drug release under electrical stimulation; conducting polymer-hydrogel hybrids offering electrically tuneable release profiles) [80, 81]. These biopolymer-based platforms not only support neuronal growth and activity but also enhance implant functionality through smart material engineering. Collectively, the research underscores the vast potential of biopolymer neural implants for next-generation neuroprosthetics and regenerative therapies.

3.5.3 Cardiac Implants

Biopolymer-based cardiac implants are in the front line for next-generation cardiovascular treatments. From conductive hydrogels [82] and smart scaffolds to drug-eluting stents [83] and calcification-resistant valves, these biomaterials provide multipurpose platforms to re-establish cardiac function and structure. Advances in surface engineering and material science—such as laser-textured stents and bioactive nanocomposite coatings—paired with spectroscopic characterisation (e.g., Raman and XPS analysis) enable implants that integrate more effectively with native tissue, reduce risks of calcification and restenosis (e.g., PLLA/ACP-coated stents show faster re-endothelialisation and lower inflammation), and even support real-time imaging applications [84, 85]. The field is incessantly moving from bench to bedside, allowing precision and personalisation into cardiac regenerative medicine.

3.5.4 Bone Implants

The bone tissue engineering field is undergoing brisk modifications through the incorporation of ceramics, biopolymers and advanced surface engineering techniques. Biopolymers like collagen, chitosan, SF and gelatin are combined with graphene oxide, bioactive ceramics (e.g., hydroxyapatite, tricalcium phosphate) and antimicrobial agents to design scaffolds that are osteoinductive, biocompatible and resistant to infection. These materials are located across applications such as orthopaedic reconstruction, dental implants, spinal repair, and fracture healing. Present-day innovations involve the use of graphene oxide for peptide-functionalised coatings for antibacterial effects [86, 87], ectopic bone formation and the application of Raman spectroscopy to assess material compatibility and implant performance [88]. The convergence of biomedical engineering and material science coupled with spectroscopic tools and powerful imaging, continues to boost the safety, efficacy and personalisation of bone implants.

3.6 Drug Delivery

Biopolymers have emerged as valuable materials for therapeutic delivery, owing to their inherent compatibility with biological systems, ability to degrade naturally, and adjustable physicochemical characteristics.

Chitin and its derivative chitosan are studied for large-scale pH-sensitive drug release, mucoadhesion, and nanoparticle formation. The chitosan-coated systems support enhanced bioavailability and cellular outlets and are often modified with other materials for tissue engineering and applications in bone repair [89]. The Alginate, with its ability to form "egg box" gels [88], is widely used for 3D-printed and colon-targeted delivery systems in the presence of divalent cations [90]. This allows sustained release with pH responsiveness for hydrophobic drugs and is frequently blended with polymers such as PVA to improve structural strength [91]. HA, known for its affinity for overexpressed CD44 receptors in tumours [92], facilitates targeted drug delivery across various routes (ocular, nasal, pulmonary, oral), and its chemical structure allows for conjugation with therapeutic molecules for multifunctional and controlled release platforms [93]. Silk fibrin is attracting interest for its capacity to produce micro- and nano formulations with adjustable crystallinity, thereby enhancing the stability of controlled drug release as well as bioactive agents.

It has been active in drug and gene delivery, diabetes treatment, bone regeneration, and as a coating material for medical implants due to its mechanical strength and excellent biocompatibility. Gelatin is extensively utilised for precise and sustained drug release through various forms such as hydrogels, microspheres, nanoparticles, and films, owing to its outstanding membrane-forming capabilities, bio adhesive properties, and potential for chemical modification to improve functionality. Its triple-helical structure and positive charge can interact with drug molecules negatively charged, improving mechanical strength and drug-efficiency.

Fibrin acts as an effective drug carrier, which enables sustainable release of chemotherapeutic agents, antibiotics and thrombolytic agents. Its versatility in hydrogel formation and strong ability to bind with nanoparticles and imaging agents render it ideal for a wide range of biomedical uses, including precise drug delivery, regenerative medicine, and repair of wounds. Collagen, recognised for its minimal immunogenic response, facilitates site-specific administration of bioactive compounds such as pharmaceuticals and growth factors, simultaneously aiding tissue repair and regeneration. It is used in nanoparticle-based delivery, wound dressings, thermosensitive hydrogels, and cartilage-targeted systems. Keratin produces a fibrous protein, biodegradable films and nanoparticles showing redox and enzyme-responsive behaviour, which are useful for pH-triggers and targeted cancer therapies. Its intrinsic cell-binding motifs (e.g., RGD, LDV) and antioxidative properties have supported the use of it in mucoadhesive drug delivery and wound healing. Cellulose, including its nanostructured forms (CNFs and CNCs), functions as a stable matrix or porous framework for sustained and targeted drug delivery because of its large surface area, adjustable porosity, and capacity to preserve drug efficacy. Derivatives such as HEC, MC, and HPMC are usually used in hydrogel-based delivery systems [94]. Finally, ELPs show unique phase transition properties, which form injectable depots or nanoparticles responsive to physiological conditions. Their biodegradability, minimal toxicity, and tuneable responsiveness make them highly attractive for in vivo drug delivery applications [95].

3.7 *Vaccine Delivery*

Vaccine delivery is a particular aspect of immunisation, requiring the development of safe, effective, and efficient methods for providing vaccines to individuals.

Biopolymers like chitosan, SF, and others have emerged as promising candidates for vaccine delivery because of their distinctive characteristics, such as compatibility with biological systems, ability to degrade naturally, and non-toxic nature.

These biopolymers offer a range of advantages, including the ability to enhance immunogenicity, protect antigens from degradation, and control the release of active agents. With their potential to improve vaccine efficacy and safety, biopolymers are being increasingly explored as carriers for drug delivery and biomedical materials [96].

Chitosan has gained attention in vaccine delivery due to its non-toxic, biodegradable, biocompatible nature and its ability to enhance permeability for peptides, proteins, and DNA vaccines. Its cationic surface facilitates loading of negatively charged protein or gene vaccines. Chitosan nanoparticles are easily formed via ionic cross-linking and spontaneous gelation in aqueous solutions through reactions between chitosan amine groups and tripolyphosphate (TPP) phosphate groups. As a vaccine delivery biopolymer, chitosan enhances immunogenicity, acts as an immune adjuvant, and protects antigens from degradation. Chitosan micro/nanoparticles, prepared by ionic gelation using TPP, offer advantages like controlled release, avoidance of toxic solvents, and mucoadhesive properties. Being a biodegradable, bio adhesive, and soft tissue-compatible polysaccharide, chitosan is widely used in pharmaceutical research and as a drug delivery carrier. It has shown promise in oral and intranasal vaccine delivery. Additionally, SF–gelatin hydrogels provide minimal adverse reactions, sustained and stimuli-responsive drug release. Other biopolymers like alginate, dextran, cellulose, poly-l-lysine, and HA are also explored for vaccine delivery. Nanoparticles, valued for their biodegradable and hydrophobic properties, serve as both adjuvants and antigen carriers in vaccine design.

3.8 Biosensing Biopolymer

Biosensing biopolymers have come up as a promising field of research, using the unique properties of biopolymers to develop innovative sensing technologies. Biopolymers such as chitosan, cellulose, and SF possess various beneficial attributes, including biocompatibility, biodegradability, and non-toxicity, which render them well-suited for biomedical uses. The integration of these biopolymers with cutting-edge detection methods, including surface-enhanced Raman spectroscopy (SERS), has facilitated the development of biosensors with high sensitivity and selectivity. Biosensing biopolymers have the capacity to transform biomedical diagnostics and monitoring, representing a fascinating research domain with significant potential to enhance human health and wellbeing [97].

Biosensing biopolymers have gained significant interest due to their promising roles in diagnostic tools, spanning medical and physiological monitoring as well as defence-related uses.

They are utilised in the detection of pesticides, E. coli, and other biomolecules using SERS-active plasmonic structures, as well as in electrochemical sensors, paper-based biosensors, and biocatalysis platforms for enzyme immobilisation and biomolecule detection. Biosensing biopolymers like chitosan and cellulose are employed in biosensors for on-site applications, such as chromatographic SERS and flexible scaffolding SERS format [98]. Additionally, they are used in the construction of biosensors and biocatalysis platforms, enabling rapid enrichment and capture of targets. The application of biosensing biopolymers has proven effective for the swift and early detection of complications caused by SAH, exhibiting high sensitivity, specificity, and comparatively low detection thresholds. Generally, the applications of biosensing biopolymers are diverse and widespread, with potential uses in medical monitoring, defence applications, and biomedical research.

3.9 3D Printing and Its Biomaterials

3D printing or additive manufacturing is a groundbreaking technology which builds physical 3D objects layer by layer by material deposition based on a digital blueprint. This revolutionary process has been referred to as the third wave of the industrial revolution due to its vast potential in a variety of fields and applications [99]. Unlike traditional subtractive manufacturing processes that remove material, 3D printing builds structures layer by layer, enabling the fabrication of complex, arbitrary geometries [100]. Its ability to rapidly produce customised and patient-specific models has been extensively explored in a broad spectrum of disciplines like medicine, dentistry, surgery, and orthopaedics [101]. Within the healthcare setting, 3D printing performs an essential function in enhancing visualisation, supporting diagnostics, enabling procedural planning and simulation, creating patient specific models and supporting the production of medical devices [100].

The prominent aspect of 3D printing particularly in biomedical and tissue engineering is utilisation of "biomaterial inks" or "bioinks" which are known to be compatible with biological systems. These biomaterial inks are broadly categorised into natural polymers and synthetic polymers [102]. Natural polymers are often favoured for their excellent biocompatibility and bioactivity, allowing them to mimic the native ECM and support cellular responses. Various natural polymers reported in the literature include alginate, collagen, gelatin, HA, chitosan and fibrin amongst others. Because of its properties, it is an excellent candidate to be used in the encapsulation of living cells and to favour their adhesion proliferation differentiation and migration [103]. However, these natural polymers may have reduced mechanical stability or rapid degradation. To overcome this, synthetic polymers, including PLA, PCL, polyethylene glycol (PEG), and polyurethane (PU), exhibit desirable features in terms of tuneable mechanical properties, degradation rates, and ease of manipulation. However, synthetic polymers can be biologically challenging in the long-run and may exert no or very little intrinsic bioactivity for direct cell communication than natural materials due mainly to their non-bio familiarity which makes them foreign substrates to host cellular activities [104]. Therefore, by combining natural and synthetic polymers we can achieve structural stability and biological compatibility. The combination of nanomaterials and 3D printing is also opening up new opportunities, including increased mechanical strength, electrical conductivity, and biocompatibility for highly functional, customisable products [105].

The application and effect of these biopolymers in 3D printing are significant, especially in areas like tissue engineering and regenerative medicine, which allows them for the development of complex scaffolds and constructs. These scaffolds can be cell seeded or encapsulate living cells in situ, with a favourable microenvironment supporting cell growth and function [106]. This ability has provided a new avenue for creating functional tissue, personalised implants, and sophisticated drug delivery systems. The constant developments of biomaterials and printing techniques are constantly pushing the industry, causing improvements in product quality, accuracy, and capacity to create complex geometries [104]. A notable development is 4D bioprinting, which involves the addition of "time" as a fourth dimension so that printed objects can alter their shape or function in response to external stimuli such as heat, light, or pH [107]. This holds potential for dynamic, adaptive structures for uses such as biosensors, biorobots, and intelligent biomedical devices. Despite impressive advancement, issues remain with bioink parameter optimisation for high cell viability, obtaining targeted mechanical properties closely resembling native tissue, and maintaining long-term stability and functionality of printed constructs in vivo [108]. With these challenges being alleviated, 3D printing and bioprinting are set to change personalised medicine and enhance patient health outcomes dramatically.

4 Artificial Intelligence

Adopting AI into the design, characterisation, structuring of biopolymers and bio-inspired composites offers a promising approach to advancing sustainability in materials science. AI, particularly through ML algorithms, excels at processing and analysing large datasets, which include molecular structures, processing conditions, and environmental impacts [109, 110]. By uncovering complex patterns and relationships within these datasets, AI facilitates accurate predictions of material properties and performance, significantly reducing the need for extensive experimental trials. In ML-driven biopolymer analysis, ML models are typically categorised into three types based on their outputs: regression models, which provide continuous numerical predictions; classification models, which assign inputs to distinct categories; and clustering models, which group data points based on shared features without supervision.

AI-based methods like Convolutional Neural Networks (CNNs) and Long Short-Term Memory (LSTM) networks are increasingly utilised for various tasks, including denoising, feature extraction and predictive modelling [111, 112]. With appropriate training data, AI models can learn to identify and interpret intricate signal patterns and features. The intrinsic structural complexity offers itself as a profound challenge and opportunity to the scope of analysis for vibrational spectroscopic methodologies [113]. Specific challenges in implementing conventional vibrational spectroscopy-based analysis in biopolymers are.

Conformation-dependent behaviour: The vibrational spectra of biopolymers can be extremely sensitive to their conformation and folding state. Also, the solvent interactions of the biopolymers can often interfere with its spectral features, adding complexity to the analysis [114].

Heterogeneity: Due to the different structural characteristics, Biopolymers often have complex, non-uniform structures with variable chain lengths and branching [115]. The heterogeneity introduces a challenge to conventional Vibrational spectra characterising methods.

Overlapping vibrational bands: Raman spectra often comprise complex and overlapping features, making it challenging to isolate meaningful information through conventional spectral interpretation methods [116].

Secondary structure sensitivity: Protein polymers have complex structures that often include defined secondary structural elements such as α-helices and β-sheets [117]. Though Raman spectroscopy is sensitive to secondary structures, minute changes are difficult to resolve unambiguously [118].

AI integration in spectroscopy can address these key challenges by streamlining data preprocessing, enhancing interpretation, enabling predictive modelling and classification, resolving spectral overlap, and aiding dynamic structure elucidation. It simplifies labour-intensive tasks and excels in complex classification and reconstructing missing spectral data. The methodologies are crucial for resolving overlapping bands and extracting information from convoluted spectra, leading to more accurate molecular interpretations. The tools can link spectral characteristics to dynamic structures, enabling the tracing of rapid conformational evolutions in features from resultant spectra.

4.1 Biopolymer Characterisation

Evaluating physicochemical characteristics of polymers and its comparison with each other is very relevant especially in the context of medical applications. It often refers to ensuring features such as molecular constitution, purity, mechanical properties, degradation rate, surface characteristics, etc. These are critical in medical settings as it affects safety and biocompatibility, efficacy and performance, it ensures reproducibility in clinical outcomes [119–121]. These properties of a biopolymer can vary over different factors such as variations in raw materials, manufacturing parameters, storage conditions and presence of impurities etc [122, 123] and is often evaluated using vibrational spectroscopic methodologies.

However, monitoring fine aspects like the degree of cross-linking and molecular composition—via spectroscopy poses a significant challenge [124, 125]. To address this, certain families of AI and ML algorithms are particularly used. Commonly used chemometric methods include Principal Component Analysis (PCA), Cluster Analysis (CA), Partial Least Square (PLS), Linear Discriminant Analysis (LDA), etc.

These methods help to improve the accuracy by addressing unmodeled variations-like instrumental and experimental artifacts-that often happens with spectroscopic data. However, these methodologies often mandate the need of preprocessing procedures like baseline correction, noise removal and scaling, multiplicative scatter correction, which may remove valuable patterns from the data, and if misused may significantly affect the material characterisation accuracy.

4.1.1 Conventional Methodologies

Of the methodologies that are being traditionally used, PCA, the unsupervised methodology, is the widely used one, specifically for dimensionality reduction. This is an especially important aspect of analysing spectroscopic data, as spectral datasets are often high-dimensional, and many classification methodologies find it extremely difficult to process it in its entirety. The methodology identifies relevant axes termed as principal components that record the maximum variance in the data, without using any labels. The methodology reduces numerous correlated wavenumber variables into a few uncorrelated principal components, achieving dimensionality reduction. However, PCA may lead to leakage of useful information in the process, and since it is label-independent, it may underperform if key classification features lie in directions of low variance [126, 127].

CA is also an unsupervised methodology, but unlike PCA, the methodology focuses on grouping similar data points into corresponding clusters based on some similarity or distance measure. Rather than transforming the data, it organises it into natural groupings. CA proves useful in scenarios where the data lacks predefined structure or labels [126].

PLS/partial least squares regression (PLSR) is a multivariate calibration technique that constructs a linear regression model to estimate the concentration of an analyte based on the spectral data of a sample, having strong capabilities in quantitatively analysing the spectroscopic data. It works especially well in cases where there are many correlated features, which is a common characteristic of spectroscopic data [113].

LDA is a supervised methodology that is useful for both classification and dimensionality reduction. The methodology projects the data in such a manner that the classes are as distinct as possible within a reduced space. The methodology is rather suitable when processed with pre-processing methods and is often preceded by PCA, leading to the PCA-LDA combination [126, 128].

Studies reliably link specific Raman peaks (e.g., amide I/III, ester/amide bands) to cross-link density or chemical modifications, which can be unveiled using these ML modalities or the combination of it [113].

4.1.2 Advanced Unified AI-Frameworks for Spectroscopic Data Interpretation

In an end-to-end model, a network learns directly from the raw spectral inputs to output predictions, eliminating any need for intermediate pre-processing steps. There are a set of algorithms that are capable of this automated learning.

CNNs, primarily designed for image processing, have emerged into powerful tools for end-to-end deep learning in the analysis of spectroscopic data. The specialised architecture is highly effective in handling grid-like data. When it comes to processing Raman data with the CNNs, an advantage is that it is highly capable of dealing with raw data. CNNs consist of multiple layers of neurons which are split into input layer, Convolutional layer, pooling layer, and fully connected layers. CNNs use convolutional filters to automatically detect localised spectral features regardless of minor shifts, noise etc., making it extremely capable of dealing with real-world spectra. Convolutional layers are made up of learnable kernels designed to identify features within the input data. These filters traverse through the input data, executing element-by-element multiplication and addition to produce an activation map. By stacking multiple convolutional and pooling layers, CNNs can capture both fine-grained, and broad spectral characteristics. The convolutional layers are verified to play a critical role in feature derivation and component identification, as models without them showed significantly lower accuracy. Thus, CNNs offer a highly scalable, interpretable, and accurate alternative to spectral analysis, especially in high throughput or real time environments, though it requires plenty of labelled data and careful design of model architecture [129, 130].

Deep Spectra, Deep-CID, Res-Net, Dense-Net etc. are few types of architectures that leverage these specialised features to attain advanced capabilities in material characterisation, quality control.

Deep-CID

Deep learning-based component identification (Deep-CID) is a novel end-to-end approach that leverages CNNs to directly facilitate component identification in mixtures from their raw Raman spectra [131], unlike many traditional methods, Deep-CID eliminates the need for explicit preprocessing steps such as smoothing, baseline correction, or peak detection. As shown in Fig. 2, The Deep-CID architecture is an assembly of four-layered-one dimensional CNN. The feed from the input layer is processed by the first convolutional layer that probes for basic patterns along the data. The Maxpooling layer extracts the most important features among the selection. The second convolution layer probes for more complex patterns from among the relevant ones, followed by another maxpooling layer that extracts the strongest signals. The final 2 convolution layers engage even more filters to capture more complex and detailed features, Pooling functions are used after these convolutions to enhance learning efficiency, improve robustness to peak shifts, and incorporate translational invariance. The ensuing flattening converts the multi-dimensional feature maps into a single long list of numbers, while the fully connected layers establish the relationship between the extracted features and the presence of a specific compound, which in turn determines the final output of the model. Dropout is applied in all convolutional and fully connected layers to prevent overfitting. ReLU

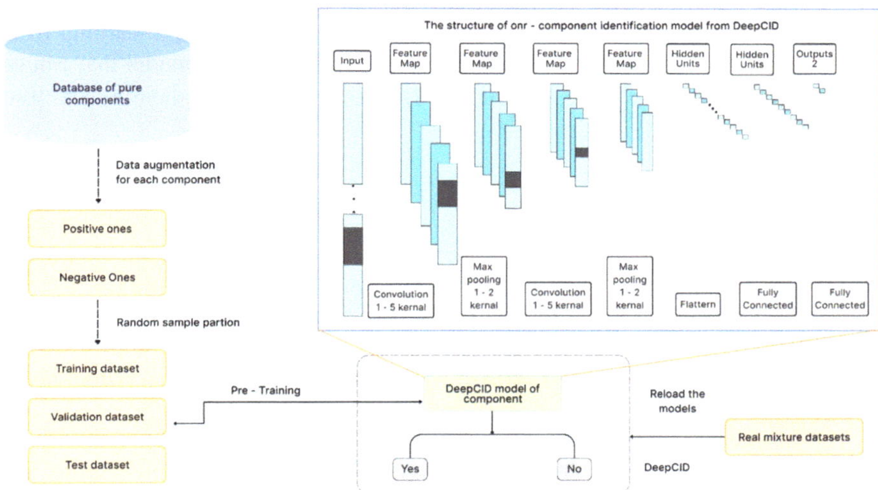

Fig. 2 Flowchart of DeepCID and the CNN model architecture. Spectra of each pure compound in the database are used for data augmentation to generate balanced positive and negative samples. The augmented dataset is split into training, validation, and testing sets. One-component identification models are trained and saved for future predictions. The CNN architecture, shown in the top right corner, consists of four layers: two convolutional layers followed by two fully connected layers. The final output layer has two nodes to determine the presence or absence of a component. This model supports pretraining with customised datasets and can be applied to evaluate real mixture datasets

serves as the activation function, with SoftMax in the final layer for output normalisation. The network is trained to engage the Adam optimizer and cross-entropy loss, and it employs ReLU and softmax activation functions.

Deep-CID models demonstrate strong stability and reproducibility, with accuracy distributions approximating Gaussian distributions with small variances over multiple training runs. In essence, Deep-CID offers an efficient, robust, and highly accurate method for identifying chemical components in complex mixtures from their spectra without manual data manipulation. However, skipping preprocessing can reduce generalisation, increasing the risk of false positives and negatives. To combat limited spectral data, data augmentation is widely used to enhance training.

DeepCID can directly extract valuable information and identify components from raw spectra, eliminating the need for preprocessing. Eventually the algorithm can potentially simplify the analysis pipeline and avoids potential errors or variability that can be introduced by preprocessing. However, non-dependence on preprocessing also avoids risk of it affecting generalisability of the model. These layers incorporate properties like shared weights (increasing learning efficiency and generalisation) and sparse connectivity (producing strong responses to local input patterns). The model effectively handles Raman shift displacement and intensity variations between instruments, noise interference, and presence of other compounds in training data. Deep-CID is proven to deal efficiently with noise interference from other compounds present in the composition. Moreover, the methodology can alleviate the displacements in peak shifts and variation in peak intensity between instruments. Due to its extensive capabilities of generalised features learning from large, augmented datasets, it can even be used as a pre-training model that can be leveraged for better precision against smaller biopolymer-based datasets and thus attain a higher accuracy.

DeepSpectra

DeepSpectra is an end-to-end deep learning framework for quantitative spectral analysis, using deep CNNs to automatically extract features and interpret raw spectroscopic data. The architecture of the DeepSpectra CNN model comprises three convolutional layers, a flattened layer, a fully connected layer, and an output layer, as depicted in Fig. 3 [127].

The speciality of the architecture is the integration of the Inception architecture into second and third convolutional layers, combining multiple convolutions and pooling operations with varying filter sizes. This allows the model to capture spectral features at different scales-narrow peaks, mid-range features, and broad or overlapping patterns. This enables the system to extract features of different sizes, i.e., the smallest convolution size may catch sharp, narrow peaks while a slightly larger convolution size may detect mid-width features, while a larger convolution layer picks up broader spectral shapes or overlapping peaks. Here, the Max pooling highlights dominant peaks, enhancing robustness to noise and peak shifts. The larger depths of the architecture can possibly extract more both high and low levels of

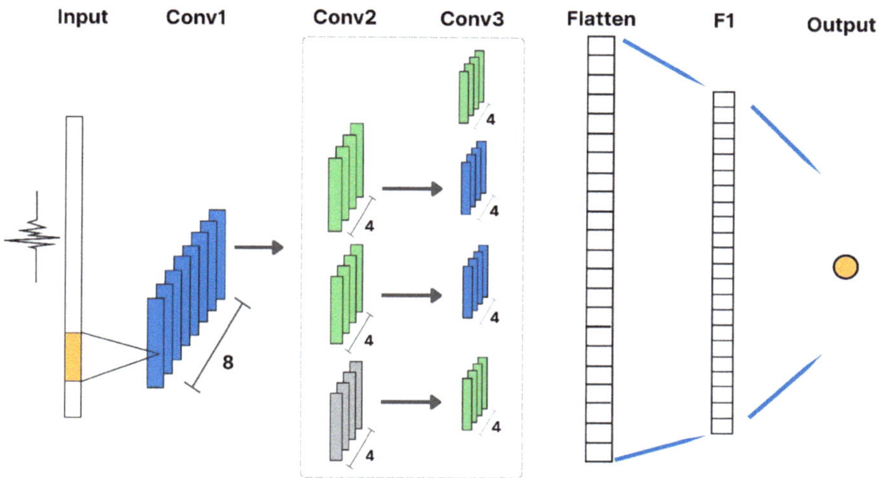

Fig. 3 Overview of the DeepSpectra model architecture, which includes three convolutional layers (Conv1, Conv2, Conv3), a flatten layer, one fully connected layer (F1), and an output layer. Blue modules denote standard convolution, green indicates 1 × 1 convolution, and gray represents max pooling. A short connection links the first 1 × 1 convolution in Conv3 directly to Conv1. The final 1 × 1 convolution in Conv3, along with two 1 × 1 convolutions in Conv2, reduces channel numbers to lower computational complexity. Small rectangles depict neurons, while larger rectangles represent feature maps in Conv1, Conv2, and Conv3

features from the raw spectra while, the enhanced width of the network by assembling parallel 1 × 1, m × 1, and n × 1 convolutions in third convolutional layer lets the architecture to have the adaptability to various scales of the local spectra features. And the model learns from concatenated outputs by the combination of these filter sizes. Thereby, the inception module enables DeepSpectra to have a large width and depth without significantly affecting the computational complexity. This design allows for an increase in both the depth and width of the network.

Advanced Feature Extraction: The CNN's convolutional layers are designed to extract spectral features. The use of the Inception module, with its parallel convolutions of different filter sizes, enables DeepSpectra to capture both local and high-level abstract features simultaneously within a single layer. This enhances the network's adaptability to various scales of local spectral features. The convolutional layers are verified to play a crucial role in spectral feature extraction and component identification, as can be seen from Fig. 4.

The architecture employs techniques such as shared weights and sparse connectivity, which increase learning efficiency and improve Generalisation ability compared to fully connected neural networks. Max pooling follows each of these convolutional layers. DeepSpectra incorporates dropout in both convolutional and fully connected layers to prevent model overfitting. Batch normalisation is also utilised after certain layers to speed up training and improve accuracy. The model is trained using the Adam optimisation algorithm with cross-entropy loss as the

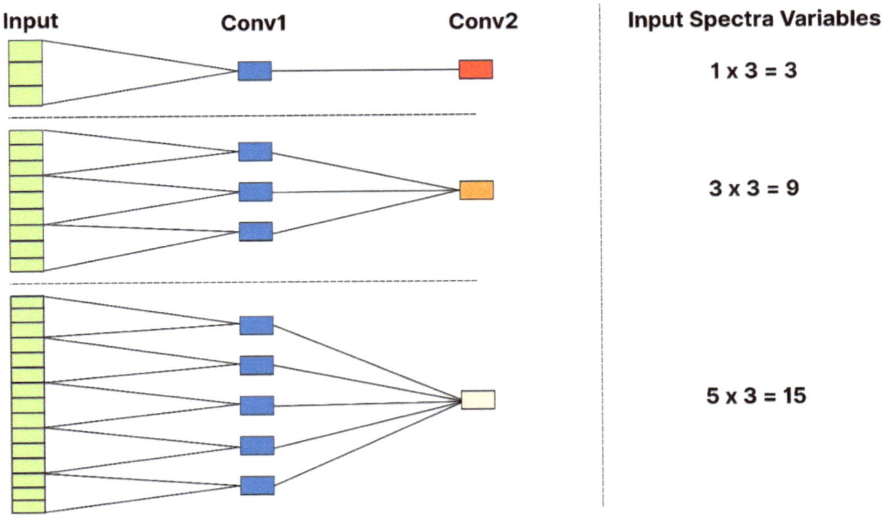

Fig. 4 Illustration of how receptive spectral regions change through two convolutional layers with varying filter sizes. The first layer uses a filter size of 3 and a stride of 3, while the second layer includes filters of sizes 1, 3, and 5. Lite-green rectangles in the input layer represent spectral variables, and rectangles in Conv1 and Conv2 layers denote individual neurons. Each neuron in the Conv2 layer corresponds to a receptive field covering 3, 9, or 15 original spectral variables, respectively

objective function. An early stopping strategy is implemented to further avoid overfitting during training.

DeepSpectra has demonstrated significantly higher accuracy compared to many other CNN models and conventional ML methods on various datasets, often achieving better results on raw data than even pre-processed data. The model is capable of alleviating interference from noise, different backgrounds, peak shift displacements, and variations in peak intensity across instruments. Stability and reproducibility: DeepSpectra exhibits great stability and reproducibility, especially with increased training samples. The system is designed for efficiency, capable of completing predictions for multiple mixture spectra within a few seconds. DeepSpectra models are structured to be easily evolved and fine-tuned with new samples to enhance their accuracy and performance over time.

This end-to-end deep learning approach is intended for quantified spectral analysis, learning patterns directly from raw data to enhance model performance. However, on the flip side, DeepSpectra models generally require longer computational time for training and mandates larger dataset size for repeatability and accuracy.

RamanNet

RamanNet aims to incorporate the principles of sparse connectivity without being restricted by temporal or spatial invariance [132]. The concept of RamanNet is that since CNNs are designed for processing image data and has got drawbacks when analysing spectroscopic data. Aspects of CNN like sparse connectivity and parameter sharing is an advantage in that aspect, However, equivariance to translation property affects the spectral processing ability of CNNs.

In RamanNet, shifted Multi-Layer Perceptrons (MLPs) were used to analyse shifted windows of the spectra. This facilitates sparse connectivity, making parameter sharing redundant and eliminates issues with translation equivariance. Hence, the convolution operation is emulated using MLPs (Fig. 5).

The fed spectrum is dismantled into overlapping dynamic windows of specific length and step size and each is handed over to a different dense block with specific number of neurons each, ensuring sparse connectivity and avoiding overfitting. This is analogous to convolution operation without translation. Further, features from all dense blocks are combined through concatenation, followed by the application of dropout, which is re-summarised using another dense layer with neurons and is regularised with dropout. The final set of features are computed from the final dense layer, termed as embedding layer. Triplet loss in the hidden layers is used to

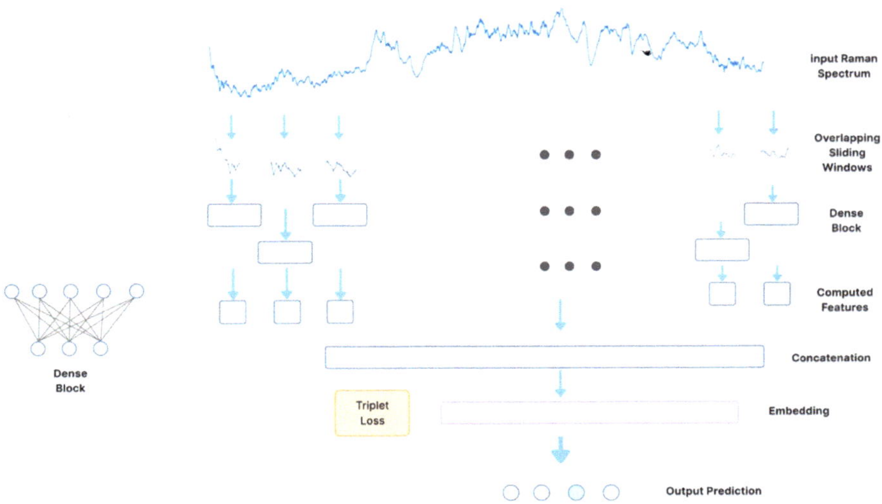

Fig. 5 Schematic of the Raman spectrum classification model using dense blocks and triplet loss. The input Raman spectrum is segmented into overlapping sliding windows, and each window is processed through dense blocks to extract local features. These computed features are then concatenated into a single feature vector and transformed into a lower-dimensional embedding space. A triplet loss function is applied during training to optimise the embedding space for discriminative representation, followed by the final output prediction (Figure Source: [132])

generate more distinct and informative representations in a reduced-dimensional space. The model has got fewer parameters and hence is better efficient computationally.

RamanNet helps overcome the confinements of the invariance property that is commonly encountered in CNNs, while leveraging sparse connectivity and triplet loss to attain good performance against Raman datasets.

Autoencoder Ridge Regression (AE Ridge) Model

The Autoencoder ridge regression is an ML regression tool designed to predict analyte concentrations from spectra, especially in the case of mixed substances. The AE-Ridge regression model combines two key ML techniques: A fully connected autoencoder for feature extraction and dimensionality reduction, and ridge regression for the final concentration prediction (Fig. 6). Hence it is a hybrid technique used for spectral data analysis [133].

Autoencoders are neural network-based models that can learn compressed representations of input data. In the case of spectroscopic data, autoencoders reduce the

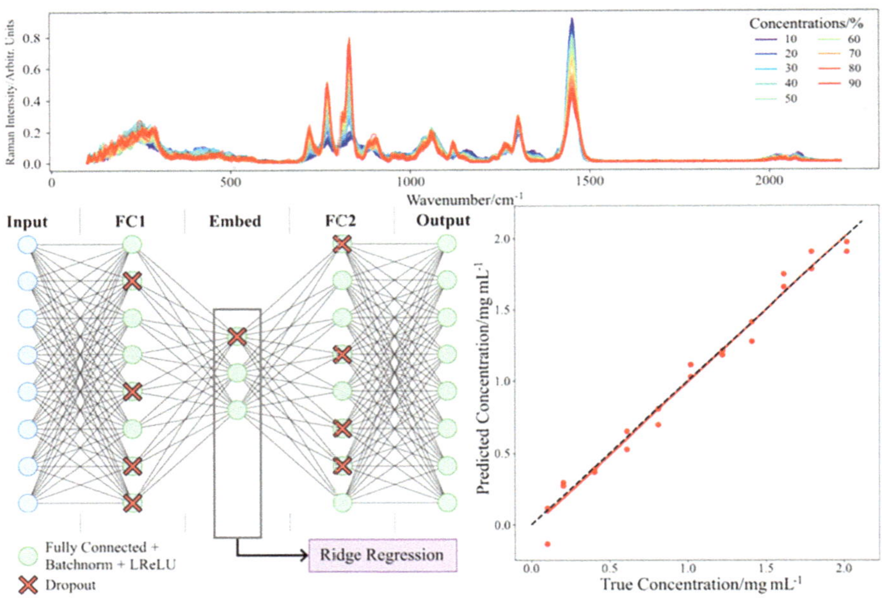

Fig. 6 Flowchart illustrating the AE-Ridge regression framework for solution concentration prediction. Salient features are first extracted from each input spectrum using a fully connected (FC) autoencoder trained to reconstruct the spectra. The resulting low-dimensional embeddings, which retain essential spectral information, are then used to train a separate ridge regression model. Ridge regression, a regularised linear model with an L2 penalty, is chosen to address potential multicollinearity. The regression model is trained on the same embedded training data used for the autoencoder and evaluated on the corresponding testing set, representing the complete AE-Ridge pipeline (Figure Source: [133])

dimensionality by learning a compact version (latent space) of the high dimensional data. Ridge Regression is a variant of linear regression that imposes a small penalty (termed as L2 regularisation or ridge penalty) to the model on instances of overfitting. This methodology is especially useful when some of the input features are closely related to each other(multicollinearity) or when the number of features exceeds the number of observations.

The autoencoder itself consists of five layers, that includes input and output layers. It features two fully connected layers in the encoder, with the final one being a 128-unit embedding layer. The decoder mirrors the encoder's architecture. Each hidden layer's output undergoes batch normalisation, followed by activation using the Leaky ReLU function. To regularise the model and reduce overfitting during training, a dropout layer is applied at the end of each hidden block. The low-dimensional embedding (feature representation) from the trained autoencoder is then used as input for a separate regression model. Ridge regression is the chosen regression model for making predictions on concentrations. The model depth and layer sizes were optimised using a grid search to minimise the mean squared error (MSE) loss. The advancements in AE-Ridge contribute to increased precision, speed, and effectiveness of analytical tools in industrial settings, aiming to reduce costs, increase product quality, and mitigate risks.

Spectroscopic Data is high-dimensional datasets and autoencoders can reduce the complexity by extracting the most significant features. In the AE Ridge hybrid model, the autoencoder reduces the dimensionality of the spectroscopic data, and the ridge regression is applied to the reduced features to predict or classify the spectra. This combination can be powerful for Noise reduction, Feature Extraction, Regularised regression on noisy, high-dimensional data.

Spectral Masked Autoencoder

SMAE is a Self-Supervised Learning (SSL) paradigm developed for Raman spectroscopy, designed to tackle the difficulty of extracting features from unlabelled spectral data. The working methodology is that it masks portions of the spectrum randomly and then attempts to reconstruct the original data, thus it learns essential spectral features that are originally crucial [134].

SMAE functions as a simple autoencoder that is intended to reconstruct partially disordered spectral signals back into their complete form, the process of which facilitates learning of crucial features. Thus, the network develops a deep recognition of the spectrum to accurately recover and generate the missing signals. The speciality of the network is that it does not need any spectral annotations during its pre-training phase. This can be highly helpful in cases where large amounts of well-annotated data are not available, as the methodology reduces the need for labelling during the pre-training phase (Fig. 7).

The workflow of the methodology can be considered into two phases. In the initial phase, the model is trained on a large amount of unlabelled data. A preparatory task is created by randomly obscuring a diverse set (around 50%) of spectral

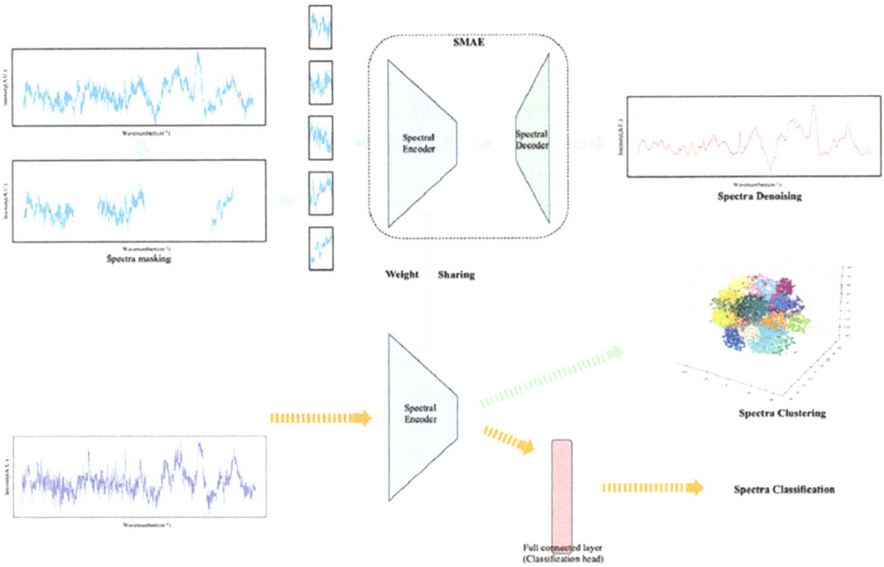

Fig. 7 Workflow of the SMAE framework. (**a**) Flowchart of the self-supervised pre-training phase using masked Raman spectra; (**b**) pathogen bacterial classifier training with limited labelled data, utilising shared autoencoder weights from the pre-training phase; (**c**) Structure of the multi-head self-attention mechanism used within the encoder and decoder blocks. (Figure Source: [134])

patches and peaks. The masked spectrum is further fed into the SMAE system, with the purpose of recovering the original input spectrum, as closely as possible. This process allows the network to learn more effectively the features beyond simple linear fitting. The encoder module in this section focuses mainly on the visible spectral features, helping the computational load and increasing the training speed. Major properties that determine the performance of the SAME algorithm are masking ratio, patch size, design of spectral encoder and decoder, positional encoding and class token are to be optimised.

Following pre-training, the encoder's weights are shared, and a fully connected classification layer is appended. This in fact is a fine-tuning phase and requires only a limited amount of labelled data to attain sublime spectral recognition, significantly saving costs and computational load compared to traditional supervised methods. Architecture components, spectral encoder, spectral decoder, loss function, optimisation.

Contrastive Learning

Contrastive learning is an advanced ML approach designed to learn robust and semantically meaningful feature representations from data. It has shown substantial promise in enhancing spectroscopic analysis, particularly in tasks such as spectrum matching and molecular structure elucidation. At its core, contrastive learning operates by organising data into positive (similar) and negative (dissimilar) pairs, encouraging the model to minimise the distance between similar instances while maximising the separation between dissimilar ones in the learned embedding space (Fig. 8). This training paradigm enables the model to extract chemically and structurally relevant information from complex spectral data [135].

In contrastive learning, the process typically begins with augmenting and transforming the original data to create multiple variations of the same sample. These variations, representing different perspectives and feature representations, form positive pairs when they share the same underlying class identity. Conversely, samples from dissimilar classes are treated as negative pairs.

The sets of positive and negative pairs are passed through an encoder-often a neural network such as a CNN or Transformer-which projects the input data into a latent feature space, where the similarities and differences between samples can be quantitatively evaluated. A projection network which maps the encoder's outputs into a lower-dimensional embedding space.

The goal of contrastive learning is to pull positive pairs closer together in the embedding space while pushing negative pairs farther apart. This is usually achieved using loss functions such as contrastive loss, triplet loss, or InfoNCE loss, chosen based on the specific application. The training process minimises the distance (maximised similarity) between positive pairs and maximises the distance (minimises similarity) between negative pairs. Through this optimisation, the model learns meaningful, discriminative data representations that can be applied to downstream

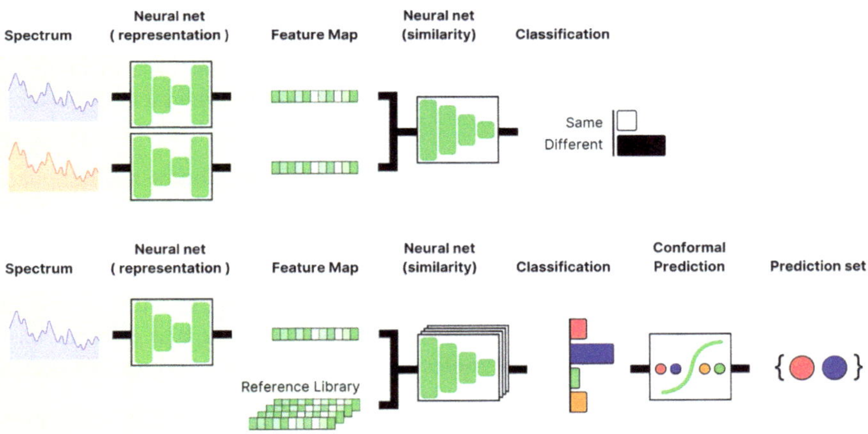

Fig. 8 Overview of the training and testing workflow for a neural network-based spectral classification model with conformal prediction. Top (training): Pairs of spectra are passed through a neural network to extract feature maps, which are then compared using a similarity network to classify whether the spectra belong to the same or different classes. Bottom (testing): a test spectrum is encoded into a feature map and compared against a reference library using the similarity network. The resulting similarity scores are used for classification and further refined using conformal prediction to generate a reliable prediction set (Figure Source: [135])

tasks such as classification, clustering, or retrieval-enabling effective generalisation to previously unseen data.

Contrastive learning is a methodology for an end-to-end learning approach, where extraction of spectral features and classification are coupled and optimised together, even from raw spectra. This results in simplification of the analysis pipeline and can lead to improved accuracy by retaining more discriminant information.

VibraCLIP

VibraCLIP is an innovative self-supervised, multi-modal contrastive learning framework specifically developed for chemical structure elucidation using vibrational spectra. VibraCLIP's main specialty is its ability to correlate spectral data (such as vibrational spectra) with other data such as molecular data. It is engineered to facilitate accurate, real-time, and scalable molecular identification, effectively overcoming key limitations of traditional vibrational spectroscopy methods—such as low throughput, complex spectral interpretation, and poor integration across analytical platforms [136].

The approach learns a unified latent space that aligns molecular graphs with their spectra using contrastive learning inspired by CLIP and operates in two stages:

- Multi-modal contrastive pre-training. The model is initially trained on large datasets, which includes diverse molecular properties and simulated spectroscopic data. This stage uses contrastive optimisation to obtain similar embeddings from the molecular graph, thereby building a shared latent space.
- Retrieval with scoring: Given an unknown molecule's IR and Raman spectra, the system computes embeddings and compares them to a database of molecular graphs using cosine similarity, ranking candidates by chemical consistency rather than exact match.

The architecture of VibraCLIP is formulated upon the CLIP (Contrastive Language-Image Pre-training) architecture, which include encoders of three distinct types. The first type is Graph Encoder which converts Simplified Molecular-input Line Entry System (SMILES) strings into graph representations. The basis of which is DimeNet++ which is a specialised graph neural network architecture that learns geometric patterns in molecules by considering both angular and distance-based features. SMILES strings describe molecular structures by encoding connectivity information in a character string representation.

FCNNs (Fully connected neural network) that process interpolated and normalised spectra where the CNNs are utilised to extract features from the spectral data. Map modality-specific features into a shared latent space, permitting direct comparison and alignment of features from different modalities.

- Independent learning rates and schedules for each modality.
- Graph-centered CLIP loss: Aligns molecular graphs separately with IR and Raman, unifying all three via the structural core.

- Mass anchoring: Incorporating molecular mass improves alignment and retrieval performance.
- Lightweight fine-tuning: Enables robust transfer from simulated to experimental data with minimal retraining.

By combining structural-spectral learning with predictive modelling, frameworks like VibraCLIP can support AI-driven biopolymer discovery, predicting key traits like strength, biodegradability, and toxicity. Generative models such as, Generative Adversarial Networks (GANs), Variational Auto Encoders (VAEs) and Transformers further contribute by designing novel architectures and evaluating full material life cycles for sustainability. AI also optimises fabrication methods (e.g., 3D printing, self-assembly), bridging lab research and industrial scalability while reducing environmental impact.

Vibrational spectroscopy remains a critical technique for probing polymer properties such as crystallinity, cross-linking, and surface features—key elements for designing advanced polysaccharide-based materials with desirable traits. These models often use design parameters as input features, enabling the predictions of critical mechanical and chemical properties such as strength, deformation, toxicity, stiffness and stability. This predictive capability accelerates the discovery of new biopolymers with optimised traits, such as improved biodegradability, mechanical strength, and thermal stability reducing both consumption and waste. Generative models form a distinct class of algorithms focused on learning from data to create new samples or design novel objects, rather than simply predicting or classifying. In the context of bio-inspired composites, GANS and VAEs can generate novel micro-architectures that mimic natural hierarchical patterns, producing materials with superior mechanical properties and reduced environmental impacts. Transformers, the foundational architecture of Large Language Models (LLMs), employ a self-attention mechanism to capture global relationships within input data, including text and images. This flexibility has made transformers one of the most widely adopted deep learning frameworks. In Biopolymer research Transformers are expected to be applied in analysing material life cycles by evaluating factors such as energy consumption, carbon emissions, and recyclability. These comprehensive analyses support the development of biopolymers that are both functional and sustainable across their entire lifespan.

AI also plays a key role in bridging the gap between laboratory discoveries and industrial production by optimising manufacturing processes such as 3D printing parameters and self-assembly techniques. This approach helps identify optimal processing conditions, reducing energy consumption, and material waste while enhancing overall material performance.

4.2 Design of Biopolymers and Composites

Designing and optimising biopolymers for targeted biomedical use is a complex challenge and traditionally reliant on trial-and-error experimentation and human intuition. The conventional process is often time consuming, costly and inefficient, with unpredictable structure-property relationships due to the influence of numerous design parameters. Modelling and designing biopolymers are a complex, interdisciplinary field that leverages a combination of spectroscopic techniques and AI. These approaches are crucial for developing biopolymers with tailored mechanical, physical and bioactive properties for diverse applications. The properties like hydrophobicity, mechanical properties, physicochemical properties, surface charge and morphology, bioavailability and biocompatibility all can be modulated by changing factors like cross-linking density and molecular weight [137]. For design application of the Biopolymers, design parameters are fed as input features, enabling the prediction of critical mechanical and chemical properties such as strength, deformation, toxicity, stiffness, and stability. This predictive capability accelerates the discovery of new biopolymers with optimised traits, such as improved biodegradability, mechanical strength, and thermal stability, reducing both resource consumption and wastage. Decision trees, random forests, neural networks, clustering algorithms, dimensionality reduction methods, self-organising maps, deep learning, fully connected neural networks, CNNs, Recurrent Neural Networks(RNNs), GANs, auto encoders, and deep belief networks.

Generative models form a distinct class of algorithms focused on learning from data to create new samples or design novel objects, rather than simply predicting or classifying. In the context of bio-inspired composites, generative models like GANs and VAEs can generate novel microarchitectures that mimic natural hierarchical patterns, producing materials with superior mechanical properties and reduced environmental impacts. Diffusion models are yet another class of generative models that can be used for generating design parameters of Biopolymers. Transformers, the foundational architecture of LLMs, employ a self-attention mechanism to capture global relationships within input data, including both text and images. This flexibility has made Transformers one of the most widely adopted deep learning frameworks. These comprehensive analyses support the development of biopolymers that are both functional and sustainable across their entire lifespan. LLMs leverage Natural Language Processing (NLP) and when finely tuned establish accurate relationships between engineering variables and requirements within their neural networks.

MeltPoolGAN, is an AI methodology that was originally developed for modelling thermal patterns in metal additive manufacturing [138], offering the possibility to repurpose for biopolymer design by leveraging its ability to learn and generate complex spatial-temporal patterns based on process parameters. MechGPT implementing Retrieval-Augmented Generation (RAG) methods [139] Multi-Modal

Mechanics Language Model (MeLM) assists engineers in solving forward and inverse design problems for various materials [140], offering multiple solutions simultaneously to optimise the conceptualisation process. LLM based methodologies like BioinspiredLLM, employs a specific RAG method to overcome typical LLM retrieval weakness and reduce the risk of hallucinations [140]. BioSpark, yet another LLM to present coherent explanations of biological mechanisms to aid designers [141].

AI also plays a key role in bridging the gap between laboratory discoveries and industrial production by optimising manufacturing processes such as 3D printing parameters and self-assembly techniques. This approach helps identify optimal processing conditions, reducing energy consumption and material waste while enhancing overall material performance.

Computational and theoretical models provide essential tools for designing and optimising biopolymer-based systems. Spatio-temporal resolutions. Multiscale modelling is categorised as Hierarchical and concurrent methods. Hierarchical methods decompose a problem into multiple levels based on temporal or spatial scales, with flow of information from fine scale to macroscopic through step-by-step analysis, allowing specialised techniques at each level.

5 Conclusion

The integration of AI with vibrational spectroscopy has marked a transformative shift in the characterisation of medical biopolymers. Conventional spectroscopic techniques such as Raman and IR spectroscopy, while powerful, often face challenges in data interpretation due to the inherent complexity and heterogeneity of biological materials. AI-driven analytical models, principally those based on ML and deep learning, now offer robust solutions for deciphering these complex spectral signatures with heightened sensitivity, specificity, and speed. New advanced, faster vibrational spectroscopy techniques are also emerging that may be applied to the medical use of biopolymers, such as stimulated Raman scattering [142].

In this chapter, we have discussed various biopolymers and its medical applications. From structure-function analysis to quality control in manufacturing, utilisation of methodologies such as generative-AI for designing novel biopolymers and the synergistic use of AI and spectroscopy. We discuss AI aspects from the conventional methodologies to advanced AI tools for spectroscopic analysis. Looking ahead, the ensuing evolution of AI models—paired with improved spectroscopic hardware and expanded biopolymer datasets—promises even greater diagnostic accuracy and material insight. As we move toward more intelligent and autonomous analytical systems, AI-aided vibrational spectroscopy is composed to become an essential tool in the next generation of medical biopolymer research and development.

References

1. Liu M, Jin J, Zhong X, Liu L, Tang C, Cai L (2024) Polysaccharide hydrogels for skin wound healing. Heliyon 10:e35014
2. Diller RB, Tabor AJ (2022) The role of the extracellular matrix (ECM) in wound healing: a review. Biomimetics 7:87
3. He S, Shi D, Han Z, Dong Z, Xie Y, Zhang F et al (2019) Heparinized silk fibroin hydrogels loading FGF1 promote the wound healing in rats with full-thickness skin excision. Biomed Eng Online:18. https://doi.org/10.1186/s12938-019-0716-4
4. Ghofrani A, Hassannejad Z (2024) Collagen-based therapies for accelerated wound healing. IntechOpen, London. https://doi.org/10.5772/intechopen.1004079
5. Askari M, Keshavarz Zarjani A, Sayyahi A, Badpa R, Naghizadeh A (2025) Chitosan nanoparticles: a promising candidate in wound healing. Int J Low Extrem Wounds. https://doi.org/10.1177/15347346251325057
6. Gardeazabal L, Izeta A (2024) Elastin and collagen fibres in cutaneous wound healing. Exp Dermatol 33:15052. https://doi.org/10.1111/exd.15052
7. Cao H, Wang J, Hao Z, Zhao D (2024) Gelatin-based biomaterials and gelatin as an additive for chronic wound repair. Front Pharmacol 15. https://doi.org/10.3389/fphar.2024.1398939
8. Ferraz MP (2025) Wound dressing materials: bridging material science and clinical practice. Appl Sci 15:1725
9. Zeng X, Sun Z, Chen L, Zhang X, Guo X, Li G (2024) Co-assembled biomimetic fibrils from collagen and chitosan for performance-enhancing hemostatic dressing. Biomater Sci 13:236–249
10. Kolodziejska B, Pajchel L, Zgadzaj A, Kolmas J (2022) A new, biomimetic collagen–apatite wound-healing composite with a potential regenerative and anti-hemorrhagic effect in dental surgery. Materials 15:8888
11. Zhong Q-K, Wu Z-Y, Qin Y-Q, Hu Z, Li S-D, Yang Z-M et al (2019) Preparation and properties of carboxymethyl chitosan/alginate/tranexamic acid composite films. Membranes 9:11
12. Zhang L, Wang M, Qiu H, Wei Y, Zhou L, Nian N et al (2023) Epicatechin gallate promotes vascularization in co-culture of human osteoblasts and outgrowth endothelial cells. Exp Biol Med 248:732–745
13. Zheng Y, Feng Z, You C, Jin Y, Hu X, Wang X et al (2013) In vitro evaluation of Panax notoginseng Rg1 released from collagen/chitosan-gelatin microsphere scaffolds for angiogenesis. Biomed Eng Online 12:134
14. Dhasmana A, Malik S, Ranjan A, Chauhan A, Tashkandi HM, Haque S et al (2022) A bioengineered quercetin-loaded 3D bio-polymeric graft for tissue regeneration and repair. Biomedicine 10:3157
15. Crane NJ, Brown TS, Evans KN, Hawksworth JS, Hussey S, Tadaki DK et al (2010) Monitoring the healing of combat wounds using Raman spectroscopic mapping. Wound Repair Regen 18:409–416
16. Crane NJ, Elster EA (2012) Vibrational spectroscopy: a tool being developed for the noninvasive monitoring of wound healing. J Biomed Opt 17:10902
17. Aramwit P, Siritientong T, Srichana T (2012) Potential applications of silk sericin, a natural protein from textile industry by-products. Waste Manag Res 30:217–224
18. Jayakumar R, Prabaharan M, Nair SV, Tamura H (2010) Novel chitin and chitosan nanofibers in biomedical applications. Biotechnol Adv 28:142–150
19. Boateng J, Catanzano O (2015) Advanced therapeutic dressings for effective wound healing—a review. J Pharm Sci 104:3653–3680
20. Yu Y, Gao Y, Zeng Y, Ge W, Tang C, Xie X et al (2024) Multifunctional hyaluronic acid/gelatin methacryloyl core-shell microneedle for comprehensively treating oral mucosal ulcers. Int J Biol Macromol 266:131221

21. Li Y, Li T, Feng J, Liu B, Wang Z, He J et al (2025) Acid-responsive contractile hyaluronic acid-based hydrogel loaded with ginsenoside Rg1 for hemostasis and promotion of gastric wound healing. Biomaterials 321:123320
22. Antoszewska M, Sokolewicz EM, Barańska-Rybak W (2024) Wide use of hyaluronic acid in the process of wound healing—a rapid review. Sci Pharm 92:23
23. Hickey RJ, Pelling AE (2019) Cellulose biomaterials for tissue engineering. Front Bioeng Biotechnol 7:45. https://doi.org/10.3389/fbioe.2019.00045
24. Barker TH, Engler AJ (2017) The provisional matrix: setting the stage for tissue repair outcomes. Matrix Biol 60:1–4
25. Zulkiflee I, Amirrah IN, Fadilah NIM, Wee MFMR, Yusop SM, Maarof M et al (2023) Characterization of dual-layer hybrid biomatrix for future use in cutaneous wound healing. Materials 16. https://doi.org/10.3390/ma16031162
26. Zhang X, Chen X, Hong H, Hu R, Liu J, Liu C (2022) Decellularized extracellular matrix scaffolds: recent trends and emerging strategies in tissue engineering. Bioact Mater 10:15–31
27. Damour O, Augustin C, Black AF (1998) Applications of reconstructed skin models in pharmaco-toxicological trials. Med Biol Eng Comput 36:825–832
28. Artal P, Ginis H, Christaras D, Villegas EA, Tabernero J, Prieto PM (2023) Inverted meniscus intraocular lens as a better optical surrogate of the crystalline lens. Biomed Opt Express 14:2129
29. Yildirim TM, Schickhardt SK, Wang Q, Friedmann E, Khoramnia R, Auffarth GU (2021) Quantitative evaluation of microvacuole formation in five intraocular lens models made of different hydrophobic materials. PLoS One 16:e0250860
30. Rusciano G, Capaccio A, Pesce G, Sasso A (2019) Experimental study of the mechanisms leading to the formation of glistenings in intraocular lenses by Raman spectroscopy. Biomed Opt Express 10:1870
31. Vacalebre M, Frison R, Corsaro C, Neri F, Santoro A, Conoci S et al (2023) Current state of the art and next generation of materials for a customized IntraOcular lens according to a patient-specific eye Power. Polymers 15:1590
32. Casey-Power S, Ryan R, Behl G, McLoughlin P, Byrne ME, Fitzhenry L (2022) Hyaluronic acid: its versatile use in ocular drug delivery with a specific focus on hyaluronic acid-based polyelectrolyte complexes. Pharmaceutics 14:1479
33. Li Q, Zhang F, Zhao H (2025) Construction and biocompatibility of penetrating corneal transplant substitute with cross-linked acellular porcine cornea and biopolymer polyurethane. Biomaterials Advances 170:214201
34. Jameson JF, Pacheco MO, Nguyen HH, Phelps EA, Stoppel WL (2021) Recent advances in natural materials for corneal tissue engineering. Bioengineering 8:161
35. Zarrintaj P, Seidi F, Youssefi Azarfam M, Khodadadi Yazdi M, Erfani A, Barani M et al (2023) Biopolymer-based composites for tissue engineering applications: a basis for future opportunities. Compos Part B Eng 258:110701
36. Suamte L, Tirkey A, Barman J, Jayasekhar BP (2023) Various manufacturing methods and ideal properties of scaffolds for tissue engineering applications. Smart Mater Manuf 1:100011
37. Nazari H, Heirani-Tabasi A, Alavijeh MS, Jeshvaghani ZS, Esmaeili E, Hosseinzadeh S et al (2019) Nanofibrous composites reinforced by MoS_2 nanosheets as a conductive scaffold for cardiac tissue engineering. ChemistrySelect 4:11557
38. Gayathri V, Khan T, Gowtham M, Balan R, Sebaey TA (2025) Functionalized conductive polymer composites for tissue engineering and biomedical applications- a mini review. Front Bioeng Biotechnol:13. https://doi.org/10.3389/fbioe.2025.1533944
39. Krishani M, Shin WY, Suhaimi H, Sambudi NS (2023) Development of scaffolds from bio-based natural materials for tissue regeneration applications: a review. Gels 9:100
40. Shirehjini LM, Sharifi F, Shojaei S, Irani S (2022) Poly-caprolactone nanofibrous coated with sol-gel alginate/mesenchymal stem cells for cartilage tissue engineering. J Drug Deliv Sci Technol 74. https://doi.org/10.1016/j.jddst.2022.103488

41. Karimi SNH, Aghdam RM, Ebrahimi SAS, Chehrehsaz Y (2022) Tri-layered alginate/poly(ε-caprolactone) electrospun scaffold for cardiac tissue engineering. Polym Int 71:1099–1108
42. Khan MUA, Razak SIA, Rehman S, Hasan A, Qureshi S, Stojanović GM (2022) Bioactive scaffold (sodium alginate)-g-(nHAp@SiO2@GO) for bone tissue engineering. Int J Biol Macromol 222:462–472
43. Dacrory S, Ali LMA, Ouahrani-Bettache S, Daurat M, El-Sakhawy M, Hesemann P et al (2025) Potential application of oxidized cellulose/alginate loaded hydroxyapatite/graphene oxide beads in bone tissue engineering. BMC Chem 19. https://doi.org/10.1186/s13065-025-01408-2
44. Guagnini B, Medagli B, Zumbo B, Cannillo V, Turco G, Porrelli D et al (2024) Alginate-Sr/Mg containing bioactive glass scaffolds: the characterization of a new 3D composite for bone tissue engineering. J Funct Biomater 15. https://doi.org/10.3390/jfb15070183
45. Salzlechner C, Walther AR, Schell S, Merrild NG, Haghighi T, Huebscher I et al (2020) Complementary techniques to analyse pericellular matrix formation by human MSC within hyaluronic acid hydrogels. Mater Adv 1:2888–2896
46. Park SH, Seo JY, Park JY, Ji YB, Kim K, Choi HS et al (2019) An injectable, click-crosslinked, cytomodulin-modified hyaluronic acid hydrogel for cartilage tissue engineering. NPG Asia Mater 11. https://doi.org/10.1038/s41427-019-0130-1
47. Yan L, Wang Y, Ren L, Wu G, Caridade SG, Fan J et al (2010) Genipin-cross-linked collagen/chitosan biomimetic scaffolds for articular cartilage tissue engineering applications. J Biomed Mater Res A 95A:465–475
48. Quade M, Schumacher M, Bernhardt A, Lode A, Kampschulte M, Voß A et al (2018) Strontium-modification of porous scaffolds from mineralized collagen for potential use in bone defect therapy. Mater Sci Eng C Mater Biol Appl 84:159–167
49. Wang Z, Kumar H, Tian Z, Jin X, Holzman JF, Menard F et al (2018) Visible light photoinitiation of cell-adhesive gelatin methacryloyl hydrogels for stereolithography 3D bioprinting. ACS Appl Mater Interfaces 10(26859–26):869
50. Noshadi I, Hong S, Sullivan KE, Shirzaei Sani E, Portillo-Lara R, Tamayol A et al (2017) In vitro and in vivo analysis of visible light crosslinkable gelatin methacryloyl (GelMA) hydrogels. Biomater Sci 5:2093–2105
51. Yu Q, Shen C, Wang X, Wang Z, Liu L, Zhang J (2023) Graphene oxide/gelatin nanofibrous scaffolds loaded with N-acetyl cysteine for promoting wound healing. Int J Nanomed 18:563–578
52. Monteiro N, Martins A, Reis RL, Neves NM (2015) Nanoparticle-based bioactive agent release systems for bone and cartilage tissue engineering. Regen Ther 1:109–118
53. Anjum S, Wang Y, Xin Y, Li X, Li T, Zhang H et al (2024) Bioinspired core-shell nanofiber drug-delivery system modulates osteogenic and osteoclast activity for bone tissue regeneration. Mater Today Bio 26:101088
54. Qu L, Dubey N, Ribeiro JS, Bordini EAF, Ferreira JA, Xu J et al (2021) Metformin-loaded nanospheres-laden photocrosslinkable gelatin hydrogel for bone tissue engineering. J Mech Behav Biomed Mater:116. https://doi.org/10.1016/j.jmbbm.2020.104293
55. AnilKumar S, Allen SC, Tasnim N, Akter T, Park S, Kumar A et al (2019) The applicability of furfuryl-gelatin as a novel bioink for tissue engineering applications. J Biomed Mater Res B Appl Biomater 107:314–323
56. Mutsenko VV, Gryshkov O, Lauterboeck L, Rogulska O, Tarusin DN, Bazhenov VV et al (2017) Novel chitin scaffolds derived from marine sponge Ianthella basta for tissue engineering approaches based on human mesenchymal stromal cells: biocompatibility and cryopreservation. Int J Biol Macromol 104:1955–1965
57. Cheng W, Ding Z, Zheng X, Lu Q, Kong X, Zhou X et al (2020) Injectable hydrogel systems with multiple biophysical and biochemical cues for bone regeneration. Biomater Sci 8:2537–2548

58. Tang H, Lu A, Li L, Zhou W, Xie Z, Zhang L (2013) Highly antibacterial materials constructed from silver molybdate nanoparticles immobilized in chitin matrix. Chem Eng J 234:124–131
59. Rajalekshmi R, Shaji AK, Joseph R, Bhatt A (2021) Scaffold for liver tissue engineering: exploring the potential of fibrin incorporated alginate dialdehyde–gelatin hydrogel. Int J Biol Macromol 166:999–1008
60. Mobarakeh ZT, Hasanzadeh E, Farzin A, Goodarzi A, Farahani MS, Shirian S et al (2023) Enhanced sciatic nerve regeneration with fibrin scaffold containing human endometrial stem cells and insulin encapsulated chitosan particles: an in vivo study. Injury 54:1462–1472
61. Periyathambi P, Vedakumari WS, Baskar SK, Bojja S, Sastry TP (2015) Osteogenic potency of magnetic fibrin nanoparticles - a novel perspective in bone tissue engineering. Mater Lett 139:108–111
62. Narayanan KB, Bhaskar R, Sudhakar K, Nam DH, Han SS (2023) Polydopamine-functionalized bacterial cellulose as hydrogel scaffolds for skin tissue engineering. Gels 9. https://doi.org/10.3390/gels9080656
63. Lv XG, Li Z, Chen SY, Xie MK, Huang JW, Peng XF et al (2016) Structural and functional evaluation of oxygenating keratin/silk fibroin scaffold and initial assessment of their potential for urethral tissue engineering. Biomaterials 84:99–110
64. Wan X, Li P, Jin X, Su F, Shen J, Yuan J (2020) Poly(ε-caprolactone)/keratin/heparin/VEGF biocomposite mats for vascular tissue engineering. J Biomed Mater Res A 108:292–300
65. Annabi N, Fathi A, Mithieux SM, Martens P, Weiss AS, Dehghani F (2011) The effect of elastin on chondrocyte adhesion and proliferation on poly (ε-caprolactone)/elastin composites. Biomaterials 32:1517–1525
66. Kuo Y-C, Ku H-F, Rajesh R (2017) Chitosan/γ-poly(glutamic acid) scaffolds with surface-modified albumin, elastin and poly-l-lysine for cartilage tissue engineering. Biomed Res Int 78:265–277
67. Castillo-Ortega MM, López-Peña IY, Rodríguez-Félix DE, Del Castillo-Castro T, Encinas-Encinas JC, Santacruz-Ortega H et al (2022) Clindamycin-loaded nanofibers of polylactic acid, elastin and gelatin for use in tissue engineering. Polym Bull 79:5495–5513
68. Ma Z, Scales RJ, Brand DD, Czernuszka JT (2024) A tri-layer tissue engineering heart valve scaffold based on atelocollagen, hyaluronic acid, and elastin. Explor BioMat-X 1:215–230
69. Janorkar AV, Rajagopalan P, Yarmush ML, Megeed Z (2008) The use of elastin-like polypeptide–polyelectrolyte complexes to control hepatocyte morphology and function in vitro. Biomaterials 29:625–632
70. Rosellini E, Madeddu D, Barbani N, Frati C, Graiani G, Falco A et al (2020) Development of biomimetic alginate/gelatin/elastin sponges with recognition properties toward bioactive peptides for cardiac tissue engineering. Biomimetics 5:1–19
71. Farokhi M, Mottaghitalab F, Ai J, Shokrgozar MA (2013) Sustained release of platelet-derived growth factor and vascular endothelial growth factor from silk/calcium phosphate/PLGA based nanocomposite scaffold. Int J Pharm 454:216–225
72. Li P, Wang Y, Jin X, Dou J, Han X, Wan X et al (2020) Fabrication of PCL/keratin composite scaffolds for vascular tissue engineering with catalytic generation of nitric oxide potential. J Mater Chem B 8:6092–6099
73. Gao J, Wang Y, Chen S, Tang D, Jiang L, Kong D et al (2017) Electrospun poly-ε-caprolactone scaffold modified with catalytic nitric oxide generation and heparin for small-diameter vascular graft. RSC Adv 7:18775
74. Zhou X, Tao Y, Chen E, Wang J, Fang W, Zhao T et al (2018) Genipin-cross-linked type II collagen scaffold promotes the differentiation of adipose-derived stem cells into nucleus pulposus-like cells. J Biomed Mater Res A 106:1258–1268
75. Wang JK, Yeo KP, Chun YY, Tan TTY, Tan NS, Angeli V et al (2017) Fish scale-derived collagen patch promotes growth of blood and lymphatic vessels in vivo. Acta Biomater 63:246–260

76. Bhuiyan MH, Clarkson AN, Ali MA (2023) Optimization of thermoresponsive chitosan/β--glycerophosphate hydrogels for injectable neural tissue engineering application. Colloids Surf B Biointerfaces 224:113193
77. Chen X, Fu Q, Jin Y, Li M, Yang R, Cui X et al (2017) In vitro studying corrosion behavior of porous titanium coating in dynamic electrolyte. Mater Sci Eng C Mater Biol Appl 70:1071–1075
78. Singh N, Chen J, Koziol KK, Hallam KR, Janas D, Patil AJ et al (2016) Chitin and carbon nanotube composites as biocompatible scaffolds for neuron growth. Nanoscale 8:8288–8299
79. Massia SP, Holecko MM, Ehteshami GR (2004) In vitro assessment of bioactive coatings for neural implant applications. J Biomed Mater Res A 68A:177–186
80. Bansal M, Dravid A, Aqrawe Z, Montgomery J, Wu Z, Svirskis D (2020) Conducting polymer hydrogels for electrically responsive drug delivery. J Control Release 328:192–209
81. Servant A, Leon V, Jasim D, Methven L, Limousin P, Fernandez-Pacheco EV et al (2014) Graphene-based electroresponsive scaffolds as polymeric implants for on-demand drug delivery. Adv Healthc Mater 3:1334–1343
82. He S, Wu J, Li S-H, Wang L, Sun Y, Xie J et al (2020) The conductive function of biopolymer corrects myocardial scar conduction blockage and resynchronizes contraction to prevent heart failure. Biomaterials 258:120285
83. Chen Y, Gao P, Huang L, Tan X, Zhou N, Yang T et al (2021) A tough nitric oxide-eluting hydrogel coating suppresses neointimal hyperplasia on vascular stent. Nat Commun 12:7079
84. Dong J, Pacella M, Liu Y, Zhao L (2022) Surface engineering and the application of laser-based processes to stents - a review of the latest development. Bioact Mater 10:159–184
85. Lach S, Jurczak P, Karska N, Kubiś A, Szymańska A, Rodziewicz-Motowidło S (2020) Spectroscopic methods used in implant material studies. Molecules 25:579
86. Lin H-Y, Chiou W-S, Shiue S-J, Cheng J-K (2020) Non-RGD peptide H-ckrwwkwirw-NH $_2$ grafting accentuates antibacterial and osteoinductive properties of biopolymer coating. Soft Mater 18:487–498
87. Ly NTK, Shin H, Gupta KC, Kang IK, Yu W (2019) Bioactive antibacterial modification of orthodontic microimplants using chitosan biopolymer. Macromol Res 27:504–510
88. Li YH, Wang YS, Zhao JS, Li ZY, Chen HH (2021) A pH-sensitive curcumin loaded microemulsion-filled alginate and porous starch composite gels: characterization, in vitro release kinetics and biological activity. Int J Biol Macromol 182:1863–1873
89. Moya-Garcia CR, Li-Jessen NYK, Tabrizian M (2023) Chitosomes loaded with docetaxel as a promising drug delivery system to laryngeal cancer cells: an in vitro cytotoxic study. Int J Mol Sci:24. https://doi.org/10.3390/ijms24129902
90. Abdellatif AAH, Ibrahim MA, Amin MA, Maswadeh H, Alwehaibi MN, Al-Harbi SN et al (2020) Cetuximab conjugated with octreotide and entrapped calcium alginate-beads for targeting somatostatin receptors. Sci Rep 10. https://doi.org/10.1038/s41598-020-61605-y
91. Malash AE, Al-esnawy AA, Ereiba KT, Bakr AM, Abdraboh AS (2025) An in vitro study of the effects of methotrexate loaded biocomposite beads on MG63 osteoblast cells. Sci Rep 15. https://doi.org/10.1038/s41598-025-85702-y
92. Lee CR, Kim GG, Park SB, Kim SW (2021) Synthesis of hyaluronic acid-conjugated fe3o4@ceo2 composite nanoparticles for a target-oriented multifunctional drug delivery system. Micromachines 12. https://doi.org/10.3390/mi12091018
93. Hsieh HY, Lin WY, Lee AL, Li YC, Chen YJ, Chen KC et al (2020) Hyaluronic acid on the urokinase sustained release with a hydrogel system composed of poloxamer 407: HA/P407 hydrogel system for drug delivery. PLoS One 15. https://doi.org/10.1371/journal.pone.0227784
94. Cai X, Yang L, Zhang LM, Wu Q (2009) Synthesis and anaerobic biodegradation of indomethacin-conjugated cellulose ethers used for colon-specific drug delivery. Bioresour Technol 100:4164–4170

95. Weeks CA, Aden B, Kilbey SM, Janorkar AV (2016) Synthesis and characterization of an array of elastin-like polypeptide-polyelectrolyte conjugates with varying chemistries and amine content for biomedical applications. ACS Biomater Sci Eng 2:2196–2206
96. Xiu X, Fu H, Zhang R, Ma S, Guo P, Li Z et al (2025) Novel antigen-presenting cell-targeted nanoparticles enhance Split vaccine immunity through microneedles inoculation. Int J Nanomed 20:5529–5549
97. Zhao P, Sun J, Zhao S, Song Y, Gao F, Wang H et al (2021) SERS-based immunoassay based on gold nanostars modified with 5,5′-dithiobis-2-nitrobenzoic acid for determination of glial fibrillary acidic protein. Microchim Acta 188. https://doi.org/10.1007/s00604-021-05081-9
98. Park M, Jung H, Jeong Y, Jeong KH (2017) Plasmonic Schirmer strip for human tear-based gouty arthritis diagnosis using surface-enhanced Raman scattering. ACS Nano 11:438–443
99. Kankala RK, Zhu K, Li J, Wang CS, Bin WS, Chen AZ (2017) Fabrication of arbitrary 3D components in cardiac surgery: from macro-, micro- to nanoscale. Biofabrication 9. https://doi.org/10.1088/1758-5090/aa8113
100. Giannopoulos AA, Mitsouras D, Yoo SJ, Liu PP, Chatzizisis YS, Rybicki FJ (2016) Applications of 3D printing in cardiovascular diseases. Nat Rev Cardiol 13:701–718
101. Theus AS, Tomov ML, Cetnar A, Lima B, Nish J, McCoy K et al (2019) Biomaterial approaches for cardiovascular tissue engineering. Emerg Mater 2:193–207
102. Pugliese R, Beltrami B, Regondi S, Lunetta C (2021) Polymeric biomaterials for 3D printing in medicine: an overview. Ann 3D Print Med 2. https://doi.org/10.1016/j.stlm.2021.100011
103. Kozaniti FK, Metsiou DN, Manara AE, Athanassiou G, Deligianni DD (2021) Recent advancements in 3d printing and bioprinting methods for cardiovascular tissue engineering. Bioengineering:8. https://doi.org/10.3390/bioengineering8100133
104. Guvendiren M, Molde J, Soares RMD, Kohn J (2016) Designing biomaterials for 3D printing. ACS Biomater Sci Eng 2:1679–1693
105. Agarwal P, Mathur V, Kasturi M, Srinivasan V, Seetharam RN, Vasanthan KS (2024) A futuristic development in 3D printing technique using nanomaterials with a step toward 4D printing. ACS Omega 9:37445
106. Shen Y, Cui J, Yu X, Song J, Cai P, Guo W et al (2024) Recent advances in three-dimensional printing in cardiovascular devices: bench and bedside applications. Smart Mater Med 5:36–51
107. Liu N, Ye X, Yao B, Zhao M, Wu P, Liu G et al (2021) Advances in 3D bioprinting technology for cardiac tissue engineering and regeneration. Bioact Mater 6:1388–1401
108. Ong CS, Fukunishi T, Zhang H, Huang CY, Nashed A, Blazeski A et al (2017) Biomaterial-free three-dimensional bioprinting of cardiac tissue using human induced pluripotent stem cell derived cardiomyocytes. Sci Rep 7:4566
109. Bin Abu Sofian ADA, Sun X, Gupta VK, Berenjian A, Xia A, Ma Z et al (2024) Advances, synergy, and perspectives of machine learning and biobased polymers for energy, fuels, and biochemicals for a sustainable future. Energy Fuel 38:1593–1617
110. Karuppusamy M, Thirumalaisamy R, Palanisamy S, Nagamalai S, El Sayed ME, Ayrilmis N (2025) A review of machine learning applications in polymer composites: advancements, challenges, and future prospects. J Mater Chem A Mater 13:16290
111. Shang L, Zhang Z, Tang F, Cao Q, Pan H, Lin Z (2023) CNN-LSTM hybrid model to promote signal processing of ultrasonic guided lamb waves for damage detection in metallic pipelines. Sensors 23:7059
112. Zhang W, Teng F, Li J, Zhang Z, Niu L, Zhang D et al (2023) Denoising method based on CNN-LSTM and CEEMD for LDV signals from accelerometer shock testing. Measurement 216:112951
113. Gieroba B, Kalisz G, Krysa M, Khalavka M, Przekora A (2023) Application of vibrational spectroscopic techniques in the study of the natural polysaccharides and their cross-linking process. Int J Mol Sci 24. https://doi.org/10.3390/ijms24032630
114. Franks F (1977) Solvation and conformational effects in aqueous solutions of biopolymer analogues. Philos Trans R Soc Lond Ser B Biol Sci 278:33–57

115. Valdman D, Lopez BJ, Valentine MT, Atzberger PJ (2013) Force spectroscopy of complex biopolymers with heterogeneous elasticity. Soft Matter 9:772–778
116. Helburn R, Nolan K (2022) Characterizing biological macromolecules with attenuated total reflectance–Fourier transform infrared spectroscopy provides hands-on spectroscopy experiences for undergraduates. Biochem Mol Biol Educ 50:381–392
117. Oberdorf R, Ferguson A, Jacobsen JL, Kondev J (2006) Secondary structures in long compact polymers. Phys Rev E 74:051801
118. Mesias VSD, Zhang J, Fu W, Dai X, Huang J (2025) Enhanced characterization of protein secondary structure transitions using Raman and SERS measurements combined with 2D correlation spectroscopy and principal component analysis. Spectrochim Acta A Mol Biomol Spectrosc 343:126607
119. Burmeister Getz E, Carroll K, Jones B, Benet L (2016) Batch-to-batch pharmacokinetic variability confounds current bioequivalence regulations: a dry powder inhaler randomized clinical trial. Clin Pharmacol Ther 100:223–231
120. Leistler B (2020) How to standardize quality requirements of raw materials. Cell Gene Ther Insights 6:1555–1560
121. Xiong H, Yu LX, Qu H (2013) Batch-to-batch quality consistency evaluation of botanical drug products using multivariate statistical analysis of the chromatographic fingerprint. AAPS PharmSciTech 14:802–810
122. Boey JY, Lee CK, Tay GS (2022) Factors affecting mechanical properties of reinforced bioplastics: a review. Polymers 14:3737
123. Shyr T-W, Ko H-C, Wu T-M, Zhu M (2021) Effect of storage conditions on the thermal stability and crystallization behaviors of poly(L-lactide)/poly(D-lactide). Polymers 13:238
124. Tinnefeld V, Sickmann A, Ahrends R (2014) Catch me if you can: challenges and applications of cross-linking approaches. Eur J Mass Spectrom 20:99–116
125. Yu C, Huang L (2018) Cross-linking mass spectrometry: an emerging technology for interactomics and structural biology. Anal Chem 90:144–165
126. Pan L, Zhang P, Daengngam C, Peng S, Chongcheawchamnan M (2022) A review of artificial intelligence methods combined with Raman spectroscopy to identify the composition of substances. J Raman Spectrosc 53:6–19
127. Zhang X, Lin T, Xu J, Luo X, Ying Y (2019) DeepSpectra: an end-to-end deep learning approach for quantitative spectral analysis. Anal Chim Acta 1058:48–57
128. Belhaouari SB, Talbi A, Elgamal M, Elmagarmid KA, Ghannoum S, Yang Y et al (2025) DeepRaman: implementing surface-enhanced Raman scattering together with cutting-edge machine learning for the differentiation and classification of bacterial endotoxins. Heliyon:11. https://doi.org/10.1016/j.heliyon.2025.e42550
129. Liu J, Osadchy M, Ashton L, Foster M, Solomon CJ, Gibson SJ (2017) Deep convolutional neural networks for Raman spectrum recognition: a unified solution. Analyst 142:4067–4074
130. Neo ERK, Low JSC, Goodship V, Debattista K (2023) Deep learning for chemometric analysis of plastic spectral data from infrared and Raman databases. Resour Conserv Recycl:188. https://doi.org/10.1016/j.resconrec.2022.106718
131. Fan X, Ming W, Zeng H, Zhang Z, Lu H (2019) Deep learning-based component identification for the Raman spectra of mixtures. Analyst 144:1789–1798
132. Ibtehaz N, Chowdhury MEH, Khandakar A, Kiranyaz S, Rahman MS, Zughaier SM (2023) RamanNet: a generalized neural network architecture for Raman spectrum analysis. Neural Comput Appl. https://doi.org/10.1007/s00521-023-08700-z
133. Poppe A, Warren C, Brooks W, Gibson S, Foster M (2025) Autoencoding Raman spectra to predict Analyte concentrations. J Raman Spectrosc. https://doi.org/10.1002/jrs.70003
134. Ren P, Zhou RG, Li Y (2025) A self-supervised learning method for Raman spectroscopy based on masked autoencoders. Expert Syst Appl:292. https://doi.org/10.1016/j.eswa.2025.128576
135. Li B, Schmidt MN, Alstrøm TS (2022) Raman spectrum matching with contrastive representation learning. Analyst 147:2238–2246

136. Rocabert-Oriols P, López N, Heras-Domingo J (2025) Multi-modal contrastive learning for chemical structure elucidation with VibraCLIP. Dig Discov. https://doi.org/10.26434/chemrxiv-2025-kwjzv
137. Jiang P, Dai Y, Hou Y, Stein J, Lin SS, Zhou C et al (2025) Artificial intelligence-assisted design, synthesis and analysis of smart biomaterials. BMEMat. https://doi.org/10.1002/bmm2.70004
138. Petrik J, Kavas B, Bambach M (2023) MeltPoolGAN: auxiliary classifier generative adversarial network for melt pool classification and generation of laser power, scan speed and scan direction in laser powder bed fusion. Addit Manuf 78. https://doi.org/10.1016/j.addma.2023.103868
139. Badini S, Regondi S, Pugliese R (2025) Enhancing mechanical and bioinspired materials through generative AI approaches. Next Mater 6. https://doi.org/10.1016/j.nxmate.2024.100275
140. Luu RK, Buehler MJ (2024) BioinspiredLLM: conversational large language model for the mechanics of biological and bio-inspired materials. Adv Sci:11. https://doi.org/10.1002/advs.202306724
141. Kang HB, Lin DCE, Chen YY, Hong MK, Martelaro N, Kittur A (2025) BioSpark: beyond analogical inspiration to LLM-augmented transfer. In: Conference on human factors in computing systems - proceedings. Association for Computing Machinery, New York. https://doi.org/10.1145/3706598.3714053
142. Krishnan Nambudiri MK, Sujadevi VG, Poornachandran P, Murali Krishna C, Kanno T, Noothalapati H (2024) Artificial intelligence-assisted stimulated Raman histology: new Frontiers in vibrational tissue imaging. Cancer 16:3917

MIX
Papier aus verantwortungsvollen Quellen
Paper from responsible sources
FSC® C105338

If you have any concerns about our products,
you can contact us on
ProductSafety@springernature.com

In case Publisher is established outside the EU,
the EU authorized representative is:
Springer Nature Customer Service Center GmbH
Europaplatz 3, 69115 Heidelberg, Germany

Printed by Libri Plureos GmbH
in Hamburg, Germany